**Man's Impact
on the Climate**

The MIT Press
Cambridge, Massachusetts,
and London, England

**Man's Impact
on the Climate**

Edited by
William H. Matthews
William W. Kellogg
G. D. Robinson

This book was designed by The MIT Press Design Department.
It was set in Linotype Baskerville
by The Colonial Press Inc.,
printed on Mohawk Neotext Offset
by The Colonial Press Inc.,
and bound in Columbia Fictionette FNV 3317 Green
by The Colonial Press Inc.
in the United States of America.

Second printing, March 1974

ISBN 0 262 13075 0 (hardcover)

Library of Congress catalog card number: 75-153295

To Carroll L. Wilson

The Director of the Study of
Critical Environmental
Problems

Contents

Preface

Over the past few years, the concept of the earth as a "spaceship" has provided many people with an awareness of the finite resources and the complex natural relationships on which man depends for his survival. These realizations have been accompanied by concerns about the impacts that man's activities are having on the global environment. Some concerned individuals, including well-known scientists, have warned of both imminent and potential global environmental catastrophes.

Theories and speculations of the global effects of pollution have included assertions that the buildup of CO_2 from fossil-fuel combustion might warm up the planet and cause the polar ice to melt, thus raising the sea level several hundred feet and submerging coastal cities. Equally foreboding has been the warning of the possibility that particles emitted into the air from industrial, energy, and transportation processes might prevent some sunlight from reaching the earth's surface, thus lowering global temperature and beginning a new ice age. Demands to ban DDT have been increasing steadily as its effects on the reproductive capabilities of birds have been determined, and as evidence is found of its accumulation in other species including man. Serious questions have been raised about the effects on oceanic and terrestrial ecosystems of systematically discharging into the environment such toxic materials as heavy metals, oil, and radioactive substances or of nutrients such as phosphorus which can overenrich lakes and coastal areas.

These and related global environmental problems were considered in depth by the one-month, interdisciplinary Study of Critical Environmental Problems (SCEP) conducted during July 1970. This Study, which was sponsored by the Massachusetts Institute of Technology, produced a report entitled *Man's Impact on the Global Environment* which was published by The M.I.T. Press in October 1970. That report presents an assessment of the existing state of scientific knowledge on several global problems and contains specific recommendations for action which would reduce the harmful effects of pollution or would provide the information required to understand more adequately the impact of man on the global environment.

The Director of SCEP was Professor Carroll L. Wilson, and the Associate Director was Professor William H. Matthews, both of M.I.T. Chairmen of major SCEP Work Groups included Drs. William W. Kellogg, National Center for Atmospheric Research; G. D. Robinson, Center for the Environment and Man, Inc.; Frederick E. Smith, Harvard University; and Edward D. Goldberg, Scripps Institution of Oceanography. These group leaders with Dr. Matthews have compiled this volume and a second volume, *Man's Impact on Terrestrial and Oceanic Ecosystems,* from the background materials prepared for SCEP, working papers written during the Study, and a few selected articles that have been previously published.

These two volumes are intended to provide more detailed scientific and technical information on global environmental problems than could adequately be summarized in the SCEP Report. It is hoped that these volumes, which reproduce, supplement, and complement material in the SCEP Report, will serve as useful reference works to researchers and students in the many disciplines involved in solving global environmental problems.

The difficulties of style and continuity which arise in any edited book have been compounded in this volume by the interdisciplinary and diverse nature of the topics and by the very nature of the SCEP exercise. Despite these problems, we have compiled these volumes as quickly as possible for we feel that the information and insights which were brought to the Study and which grew out of it should be made widely available while they are timely.

This volume contains forty-eight papers of varying length, scope, depth, complexity, and style. Some are general reviews; others are detailed analyses of one component of one problem. Some were prepared during the Study, and others have been polished for previous publications. Despite these variations, we believe that this volume represents a unique collection of ideas, facts, techniques, wisdom, and judgment.

Such a collection could have been compiled only through the cooperation of virtually all federal agencies with responsibilities in these areas and the active participation of scientists and professionals from over a dozen universities and research institutions. To all these and many others, we owe sincere thanks. We also express our deep appreciation for the extraordinary assistance of Miss Ada Demb of M.I.T. in converting this collection of papers into a publishable manuscript.

The subject of this volume is the impact of man's activities on the climate. Even a general understanding of the nature and implications of that impact requires familiarity with the present state of knowledge about the climate, the atmospheric processes that produce climate and climate change, and the interaction of pollutants with these processes. It is also necessary to have an appreciation of the modeling and monitoring tools available for learning more about these areas. Finally, the concerned citizen must be able to look beyond the technical facts of any given situation and grasp the implications of actions that might be taken to ameliorate problems which are understood. The papers brought together here provide this broad overview.

This volume is divided into eleven parts, each preceded by an introduction to the papers in that section. These introductions serve to alert the reader to the content of the papers and to relate the papers to one another. Part I reproduces the SCEP Work Group reports dealing with climate and with atmospheric monitoring and also the summary of those reports. These present the distillation of the papers in this volume and of numerous other works by many of the participants of the month-long Study.

Part II provides a broad semitechnical view of the factors that are involved in determining climate and in changing climate, and also outlines how man might affect these processes through the introduction of pollutants. The understanding of these processes and interactions can be achieved only through mathematical modeling and through monitoring. The reader is introduced to the concepts of these two areas in Parts III and IV, respectively.

Parts V through IX are devoted to discussions of specific pollutants that may affect the climate. Each section treats the theoretical and empirical evidence available on predicted or observed effects and also includes discussions of the monitoring and measurement methods that can be used to increase knowledge in these areas and/or alert man to the impact which he is having on the climate. These five sections are entitled Carbon Dioxide and Atmospheric Heating, Particles and Turbidity, Particles and Clouds, Contaminants of the Upper Atmosphere, and A Nonproblem and a Potential Problem (oxygen depletion and clearing of the Amazon forest). Some monitoring techniques which are applicable to most of these problems are discussed in detail in Part X.

The final section of this volume, Part XI, contains several papers which illustrate the complex social and political issues which must be addressed if decisions which could avert potential disasters are ever to be made and implemented. These are issues which cannot simply be resolved by "more research"—the scientist and his fellow citizens in other fields must confront them together. We hope that this volume will contribute to those efforts.

William H. Matthews
William W. Kellogg
G. D. Robinson

Contributors

A. P. Altshuller, Director
DIVISION OF CHEMISTRY AND
PHYSICS
AIR POLLUTION CONTROL
OFFICE
ENVIRONMENTAL PROTECTION
AGENCY

Walter Bischof
INSTITUTE OF METEOROLOGY
UNIVERSITY OF STOCKHOLM
SWEDEN

Richard D. Cadle, Head
CHEMISTRY DEPARTMENT
NATIONAL CENTER FOR
ATMOSPHERIC RESEARCH

Richard A. Carpenter, Chief
ENVIRONMENTAL POLICY DIVISION
CONGRESSIONAL RESEARCH
SERVICE
LIBRARY OF CONGRESS

T. Carpenter
AIR RESOURCES LABORATORY
NATIONAL OCEANIC AND
ATMOSPHERIC ADMINISTRATION

Willard A. Crandall
CHIEF CHEMICAL ENGINEER
CONSOLIDATED EDISON COMPANY
OF NEW YORK, INC.

T. G. Dopplick
DEPARTMENT OF METEOROLOGY
MASSACHUSETTS INSTITUTE OF
TECHNOLOGY

A. J. Drummond
CHIEF SCIENTIST
THE EPPLEY LABORATORY, INC.

Seymour Edelberg
LINCOLN LABORATORY
MASSACHUSETTS INSTITUTE OF
TECHNOLOGY

Philip Hanst
NATIONAL AERONAUTICS AND
SPACE ADMINISTRATION

E. Hughes
ANALYTICAL CHEMISTRY DIVISION
NATIONAL BUREAU OF STANDARDS

Christian E. Junge, Director
MAX-PLANCK INSTITUT FÜR
CHEMIE
MAINZ, GERMANY

William W. Kellogg, Associate
Director
NATIONAL CENTER FOR
ATMOSPHERIC RESEARCH

Jules Lehmann
EARTH OBSERVATION PROGRAM
NATIONAL AERONAUTICS AND
SPACE ADMINISTRATION

Julius London
DEPARTMENT OF
ASTROGEOPHYSICS
UNIVERSITY OF COLORADO

Edward N. Lorenz
DEPARTMENT OF METEOROLOGY
MASSACHUSETTS INSTITUTE OF
TECHNOLOGY

John H. Ludwig
ASSISTANT COMMISSIONER FOR
SCIENCE AND TECHNOLOGY
AIR POLLUTION CONTROL OFFICE
ENVIRONMENTAL PROTECTION
AGENCY

Thomas B. McMullen
DIVISION OF AIR QUALITY AND
EMISSION DATA
AIR POLLUTION CONTROL OFFICE
ENVIRONMENTAL PROTECTION
AGENCY

Lester Machta, Director
AIR RESOURCES LABORATORY
NATIONAL OCEANIC AND
ATMOSPHERIC ADMINISTRATION

Syukuro Manabe
GEOPHYSICAL FLUID DYNAMICS
LABORATORY
NATIONAL OCEANIC AND
ATMOSPHERIC ADMINISTRATION

Edward A. Martell
NATIONAL CENTER FOR
ATMOSPHERIC RESEARCH

William H. Matthews
DEPARTMENT OF CIVIL
ENGINEERING
MASSACHUSETTS INSTITUTE OF
TECHNOLOGY

J. Murray Mitchell, Jr.
CLIMATIC CHANGE PROJECT
NATIONAL OCEANIC AND
ATMOSPHERIC ADMINISTRATION

George B. Morgan, Director
DIVISION OF AIR QUALITY
EMISSION DATA
AIR POLLUTION CONTROL OFFICE
ENVIRONMENTAL PROTECTION
AGENCY

Reginald E. Newell
DEPARTMENT OF METEOROLOGY
MASSACHUSETTS INSTITUTE OF
TECHNOLOGY

Guntis Ozolins
AIR POLLUTION CONTROL OFFICE
ENVIRONMENTAL PROTECTION
AGENCY

Hans A. Panofsky
PROFESSOR OF ATMOSPHERIC
SCIENCES
PENNSYLVANIA STATE UNIVERSITY

John B. Pate
NATIONAL CENTER FOR
ATMOSPHERIC RESEARCH

James T. Peterson
AIR POLLUTION CONTROL OFFICE
ENVIRONMENTAL PROTECTION
AGENCY

Henry Reichle
LANGLEY RESEARCH CENTER
NATIONAL AERONAUTICS AND
SPACE ADMINISTRATION

Roger Revelle, Director
CENTER FOR POPULATION STUDIES
HARVARD UNIVERSITY

G. D. Robinson
CENTER FOR THE ENVIRONMENT
AND MAN, INC.

Bernard E. Saltzman
DEPARTMENT OF
ENVIRONMENTAL HEALTH
UNIVERSITY OF CINCINNATI

T. Sasamori
NATIONAL CENTER FOR
ATMOSPHERIC RESEARCH

Vincent J. Schaefer
ATMOSPHERIC SCIENCES
RESEARCH CENTER
STATE UNIVERSITY OF
NEW YORK AT ALBANY

Joseph Smagorinsky
GEOPHYSICAL FLUID DYNAMICS
LABORATORY
NATIONAL OCEANIC AND
ATMOSPHERIC ADMINISTRATION

Walter O. Spofford, Jr.
QUALITY OF THE ENVIRONMENT
PROGRAM
RESOURCES FOR THE FUTURE, INC.

Morris Tepper
DEPUTY DIRECTOR OF EARTH
OBSERVATION PROGRAM AND
DIRECTOR OF METEOROLOGY
NATIONAL AERONAUTICS AND
SPACE ADMINISTRATION

Warren M. Washington
NATIONAL CENTER FOR
ATMOSPHERIC RESEARCH

The three papers in this section are products of the Study of Critical Environmental Problems (SCEP) which was sponsored by the Massachusetts Institute of Technology and conducted during July 1970 at Williamstown, Massachusetts. The papers are reprinted from the SCEP Report, *Man's Impact on the Global Environment,* which was published shortly after the Study by The M.I.T. Press.

SCEP was a one-month, interdisciplinary examination of the global, climatic, and ecological effects of man's activities. The disciplines represented by the more than fifty participants included meteorology, oceanography, ecology, chemistry, physics, biology, geology, engineering, economics, social sciences, and law. The participants were drawn from seventeen universities, thirteen federal agencies, three national laboratories, and eleven nonprofit and industrial corporations. The Study was supported professionally and/or financially by twelve federal agencies, four foundations, six industrial corporations, a research institution, and an academic institution.

The SCEP Report contains a summary of the major findings and recommendations of the Study and also the reports of the seven SCEP Work Groups. These Work Group reports were developed through intensive, full-time discussion and study by the group members. The report of the Work Group on Climatic Effects which appears in this section is the product of four weeks of effort. The participants were William W. Kellogg, National Center for Atmospheric Research, Chairman; Richard D. Cadle, National Center for Atmospheric Research; Robert G. Fleagle, University of Washington; Stanley Greenfield, RAND Corporation; Christian E. Junge, Max-Planck Institut Für Chemie; Charles D. Keeling, Scripps Institution of Oceanography; Julius London, University of Colorado; Lester Machta, National Oceanic and Atmospheric Administration; J. Murray Mitchell, Jr., National Oceanic and Atmospheric Administration; Reginald E. Newell, Massachusetts Institute of Technology; Hans A. Panofsky, Pennsylvania State University; James T. Peterson, Environmental Protection Agency; Walter Ramberg, Department of State; G. D. Robinson, Center for the Environment and Man, Inc.; Silvio G. Simplicio, Weather Bureau; Joseph Smagorinsky, National Oceanic and Atmospheric Administration; Howard J.

Taubenfeld, Southern Methodist University School of Law; Morris Tepper, National Aeronautics and Space Administration; F. Joachim Weyl, Hunter College of the City University of New York; and William G. Anderson, Harvard Law School, Rapporteur.

The report of the Work Group on Monitoring reprinted here (with sections on ecological monitoring omitted) is the result of three weeks of study. The members of this study were G. D. Robinson, Center for the Environment and Man, Inc., Chairman; A. P. Altshuller, Environmental Protection Agency; Richard D. Cadle, National Center for Atmospheric Research; Robert Citron, The Smithsonian Institution; Seymour Edelberg, Massachusetts Institute of Technology; Gifford Ewing, Woods Hole Oceanographic Institution; Dale W. Jenkins, The Smithsonian Institution; Jules Lehmann, National Aeronautics and Space Administration; Henry Reichle, National Aeronautics and Space Administration; Morris Tepper, National Aeronautics and Space Administration; and Jonathan Marks, Harvard Law School, Rapporteur.

The summary preceding these two reports is from the SCEP Report and presents a distillation of the major findings and recommendations contained in them.

1

Climatic Effects of Man's Activities

Summary of SCEP Report

Introduction

There is geological evidence that there have been five or six glacial periods (ice ages); the most recent (the Pleistocene) lasted 1 to 1.5 million years. In the past century there has been a general warming of the atmosphere of about 0.4°C up to 1940, followed by a few tenths degree cooling. It seems clear that our climate is subject to a wide variety of fluctuations, with periods ranging from decades to millennia, and that it is changing now.

We know that the atmosphere is a relatively stable system. The solar radiation that is absorbed by the planet and heats it must be almost exactly balanced by the emitted terrestrial infrared radiation that cools it; otherwise the mean temperature would change much more rapidly than just noted. This nearly perfect balance is the key to the changes that do occur, since a reduction of only about 2 percent in the available energy can, in theory, lower the mean temperature by 2°C and produce an ice age.

That there have not been wider fluctuations in climate is our best evidence that the complex system of ocean and air currents, evaporation and precipitation, surface and cloud reflection and absorption form a complex feedback system for keeping the global energy balance nearly constant. Nonetheless, the delicacy of this balance and the consequences of disturbing it make it very important that we attempt to assess the present and prospective impact of man's activities on this system.

The total mass of the atmosphere and the energy involved in even such a minor disturbance as a thunderstorm (releasing the energy equivalent to many hydrogen bombs) should convince us immediately that man cannot possibly hope to intervene in such a gigantic arena. However, in reality man does intervene, because he can—without intending to do so—reach some leverage points in the system.

All the important leverage points that this Study has identified control the radiation balance of the atmosphere in one way or another, and most of them control it by changing the compo-

Reprinted from Study of Critical Environmental Problems (SCEP), 1970. *Man's Impact on the Global Environment* (Cambridge, Massachusetts: The M.I.T. Press), pp. 10–19.

sition of the atmosphere. For example, man can change the temperature of the atmosphere by introducing a gas such as CO_2 or a cloud of particles that absorbs and emits solar and terrestrial infrared radiation, thereby altering the delicate balance we have described. He can also affect the heat balance by changing the face of the earth or by adding heat as a result of rising energy demands.

A thorough understanding and reliable prediction of the influence of atmospheric pollutants on climate requires the mathematical simulation of atmosphere-ocean systems, including the pollutants. At present, computer models successfully simulate many observed characteristics of the climate and have significantly advanced our knowledge of atmospheric phenomena. They have, however, a number of drawbacks that become serious when modeling new states of equilibrium or changes of climate in its transition toward these new states. Unless these limitations are overcome, it will be difficult, if not impossible, to predict inadvertent climate modifications that might be caused by man.

Recommendations

1.   We recommend that current computer models be improved by including more realistic simulations of clouds and air-sea interaction and that attempts be made to include particles when their properties become better known. Such models should be run for periods of at least several simulated years. The effects of potential global pollutants on the climate and on phenomena such as cloud formation should be studied with these models.

2.   We recommend that possibilities be investigated for simplifying existing models to provide a better understanding of climatic changes. Simultaneously, a search should be made for alternative types of models which are more suitable for handling problems of climatic change.

Carbon Dioxide from Fossil Fuels

All combustion of fossil fuels produces carbon dioxide (CO_2), which has been steadily increasing in the atmosphere at 0.2 percent per year since 1958. Half of the amount man puts into the atmosphere stays and produces this rise in concentration. The other half goes into the biosphere and the oceans, but we are not certain how it is divided between these two reservoirs. CO_2 from fossil fuels is a small part of the natural CO_2 that is constantly being exchanged between the atmosphere/oceans and the atmosphere/ forests.

A projected 18 percent increase resulting from fossil fuel combustion to the year 2000 (from 320 ppm to 379 ppm) might increase the surface temperature of the earth 0.5°C; a doubling of the CO_2 might increase mean annual surface temperatures 2°C. This latter change could lead to long-term warming of the planet. These estimates are based on a relatively primitive computer model, with no consideration of important motions in the atmosphere, and hence are very uncertain. However, these are the only estimates available today.

Should man ever be compelled to stop producing CO_2, no coal, oil, or gas could be burned and all industrial societies would be drastically affected. The only possible alternative for energy for industrial and commercial use is nuclear energy, whose by-products may also cause serious environmental effects. There are at present no electric motor vehicles that could be used on the wide scale our society demands.

Although we conclude that the probability of direct climate change in this century resulting from CO_2 is small, we stress that the long-term potential consequences of CO_2 effects on the climate or of societal reaction to such threats are so serious that much more must be learned about future trends of climate change. Only through these measures can societies hope to have time to adjust to changes that may ultimately be necessary.

Recommendations
1.   We recommend the improvement of present estimates of future combustion of fossil fuels and the resulting emissions.
2.   We recommend study of changes in the mass of living matter and decaying products.
3.   We recommend continuous measurement and study of the carbon dioxide content of the atmosphere in a few areas remote from known sources for the purpose of determining trends. Specifically, four stations and some aircraft flights are required.

4.   We recommend systematic scientific study of the partition of carbon dioxide among the atmosphere, the oceans, and the biomass. Such research might require up to 12 stations.

Particles in the Atmosphere
Fine particles change the heat balance of the earth because they both reflect and absorb radiation from the sun and the earth. Large amounts of such particles enter the troposphere (the zone up to about 12 km or 40,000 feet) from natural sources such as sea spray, windblown dust, volcanoes, and from the conversion of naturally occurring gases into particles.

Man introduces fewer particles into the atmosphere than enter from natural sources; however, he does introduce significant quantities of sulfates, nitrates, and hydrocarbons. The largest single artificial source is the production of sulfur dioxide from the burning of fossil fuel that subsequently is converted to sulfates by oxidation. Particle levels have been increasing over the years as observed at stations in Europe, North America, and the North Atlantic but not over the Central Pacific.

In the troposphere, the residence times of particles range from 6 days to 2 weeks, but in the lower stratosphere micron-size particles or smaller may remain for 1 to 3 years. This long residence time in the stratosphere and also the photochemical processes occurring there make the stratosphere more sensitive to injection of particles than the troposphere.

Particles in the troposphere can produce changes in the earth's reflectivity, cloud reflectivity, and cloud formation. The magnitudes of these effects are unknown, and in general it is not possible to determine whether such changes would result in a warming or cooling of the earth's surface. The area of greatest uncertainty in connection with the effects of particles on the heat balance of the atmosphere is our current lack of knowledge of their optical properties in scattering or absorbing radiation from the sun or the earth.

Particles also act as nuclei for condensation or freezing of water vapor. Precipitation processes can certainly be affected by changing nuclei concentrations, but we do not believe that the effect of man-made nuclei will be significant on a global scale.

Recommendations

1.   We recommend studies to determine optical properties of fine particles, their sources, transport processes, nature, size distributions, and concentrations in both the troposphere and stratosphere, and their effects on cloud reflectivity.

2.   We recommend that the effects of particles on radiative transfer be studied and that the results be incorporated in mathematical models to determine the influence of particles on planetary circulation patterns.

3.   We recommend extending and improving solar radiation measurements.

4.   We recommend beginning measurements by lidar (optical radar) methods of the vertical distribution of particles in the atmosphere.

5.   We recommend the study of the scientific and economic feasibility of initiating satellite measurements of the albedo (reflectivity) of the whole earth, capable of detecting trends of the order of 1 percent per 10 years.

6.   We recommend beginning a continuing survey, with ground and aircraft sampling, of the atmosphere's content of particles and of those trace gases that form particles by chemical reactions in the atmosphere. For relatively long-lived constituents about 10 fixed stations will be required, for short-lived constituents, about 100.

7.   We recommend monitoring several specific particles and gases by chemical means. About 100 measurement sites will be required.

The Role of Clouds

The importance of clouds in the atmosphere stems from their relatively high reflectivity for solar radiation and their central role in the various process involved in the heat budget of the earth-atmosphere system.

Recommendations

1.   We recommend that there be global observations of cloud distribution and temporal variations. High spatial resolution satellite observations are required to give "correct" cloud population

counts and to establish the existence of long-term trends in cloud-iness (if there are any).

2.   We recommend studies of the optical (visible and infrared) properties of clouds as functions of the various relevant cloud and impinging radiation parameters. These studies should include the effect of particles on the reflectivity of clouds and a determination of the infrared "blackness" of clouds.

Cirrus Clouds from Jet Aircraft

Contrail (condensation trail) formation, which is common near the world's air routes, is more likely to occur when jets fly in the upper troposphere than in the lower troposphere because of the different meteorological conditions in these two regions.

There are very few, if any, statistics that permit us to determine whether the advent of commercial jet aircraft has altered the frequency of occurrence or the properties of cirrus clouds. We do not know whether the projected increase in the operation of subsonic jets will have any climate effects.

Two weather effects from enhanced cirrus cloudiness are possible. First, the radiation balance may be slightly upset, and, second, cloud seeding by falling ice crystals might initiate precipitation sooner than it would otherwise occur.

Recommendations

1.   We recommend that the magnitude and distribution of increased cirrus cloudiness from subsonic jet operations in the upper troposphere be determined. A study of the phenomenon should be conducted by examining cloud observations at many weather stations, both near and remote from air routes.

2.   We recommend that the radiative properties of representative contrails and contrail-produced cirrus clouds be determined.

3.   We recommend that the significance, if any, of ice crystals falling from contrail clouds as a source of freezing nuclei for lower clouds be determined.

Supersonic Transports (SSTs) in the Stratosphere

The stratosphere where SSTs will fly at 20 km (65,000 feet) is a very rarefied region with little vertical mixing. Gases and particles produced by jet exhausts may remain for 1 to 3 years before disappearing.

We have estimated the steady-state amounts of combustion products that would be introduced into the stratosphere by the Federal Aviation Agency projection of 500 SSTs operating in 1985–1990 mostly in the Northern Hemisphere, flying 7 hours a day, at 20 km (65,000 feet), at a speed of Mach 2.7, propelled by 1,700 engines like the GE-4 being developed for the Boeing 2707-300. We have used General Electric (GE) calculations of the amount of combustion products because no test measurements exist. In our calculations we used jet fuel of 0.05 percent sulfur. We have been told that a specification of 0.01 percent sulfur could be met in the future at higher cost.

We have compared the amounts that would be introduced on a steady state basis with the natural levels of water vapor, sulfates, nitrates, hydrocarbons, and soot in the stratosphere. We have also compared these levels with the amounts of particles put into the atmosphere by the volcano eruption of Mount Agung in Bali in 1963.

Based on these calculations, we have concluded that no problems should arise from the introduction of carbon dioxide and that the reduction of ozone due to interaction with water vapor or other exhaust gases should be insignificant. Global water vapor in the stratosphere may increase 10 percent, and increases in regions of dense traffic may be 60 percent.

Very little is known about the way particles will form from SST-exhaust products. Depending upon the actual particle formation, particles from these 500 SSTs (from SO_2, hydrocarbons, and soot) could double the pre-Agung eruption global averages and peak at ten times those levels where there is dense traffic. The effects of these particles could range from a small, widespread, continuous "Agung effect" to one as big as that which followed the Agung eruption. (The analogy between the SST input and that by the Mount Agung eruption is not exact.) The temperature of the equatorial stratosphere (a belt around the earth) increased 6° to 7°C after the eruption and remained at 2° to 3°C above the pre-Agung level for several years. No apparent temperature change was found in the lower troposphere.

Clouds are known to form in the winter polar stratosphere.

Two factors will increase the future likelihood of greater cloudiness in the stratosphere because of moisture added by the SSTs: the increased stratospheric cooling due to the increasing CO_2 content of the atmosphere and the closer approach to saturation indicated by the observed increase of stratospheric moisture. Such an increase in cloudiness could affect the climate. The introduction of particles into the stratosphere could also produce climatic effects by increasing temperatures in the stratosphere, with possible changes in surface temperatures.

A feeling of genuine concern has emerged from these conclusions. The projected SSTs can have a clearly measurable effect in a large region of the world and quite possibly on a global scale. We must, however, emphasize that we cannot be certain about the magnitude of the various consequences.

Recommendations

1.   We recommend that uncertainties about SST contamination and its effects be resolved before large-scale operation of SSTs begins.

2.   We recommend that the following program of action be initiated as soon as possible:

a. Begin now to monitor the lower stratosphere for water vapor, cloudiness, oxides of nitrogen and sulfur, hydrocarbons, and particles (including the latter's composition and size distribution).

b. Determine whether additional cloudiness or persistent contrails will occur in the stratosphere as a result of SST operations, particularly in certain cold areas, *and* the consequences of such changes.

c. Obtain better estimates of contaminant emissions, especially those leading to particles, under simulated flight conditions and under real flight conditions, at the earliest opportunity.

d. Using the data obtained in carrying out the preceding three recommendations, estimate the change in particle concentration in the stratosphere attributable to future SSTs *and* its impact on weather and climate.

3.   We recommend implementation now of a special monitoring program for the lower stratosphere (about 20 km or 60,000 to 70,000 feet) to include the following activities:

a. Measurement by aircraft and balloon of the water vapor content of the lower stratosphere. The area coverage required is global, but with special emphasis on areas where it is proposed that the SST should fly.

b. Sampling by aircraft of stratospheric particles, with subsequent physical and chemical analysis.

c. Monitoring by lidar (optical radar) of optical scattering in the lower stratosphere, again with emphasis on the region in which heavy traffic is planned.

d. Monitoring of tropospheric carbon monoxide concentration because of its potential effects on the chemical composition of the lower stratosphere.

Atmospheric Oxygen: Nonproblem

Atmospheric oxygen is practically constant. It varies neither over time (since 1910) nor regionally and is always very close to 20.946 percent. Calculations show that depletion of oxygen by burning all the recoverable fossil fuels in the world would reduce it only to 20.800 percent. It should probably be measured every 10 years to be certain that it is remaining constant.

Surface Changes and the Climate

The most important properties of the earth's surface that have a bearing on climate and are likely to be affected by human activity are reflectivity, heat capacity and conductivity, availability of water and dust, aerodynamic roughness, emissivity in the infrared band, and heat released to the ground.

Since the amount of carbon dioxide in the atmosphere is dependent on the biomass of forest lands which serves as a reservoir, widespread destruction of forests could have serious climatic effects. Population growth or overgrazing that increases the arid or desert areas of the earth creates conditions that allow the introduction of dust particles to the atmosphere.

Other important surface changes are from man's activities that modify snow and ice cover, particularly in polar regions, and from some possible projects involving the production of new, very large water bodies. Increased urbanization is of possible global importance only as it produces extended areas of contiguous cities. Still, it is not certain whether effects of urbanization extend far beyond the general region occupied by the cities.

Recommendation

We recommend that before actions are taken which result in some of the very extensive surface changes described mathematical models be constructed which simulate their effects on the climate of a region or, possibly, of the earth.

Thermal Pollution

Although by the year 2000 global thermal power output may be as much as six times the present level, we do not expect it to affect global climate. Over cities it does already create "heat islands," and as these grow larger they may have regional climatic effects. We recommend that these potential effects be studied with computer models.

2

Work Group on Climatic   SCEP Report
Effects

Introduction

Our Frame of Reference

We have chosen to restrict ourselves as far as possible to atmospheric problems that are global in scale and critical in the sense that they may affect the environment. We have emphasized those aspects where man's activities seem to have an influence on the atmosphere, but we are keenly aware that one cannot judge this impact without an understanding of the natural background or the competition. Thus, we have dealt with a wide variety of atmospheric features and their man-made alterations, but we have had to ignore some also, such as local weather modification and urban air pollution.

There is another self-imposed limitation to our frame of reference, in that we have not considered any ecological effects of atmospheric change or pollutants. Actually, we are not aware of any global effects of air pollution on living things, though of course the air is a carrier of some persistent compounds that can build up to harmful levels in the oceans. This was of great concern to the Work Group 2, which dealt with the matter of accumulation of persistent pesticides and heavy metals.

Perhaps the most substantial contribution we can make to the matter of global atmospheric changes by man is the combination of meteorological, chemical, and economic expertise. The atmospheric scientists can, for the first time in some instances, work with quantitative inputs concerning the amounts and distributions of man-made contaminants of various kinds and can weigh these against natural sources and sinks. Even though we cannot trace all the implications now, the material will be available and in a form such that we and others can continue to work with it.

We have organized our report in such a way that it deals first with the way various contaminants are introduced, carried, dispersed, and removed from the atmosphere; then we deal with

Reprinted from Study of Critical Environmental Problems (SCEP), 1970. *Man's Impact on the Global Environment* (Cambridge, Massachusetts: The M.I.T. Press), pp. 40–112.

changes that man may produce through his use of the earth's solid and watery surface; and finally we deal with the effects of each of these elements on the climate.

The Atmosphere

Planet Earth has a compressible atmosphere made up of nitrogen (N_2) and oxygen (O_2) in amounts that stay remarkably constant in time. In addition, there are a number of other minor gases that are reasonably constant atmospheric constituents. Also present, however, are gases found in relatively small and in some cases highly variable concentrations such as water vapor, carbon dioxide, and ozone, each of which plays a significant role in determining the temperature structure and the energy transfer of the atmosphere.

It is important that water can exist in all three phases (ice, liquid, vapor) in the earth-atmosphere system—and that transformations take place among all three phases. Some of these changes, such as condensation from water vapor to liquid in the formation of clouds, involve the release of latent heat and therefore represent an important physical link in the energy cycle of the atmosphere.

The amount of water vapor present in the atmosphere depends on many factors and in the lower atmosphere is highly variable. In some cases, as over tropical areas, the concentration relative to air (mixing ratio) can be as high as 0.04 by mass. Water vapor mixing ratio decreases with height to a constant value in the stratosphere of about 3×10^{-6} (3 ppm by mass).

The concentration of carbon dioxide is nearly constant in the troposphere and stratosphere at a mixing ratio of about 320 ppm by volume, whereas ozone exists in relatively small amounts in the troposphere (except for very local areas) and has its maximum concentration at about 25 km. Its mixing ratio at that altitude can be of the order of 10 ppm by volume.

In addition to these gases, the atmosphere acquires from natural and man-made sources a number of variable trace gases (such as SO_2, oxides of nitrogen) and solid particles such as dust, sea spray, and sulfates. The latter are important in the atmosphere, primarily because they provide nuclei for condensation and freezing but also because they are involved with radiative scattering and absorption processes and therefore have an influence on the temperature.

The temperature distribution in the atmosphere is the result of many interacting processes involving solar and terrestrial radiation and motions within the atmosphere itself. On the average the earth receives solar radiation in the amount of about 2 cal cm^{-2} min^{-1} at the top of the atmosphere on a unit surface perpendicular to the sun's rays (or about 0.5 cal cm^{-2} min^{-1} when spread over the entire globe. Although it is not certain, it is believed that there is very little (if any) variation in this "solar constant."

As the solar beam passes through the high atmosphere, a very small amount is absorbed. In the stratosphere, ozone absorbs about 3 percent of the incoming solar radiation, and this absorption is responsible for the relatively high temperature found in the stratospheric region from 25 to 50 km. As the solar beam penetrates the lower atmosphere, it is reduced still further by absorption (mainly by water vapor) and by back reflection (mainly from clouds, but also from air molecules and large dust-type particles). The average total absorption of solar radiation by atmospheric gases, clouds, and dust amounts to a little more than 20 percent of the incoming solar radiation.

The average reflectivity (planetary albedo) of the earth-atmosphere system is approximately 30 to 35 percent, with the clouds being responsible for over three-fourths of this amount. Thus, just under 50 percent of the incoming radiation is absorbed at the earth's surface. It is important to note, however, that there are large variations around the earth in this absorption, due to the varying nature of the underlying surface. In the polar regions, for instance, where there is ice and snow, the surface reflectivity is very high, and little of the solar energy received at the surface is absorbed. On the other hand, in the tropics, where the oceans represent a large part of the surface area, the reflectivity is relatively low, and the absorbed energy is used for the most part in evaporating water from the ocean surface.

The earth emits infrared radiation. This radiation is almost completely absorbed by the principal polyatomic gases in the atmosphere (H_2O, CO_2, and O_3), which in turn reemit the radiation both upward (eventually out to space) and downward (back to the ground). Clouds absorb and emit infrared radiation also and, in general, do so as black bodies (that is, very efficiently). The amount of radiation returned to the ground, therefore, depends principally on the water vapor content, carbon dioxide content, and cloudiness of the atmosphere. Changes in these variables would, of course, affect the radiation budget of the atmosphere.

On the average the atmosphere continuously loses energy as a net result of these radiative processes. The energy is replenished in two ways. First, there is conduction of heat from the ground to the atmosphere. Second, and more important, solar radiation absorbed at the ocean surface evaporates water, and, when this water vapor finally condenses in the atmosphere, its latent heat is released to the air. Approximately two-thirds of the energy transfer from the surface to the air is accomplished by this latter mechanism.

Since incoming solar radiation on a unit horizontal surface is more direct in equatorial than in polar regions, more than twice as much solar radiation on an annual average is available at the equator as is available at the poles. Outgoing radiation depends strongly on the temperature of the radiating substance, but, since the temperature in the atmosphere does not vary much (on an absolute scale) from equator to pole, there is only a slight decrease with latitude of emitted radiation. As a result there is an excess of incoming solar radiation over emitted infrared radiation in equatorial regions and a deficit in polar regions. If the earth did not rotate, this would be somewhat analogous to heating a fish tank at one end and cooling it at the other. In the fish tank this would give rise to a direct circulation of the fluid—rising at the heated end, flowing toward the cooled end, descending and returning to the heated end.

The large-scale circulation of the real atmosphere, however, although forced initially in response to the unequal geographic heating and cooling, is much more complicated, due primarily to the effect of the earth's rotation. There are many additional perturbing factors affecting the circulation, among which are the uneven distribution of land and water and the roughness of the ground.

One obvious consequence of large-scale motions for the general structure of the atmosphere is that the average latitudinal variation of temperature that would obtain, if the temperature distribution resulted solely from a balance of radiative factors, is decreased. Atmospheric motions are also instrumental in adjusting the average vertical temperature distribution and, together with radiative processes, help to produce the different characteristics of the troposphere and stratosphere. The qualitative difference in lapse rate (vertical temperature change) between troposphere (decreasing temperature with height) and stratosphere (constant or increasing temperature with height) and the action of scavenging processes by rain and snow lead to an important difference in the residence times of various atmospheric constituents in these two regions. Gases and particles in the troposphere are fairly well mixed vertically within periods of a few days to about a month and at the same time are removed by scavenging and by direct contact with the surface. In the lower part of the stratosphere, however, their residence times are of the order of two years.

The net result of the various highly interactive processes and features of the atmosphere is a long-term average weather and climate pattern which has some general features that are reasonably well understood, for example, the mean latitudinal and vertical temperature distribution, the existence of trade winds in the ocean areas of the tropics, and so forth. Many details of these patterns, however, involving both subtle and very obvious variations from year to year in circulation, temperature, and rainfall, cannot be satisfactorily explained at present. These irregular variations exist even with the basic features of the earth-atmosphere system and of the sun unchanged. But if some of these features were to change, the statistical properties of the atmosphere would probably undergo additional variations.

We know that in the earth's history the climate has changed on many times scales, and many different theories have been invoked to explain these changes. These theories involve variations of the sun's radiation, the earth's orbit, the earth's surface, the composition of the atmosphere, or the circulation of the oceans (Mitchell, 1965, 1968; Sellers, 1965). Some of these variations occur naturally, as, for example, changes in the earth's orbital elements (which can be calculated and forecast quite precisely) and changes in the number of particles due to volcanic eruptions (which cannot at present be forecast). Other variations, so far fairly modest, are the result of man's activity; for example, the recent change in the concentration of carbon dioxide. One thing appears clear: climate has changed over decades, centuries, and millennia without the influence of human activity.

Between 1900 and approximately 1945, global temperatures rose at a rate of about 0.8°C per century, and glaciers receded. Since then a cooling trend has set in, at a somewhat slower rate than the previous warming. There is a tendency to blame these changes on some effect of increasing population and industrialization, but the magnitudes of those climatic changes are no larger than earlier changes that cannot be attributed to man. This does not mean that man's activities may not have influenced climatic changes during the twentieth century; merely that we do not know how the climate would have changed if man had not been present. Man-made climatic changes could not have been larger than natural changes—and they may have been a good deal smaller.

Still, we suspect that this situation may not last; eventually man can become so numerous and powerful that his activities will change the composition of the atmosphere or the character of the earth's surface enough to produce effects larger than those "naturally" expected. Ice advances and retreats, widespread droughts, changes of the ocean level, and so forth, were accompanied by only slight shifts in the mean circulation pattern and only small changes in the average temperature over large parts of the earth. Year-to-year variations in the rainfall, temperature, and circulation characteristics of the atmosphere are much larger than the mean atmospheric properties that are found when the climate changes. It is likely that there exists in the atmosphere an almost continuous set of possible climatic regimes; that a small change in a mean condition of the atmosphere may be accomplished by a relatively small perturbation in one or more of the relevant parameters; and that small changes in the mean conditions of the atmosphere or the surface can then produce a change from one climatic regime to another (see, for example, Lorenz, 1970).

This thought will emerge repeatedly in the sections that follow, as we try to assess the influence of man on the climate. We will be seeking to identify those "leverage points" that man can reach, points where his relatively subtle alterations of the environment could influence significantly the global climate. It is in the interest of a rational society to be on the lookout for any such changes and to develop theories of atmospheric behavior sufficient to allow us to forecast the atmosphere's future course, given a knowledge of what man will be doing. The effort expended will certainly be trivial compared to the possible return (Hess, 1959; Lamb, 1966; Fletcher, 1969; Flohn, 1969).

Man-Made Atmospheric Contaminants

When we consider the ways that man can modify the envelope of air that covers his planet, his introduction to it of contaminants in ever greater amounts comes to mind first. Therefore, our first purpose will be to point to the magnitudes of the changes in the atmosphere due to man's contaminants (including waste heat) and to say as much as we can about trends and future levels of these contaminants.

Carbon Dioxide and Other Trace Gases
That May Affect Climate

CARBON DIOXIDE

ATMOSPHERIC CONTENT OF CO_2

Carbon dioxide (CO_2) is a trace gas in the earth's atmosphere with a concentration of a little over 0.03 percent by volume (about 320 ppm). In spite of its relatively small concentration it plays an important role in determining the temperature of the planet. It absorbs sunlight to a modest degree; more importantly, it absorbs (and emits) infrared radiation. By intercepting a part of the infrared radiation that is emitted by the earth's surface and reradiating it back toward the earth, it cuts down the rate of surface cooling, and at the same time it acts to cool the upper atmosphere.

It is this influence on the heat balance of the earth-atmosphere system that arouses our concern over any change in its concentration. The idea that climatic change could result from changes in atmospheric CO_2 content was suggested independently by the American geologist T. C. Chamberlain in 1899, and the Swedish chemist S. Arrhenius in 1903. It was rather easy to show that the increasing production of CO_2 by the burning of fossil fuels (coal, oil, and natural gas) *ought* to influence the amount of CO_2 in the air. Nevertheless, it was not until the period of the

International Geophysical Year in 1958 that the first systematic and accurate observations of its concentration were begun.

There are now observations, some for only limited periods, from Swedish aircraft (Bolin and Bischof, 1969), Point Barrow (Kelley, 1969), the Antarctic (Brown and Keeling, 1965), and Mauna Loa (Pales and Keeling, 1965), the latter being the most continuous and frequent. Figure 2.1 shows the trends of yearly mean values of CO_2 concentration at these four places, and Figure 2.2 shows the changes in monthly mean values for Mauna Loa.

From these observations the following generalizations can be made:

1. CO_2 seems to have been increasing throughout the world at about 0.2 percent per year, or 0.7 ppm out of 320 ppm. The dashed "best fit" curve in Figure 2.2 shows that this rate is not constant, but the length of record is too short to place much emphasis on the deviations from a linear trend.

2.   There is somewhat more CO_2 at the most northerly station and the least is at the Antarctic station, but the total difference

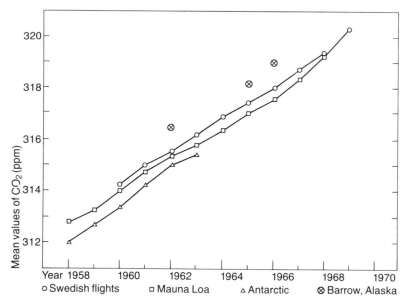

Figure 2.1 Annual mean values of CO_2

Sources: Swedish flights (Bolin and Bischof, 1969); Mauna Loa (Pales and Keeling, 1965, Bainbridge, 1970); Antarctic (Brown and Keeling, 1965); Barrow, Alaska (Kelley, 1969).

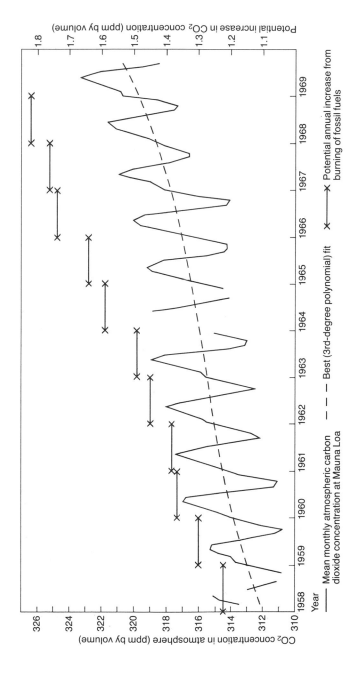

Figure 2.2 CO₂ concentration from burning of fossil fuels

Sources: Monthly (Pales and Keeling, 1965), (Bainbridge, 1970); Best fit (Cotten, 1970); Annual increase (United Nations, *World Energy Supplies*).

is only about 1 ppm, corresponding to a lag from north polar regions to south polar regions of about 18 months.

3.   In the Northern Hemisphere there is a distinct seasonal oscillation in CO_2, with an amplitude at the surface at high latitude in the Northern Hemisphere of about 9 ppm and at Mauna Loa (altitude 3,398 meters, latitude 19°34'N) of 6 or 7 ppm. This oscillation decreases in amplitude with increasing altitude, according to the Swedish observations with aircraft.

4.   It is estimated that, in the interval 1850–1950, fossil fuel consumption produced an amount of CO_2 equal to 10 percent of the amount estimated to be in the atmosphere in 1950 (United Nations, 1956). Analysis of tree rings and other biological specimens of known age show that there was a 1 to 2 percent decrease of radioactive carbon-14 (a 1 to 2 percent "Suess effect") over the same period (Suess, 1965).

These are the main facts regarding the changing atmospheric CO_2 content. What are some of the related facts that help to explain this change that has been taking place? And what is our best guess about the future? The picture is far from clear, but we do have a fairly good idea of what is happening—even if we sometimes have difficulty assigning the right numbers.

In 1950, the concentration of CO_2 in the atmosphere was probably a little over 300 parts per million (ppm) by volume, or 445 ppm by mass. (It was not measured at that time, so this is an estimate.) Since the mass of the entire atmosphere is 5.14×10^{21} grams, there was about 2.3×10^{18} grams of CO_2 in the atmosphere. In that year about 0.67×10^{16} grams of CO_2 were released into the atmosphere by burning fossil fuel, and this was 0.28 percent of the amount already in the atmosphere in 1950 (President's Science Advisory Committee [PSAC], 1965). The amounts added each year have been increasing ever since, and are expected to rise further. The amounts are shown in abbreviated form in Table 2.1.

RESERVOIRS FOR CO_2

In order to predict the future CO_2 concentration, we could, as some have done, assume that the fraction of CO_2 produced by combustion of fossil fuels that has stayed in the atmosphere in the past decade (about 50 percent) will remain the same in the future. This is certainly a rational first assumption, but we have

Table 2.1 Carbon Dioxide Production and Accumulation

Year	Amount Added from Fossil Fuel (Mt/yr)[a]	Cumulative Amount Added over Previous Decade[a]	Concentration by Volume (ppm)[b]	Total Amount in Atmosphere (Mt)[c]
1950	6,700	52,200	306	2.39×10^6
1960	10,800	82,400	313	2.44×10^6

Note: The percentage of CO_2 remaining in the atmosphere of that introduced in 1960 can be calculated from the data in this table and by noting from Figure 2.1 that the increase of CO_2 concentration by volume was approximately 0.7 ppm. This corresponds to 5,450 Mt increase in 1960 (0.7 times 2.44×10^6 divided by 313). The first column notes that 10,800 Mt were introduced in 1960, thus the percentage increase was 51.

[a] Source: President's Science Advisory Committee (PSAC), 1965.

[b] The 1960 number is the average of the observed values in Figure 2.1. The 1950 value was obtained by a linear extrapolation of the curves in Figure 2.1.

[c] This column is computed from data in the preceding column by taking the mass of the atmosphere to be 5.14×10^{21} grams. Note that 1 mole of air weighs 29 grams.

the uneasy feeling that the real world is not likely to remain so simple and linear. The other 50 percent had to go into some other reservoirs, and this leads us to inquire about them and how they are likely to behave.

The two reservoirs that must be involved are, first, the oceans, and second, what we will call the biosphere, the mass of living things and nonliving organic matter on land and in the oceans, of which the forests represent the largest portion by far.

It is difficult to estimate the total mass of living things and the organic material on which they feed. Attempts to do this have arrived at numbers that are roughly one to two times the amount of carbon in atmospheric CO_2 (2.5×10^6 megatons [Mt] of CO_2 or 0.68×10^6 Mt of carbon). The biosphere would respond to an increase in atmospheric CO_2 by growing faster, since a 10 percent increase in CO_2 corresponds to an increase in the rate of photosynthesis (if living things were not nutrient or water limited) of 5 to 8 percent (Keeling, 1970). At the same time, man, by cutting down forests and removing the wood from the carbon cycle, can decrease the mass of the biosphere. Actually, however, we do not know quantitatively how the biosphere responds to a change in atmospheric CO_2 or to man's temporary removal of part of it.

Turning to the oceans, it must be made clear at this point that, when we speak of the ocean reservoir, we are referring to that part of the ocean that can exchange CO_2 with the atmosphere and remain more or less in equilibrium with the atmosphere, in the course of a few years. This must be the surface layer that mixes fairly rapidly. The total ocean contains nearly 60 times as much CO_2 as the atmosphere, but the time needed to exchange that part of the water below 1 km with the surface water is probably 500 to 1,000 years, perhaps longer (Keeling, 1970). Thus, except for a few limited parts of the ocean, such as a limited portion of the North Atlantic and the Weddell Sea, where there can be an overturning of the deep waters, we are only concerned with the first few hundred meters of near surface water.

There are two kinds of oceanic variations that could have an effect on atmospheric CO_2. The overturning of the deep waters just referred to, which would bring CO_2-rich water to the surface, could release large amounts of CO_2 in a short time. Furthermore, a change of the surface water temperature over a large area would influence the CO_2 balance between ocean and atmosphere, because an increase in water temperature lowers the ability of water to hold CO_2. Therefore, though we have not taken these factors into account, changes in the ocean could have an important bearing on future CO_2 concentrations.

The inorganic carbon in ocean water is mostly in the form of carbonate and bicarbonate, balanced against the available metallic cations, but about 1 percent is in solution as dissolved CO_2. Taking into account the thermodynamic relationships between the chemical species, we can calculate that, although a layer of ocean water 60 m deep contains as much inorganic carbon as the atmosphere, the layer will take up only one-tenth as much CO_2 as the atmosphere will if a process such as fossil fuel combustion injects more CO_2 into the air. Based on evidence from the rate of mixing of radioactive tracers from nuclear tests, we expect between 200 and 400 m of ocean water to mix with surface water in a period of a few years, but our knowledge of this kind of mixing is very sketchy indeed, and the process must vary widely in different parts of the ocean.

This limited ability of the oceans to take up additional CO_2 in a short time, and the apparent inability of the biosphere to increase its uptake by more than a few percent, accounts for the fact that about half of the added CO_2 from fossil fuels has remained in the atmosphere (see notes on Table 2.1 and Table 2.2). This does not tell us what the ultimate capacity of the ocean and biosphere reservoirs is, but merely that together they respond by taking up about half of each new injection into the atmosphere.

In order to estimate the size of these reservoirs, it is necessary to consider the Suess effect. In general terms this effect is the percent dilution of natural carbon-14 caused by the release (as CO_2) of carbon-12 and carbon-13 to the atmosphere from the combustion of fossil fuels. Fossil fuels are very low in radioactive carbon-14, since the half-life of carbon-14 is about 6,000 years. As we have pointed out, during the century prior to 1950 the estimated additional CO_2 available to the atmosphere (based on fossil fuel usage) was 10 percent of the amount present in the atmosphere in 1950, and the observed decrease in carbon-14 was 1 to 2 percent. Taking into account that the ocean and biosphere reservoirs, plus the atmosphere, share in the process of diluting the carbon-14, the conclusion is reached that a 1 percent Suess effect implies that the total reservoir (ocean, biosphere, and atmosphere) is 10 percent/1 percent or 10 times larger than the atmospheric content alone and that a 2 percent Suess effect implies a total reservoir that is 5 times larger. This gives us an indication of the size of the ocean-plus-biosphere reservoir for $C^{14}O_2$, relative to the atmosphere, but does not determine the "sizes" of the ocean and the biosphere relative to each other.

We must emphasize that this conclusion only describes the size of the reservoir for $C^{14}O_2$ and that our concern is with the size of the reservoir for atmospheric CO_2 (of which $C^{14}O_2$ is but a very small fraction). Assuming that the portion of the biosphere that exchanges $C^{14}O_2$ in a few years time is about the same size as the atmospheric reservoir (recall the estimates of the mass of the biosphere in terms of the mass of atmospheric carbon), the biosphere plus atmosphere would represent twice the atmospheric reservoir, and the size of the oceanic reservoir, as deduced for a 2 to 1 percent Suess effect, would be three to eight times larger, respectively. Such an ocean reservoir would be equivalent to a layer of ocean water 180 to 480 m in average depth. This agrees with the statement made earlier that 60 m of ocean surface water will take up one-tenth as much injected CO_2 as will the atmosphere. We may conclude that the accessible ocean reservoir for atmospheric CO_2 is 0.3 to 0.8 times that of the atmosphere itself. The previous argument about mixing times in the surface of the ocean suggests that the smaller depth would be more likely but does not prove conclusively that the larger depth is unreasonable.

FUTURE TRENDS

If the ocean reservoir is dominant, then it will take up its fraction of the added CO_2 in a regular way, in proportion to the partial pressure of atmospheric CO_2, and it will store it indefinitely. Further, since the mixed layer is slowly exchanged with the deep ocean water over a period of decades, the ocean will take up still more. If, on the other hand, the land plants of the biosphere are the main reservoir, we should expect to see the CO_2 partly returned to the atmosphere within a few decades, as the previously added growth of trees falls and begins to rot. The biosphere can take up the increased CO_2, but only temporarily. The trend toward depleting the major remaining stands of virgin forests, such as those in tropical Brazil, Indonesia, and the Congo, will further reduce this possible reservoir and may release even more CO_2 to the atmosphere.

A model has been devised by Machta, Machta, and Olson, at this Study, that takes into account the best estimate of the ocean and biosphere capacities to take up additional CO_2. The model, which resembles others that have been used in this area, was used to estimate the buildup of CO_2 until the year 2000. Its results are shown in Table 2.2.

One further point should be made concerning the longer view (beyond the year 2000). If by the year 2000 we have consumed from 2 to 12 percent of the total recoverable fossil fuel reserves of the globe and the remainder were burned, then 8 to 50 times the cumulative amount released before the year 2000 would go into the atmosphere. If half of that remained in the atmosphere, it would be within man's power in the next century to increase the CO_2 of the atmosphere by a factor of 4 or more. The fact that it is possible leads us to urge that its implications be investigated.

OTHER TRACE GASES

The list of trace gases that exist in the atmosphere is very long and includes ozone and water vapor (which man does not produce enough of to compete with nature globally); methane, carbon

Table 2.2 Possible Atmospheric Carbon Dioxide Concentrations[a]

Year	Amount Added from Fossil Fuel (Mt/yr)[b]	Cumulative Amount Added over Previous Decade (Mt)	Concentration by Volume (ppm)	Total Amount in Atmosphere (Mt)	Percentage of Annual Addition Remaining in Atmosphere
1970	15,400	126,500	321	2.50×10^6	52
1980	22,800	185,000	334	2.61×10^6	52
1990	32,200	268,000	353	2.75×10^6	52
2000	45,500	378,000	379	2.95×10^6	51

Note: This table is not a projection or a prediction. These calculations have been developed to provide insight into the nature of problems that may exist over the next several decades.

[a] The data in this table are taken from a model of the carbon cycle developed at the Study by Jon Machta, Lester Machta, and Jerry Olson, which includes biospheric and oceanic uptake of CO_2 (see Appendix to the report of Work Group 2). The model calculates yearly values for atmospheric CO_2, using past fossil fuel CO_2 production data from the United Nations. It was assumed that the atmosphere and the oceans were in equilibrium in 1860 with a concentration of 290 ppm by volume. The mechanisms used for oceanic uptake were gaseous concentration difference and the chemical mixing of CO_2 into a larger pool of carbonates. The mechanism used for continental biospheric uptake was CO_2 fertilization, that is, an increase in atmospheric CO_2 concentration, causing a proportional increase in photosynthetic rate. According to Keeling (1970) the mass of CO_2 in that layer of the ocean which is in communication with the atmosphere is between 5 and 8 times the mass of CO_2 in the atmosphere. The model assumed a mixing layer in the ocean that contains 8 times the mass of CO_2 that is in the atmosphere. The fertilization coefficient was adjusted to agree roughly with the Mauna Loa measurements (see Figure 2.2).

[b] For illustrative purposes only, the numbers in the first column are based on 4 percent annual growth rate until 1980 and 3.5 percent thereafter. These numbers are slightly larger than those compiled by Work Group 7.

Note: One million metric tons equals one megaton (Mt) or 10^{12} grams (g).

monoxide, formaldehyde, and a host of other organic compounds (which do not have a significant effect on the atmospheric heat balance by themselves); and SO_2, oxides of nitrogen, and hydrocarbons (which do not have a significant effect on radiation by themselves, but which can form particles that do). We do not consider these gases to have a global significance except in so far as they form particles.

Conclusions

1. The atmospheric CO_2 concentration has increased at the average rate of 0.2 percent per year during the period 1958–1969.
2. For this period, approximately one-half of the input has remained in the atmosphere. The other portion has gone into the

ocean and the biosphere. The partitioning of this other portion is uncertain.

3.   Predictions of future atmospheric CO_2 concentrations depend both on the growth rate of fossil fuel combustion and on the partitioning of man-made CO_2 among the atmospheric, oceanic, and biospheric reservoirs. This partitioning is too uncertain to allow accurate predictions. However, the assumption of about a 4 percent annual growth in fossil fuel combustion, and of a continuation of the partitioning found during the 1958–1969 period, leads to an increase in atmospheric CO_2 over present levels of almost 20 percent in the year 2000, from 320 to 379 ppm.

Recommendation

1.   We recommend that there be developed a station net with continuous analysis and with sufficiently high accuracy and resolution to determine the global trend and to analyze the partitioning of CO_2 uptake between the ocean and the biosphere (see report of Work Group on Monitoring).

In order to obtain the full benefit of this station net, supporting programs are needed, especially addressed to the gathering of pertinent oceanographic data, the procurement of stratospheric observations of trends of CO_2 concentration, and the supplementation of station data by satellite and aircraft measurements to improve geographical and vertical coverage.

Particles and Turbidity

ATMOSPHERIC CONTENT OF PARTICLES

The visible effect of particles in the air (or aerosols—we shall use these terms interchangeably) is to scatter sunlight. Air molecules also scatter sunlight, but they are much smaller than the wavelengths of visible light and scatter light in a predictable and isotropic way (Rayleigh scattering). Particles of more than about 0.1-micron (μ) radius scatter in a complicated manner, with a preponderance of scattered light in the forward direction (Mie scattering) and a relatively weak wavelength dependence. They can also absorb sunlight to some extent.

The solar beam, on its way to the earth's surface, is attenuated by absorption in ozone and other trace gases, is scattered isotropically by the air molecules, and is also scattered and absorbed by aerosols. This effect can be summarized by the formula for the intensity of the solar beam at a given wavelength:

$$I = I_0 e^{-(\alpha_a + \alpha_s + \alpha_d)m}$$

Where I_0 is the intensity outside the atmosphere and where α_a is the total absorption coefficient by gases, α_s is the gaseous scattering coefficient, α_d is the "turbidity coefficient" and accounts for aerosol scattering and absorption, and m is the path length relative to the vertical path length through the atmosphere (the cosecant of the solar zenith angle). When I is measured, the turbidity coefficient, α_d, can be calculated, since the other parameters are fairly well known. Measurement on cloudless days by standardized methods of such a turbidity index is probably the simplest and soundest way to obtain a continuous record, in relative terms, of the particulate load of the atmosphere.

Unfortunately, this method of turbidity determination lumps together the loss from the beam due to both scattering and absorption by particles and makes it difficult to assess the effect of particles on the radiation balance of the earth (Bryson, 1968).

SOURCES

Particles in the troposphere can be most generally characterized by their sources:

1.   Natural continental aerosols:

From dust storms and desert areas, with size range above about 0.3-μ radius.

From photochemical gas reactions between ozone and hydrocarbons from plants, resulting in very small particles of less than 0.2-μ radius.

From photochemical reactions between trace gases such as SO_2, H_2S, NH_3, and O_3 or atomic O. Such reactions are strongly influenced by humidity or the presence of cloud droplets.

From volcanic eruptions, which emit particles of all sizes and trace gases (especially SO_2) that subsequently can become particles in the stratosphere.

2.   Natural oceanic aerosols:

From the evaporation of ocean spray. These have essentially the same composition as sea salt, with size range above about 0.3-μ radius.

3.   Man-made aerosols:

From "smoke," that is, solid particles formed by combustion.

From photochemical gas reactions between unburned or partially burned organic fuel (for example, gasoline) and oxides of nitrogen ("smog"). This results in small particles, initially, of less than 0.2-μ radius.

From photochemical reactions between SO_2 and O_3 or atomic oxygen, such reactions being essentially the same as those of natural SO_2.

While the number concentration of small particles (less than 0.1-μ radius) falls off with increasing altitude, consistent with a terrestrial origin, the concentration of the large (0.1- to 1-μ radius) particles actually shows a maximum at about 18 km, suggesting that these particles (predominantly sulfates with evidence of many nitrates) are formed in the stratosphere, as suggested by Junge (1963), who first identified this stratospheric phenomenon. The most generally accepted hypothesis for their origin is that they are formed by the oxidation of gaseous SO_2 and then grow, at least initially, as very hygroscopic sulfuric acid droplets (Cadle et al., 1970). Any cations would, of course, combine with the acid, but to what extent the number of droplets is reduced by such combination is not known.

The origin of the stratospheric SO_2 is of considerable interest. We see very marked increases of the sulfate layer following large volcanic eruptions in the tropics, the most celebrated ones of recent history being the eruptions of Mont Tambora (1815), Krakatoa (1883), and Mount Agung (1963). Of course, there were no opportunities to sample the earlier stratospheric layers, but the characteristic height (about 18 km), persistence (several years), and twilight displays of colors leave little doubt that all major tropical eruptions have been alike in injecting large amounts of SO_2 (or possibly sulfates) into the lower stratosphere and that the particles that were subsequently formed spread worldwide in the surprisingly short time of about six months. The Agung particles had a measurable effect on the temperature of the atmosphere, as discussed in a later section.

Typical concentrations of such stratospheric particles, measured several years after the Agung eruption and possibly representing a partial return to "normal," are as follows (Cadle et al., 1969; Cadle et al., 1970):

18 km in tropics   0.25 μ g/m^3 ambient
18 km at mid-latitudes   0.40 μ g/m^3 ambient
13 km at mid-latitudes   0.15 μ g/m^3 ambient

(At 50 mb, or about 20 km, 1 μ g/m^3 ambient represents approximately 20 ppb by weight of sulfate particles.)

There is a first suspicion that man-made particles have begun to contaminate the stratosphere: the Cl/Br ratio for stratospheric particles is about 1/20 of its value for seawater, raising the possibility of a significant uptake of automobile exhaust particles that contain bromine (Cadle et al., 1970). The discovery of stable lead in the stratospheric population of particles would go a long way toward confirming this conjecture, and a systematic search for it is therefore in progress (Brookhaven National Laboratory, 1969).

Estimates of the present and future, natural and man-made production of particles is contained in reports of Work Groups 5, 6, and 7.

LIFETIMES AND BEHAVIOR OF PARTICLES

The removal of particles, both those that are soluble in water and those that are insoluble, is accomplished primarily by rain or snow. Very small particles are collected by small cloud droplets that are under the influence of Brownian motion, and subsequently the cloud droplets are rained out. Larger particles can be removed by the same two-stage process or by direct washout by falling raindrops.

The average lifetimes of particles in the lower atmosphere depend on the rainfall (or snowfall) regime in which they reside. Studies using various radioactive tracers have given lifetimes ranging from 6 days to 2 or more weeks in the lower troposphere. At mid-latitudes the shorter lifetime seems to be more accurate. In the upper troposphere the residence time is probably 2 to 4 weeks, while in the lower stratosphere the residence time varies from 6 months at high latitudes to about a year just above the tropical tropopause. The residence time continues to increase with altitude and is about 3 to 5 years in the upper stratosphere, and 5 to 10 years in the mesosphere, based on experience with specific radioactive nuclides that were injected as debris at high altitude by various nuclear tests (Junge, 1963; Telegadas and List, 1964; Bhandari, Lal, and Rama, 1966; Leipunskii et al., 1970; Martell, 1970).

Because particles stay in the air for some time, and because there are a variety of different substances in the mix of aerosols, particularly over land, it is unusual for a particle to retain its physical identity. The accretion of foreign substances by a particle takes place by coagulation of particles with each other, by cycles of condensation and evaporation in water droplets, and by various gas reactions on a particle's surface.

One particularly important physical characteristic of particles is their ability to serve as nuclei for condensation of water vapor. Generally the larger particles (0.1 to 1 μ or larger) are the most effective condensation nuclei, but hygroscopic salts and acid droplets also serve in the same role. As the relative humidity increases in an updraft and approaches saturation, the more effective condensation nuclei start to grow first until from about 50 to several hundred droplets per cubic centimeters are formed, depending on the rate of cooling and the availability of nuclei. This is the mechanism by which warm clouds (most clouds, in fact) are formed. Whether they will precipitate depends very largely on the number of nuclei that were initially available in the updraft.

Another important physical characteristic of particles is their ability to act as freezing nuclei that initiate the freezing of super-cooled water droplets. A number of kinds of clay commonly serve as freezing nuclei, and there are other substances that are equally effective—including the celebrated silver iodide that is so popular with cloud seeders. Artificial sources of freezing nuclei that have been identified are steel foundries and some other plants that put out particles. The lead from automobile exhaust, when combined with any iodine that happens to be in the air, can be another large source in urban areas (Schaefer, 1966).

TRENDS

There is substantial evidence of an increasing trend in turbidity. The two longest records of turbidity, using standardized techniques at Washington, D.C., and Davos, Switzerland (McCormick and Ludwig, 1967), both show an upward trend in the past three or more decades. The Washington trend may be ascribed to the urban pollution nearby, but the Davos change, since it is recorded at a mountain observatory, may represent a more general increase in aerosol content.

Ludwig, Morgan, and McMullen (1969) showed that at 20 nonurban U.S. sites average ambient particulate concentrations increased by about 12 percent during the period 1962–1966, and suggested that this increase was the result of particles generated by man's activity. Schaefer (1969) found that at Yellowstone National Park and at Flagstaff, Arizona, Aitken nuclei concentrations increased by a factor of 10 within a five-year span during the mid-1960s.

Observations of direct solar radiation in North America and Europe (Budyko, 1969) show a radiation decrease of about 4 percent from the 1930s and early 40s to 1960. Budyko suggests that this decrease is due to atmospheric aerosols resulting from man's activity. The standardized network organized by the National Air Pollution Control Administration (NAPCA) has been in existence long enough to establish geographical patterns of total particulate loading, but not long enough to establish secular trends definitely.

Although very little is known about trends in turbidity far from sources of man-made particles, there are two sets of data of atmospheric electrical conductivity over the oceans that do give an indication. The conductivity of the lower atmosphere can be measured with good absolute accuracy from ships and is roughly inversely proportional to the particulate loading. Series of such measurements were made by the research vessels *Carnegie,* starting in 1907, and *Oceanographer* in 1967. Comparison of the two series (Cobb and Wells, 1970) showed no change in the South Pacific data, but the North Atlantic data indicated an increase in particle concentration by about a factor of 2, presumably due to pollution from the North American continent which persisted after it had been carried far out to sea.

Conclusions

1.   The stratosphere, because of the long residence times of contaminants and the photochemical processes there, is more sensitive to the injection of particles than is the troposphere. There is some evidence that man-made contaminants (for example, bromine) released at the ground are accumulating in the stratosphere

in trace amounts. The major source of stratospheric particles at present is volcanic, consisting primarily of sulfates and nitrates, the former usually predominating. It is not known what the non-volcanic fraction is, or the amounts of cations (ammonia, metallic ions) that are available to combine with the sulfates and nitrates. There is still some question about the magnitude of the increase in stratospheric particles following the last major eruption (Mount Agung in 1963), but it was probably more than ten times on a worldwide basis.

2. The current quantitative yields of the principal sources of tropospheric particles that affect turbidity ($r \leq 2.5\ \mu$) have been estimated. The man-made contribution averaged over the globe amounts to about one part in five by weight, and the ratio of these number concentrations is about the same. The ratio is expected to rise. The overall trend in particles has been found:

a. To be increasing as shown by European and American solar radiation data (Budyko, 1969).

b. To be increasing over the North Atlantic, as shown by conductivity measurements (Cobb and Wells, 1970).

c. To be increasing at nonurban U.S. stations, as shown by high-volume filter samples (Spirtas and Levin, 1970).

d. To be increasing at Yellowstone National Park and at Flagstaff, Arizona, as shown by Aitken nuclei measurements (Schaefer, 1969).

e. Not to be increasing significantly in data collected in the central Pacific (Cobb and Wells, 1970).

3. In view of their relatively short residence times, the population of tropospheric particles can maintain large inhomogeneities. Problems of climate modification due to changes in this population will therefore tend to become manifest on a local level long before they do so on a global scale. Furthermore, the short residence time permits man to reverse the trend of growing atmospheric particle burdens within a few months if control measures are employed.

Recommendations

We recommend as priority efforts in order to identify particles and their trends (in addition to those recommendations concerning *effects* of turbidity contained in a later section—also see report of Work Group on Monitoring):

1.   Monitoring at sites not biased by proximity to pollution sources of:

—Atmospheric turbidity; the minimum net of coverage being the same as that for CO_2. (For economic reasons they should as far as possible be the same set of stations.)

—Direct and global (all-sky) solar radiation, using standardized observations, with accuracy consistent with determining long-term trends. These should include measurements at several wavelengths when possible.

—Nature and concentration of particles (a) *in the stratosphere,* to support source identification, (b) *in the troposphere,* to support, in addition, an understanding of radiation and cloud physics effects.

—Trace gases that are the precursors of particles.

2.   Laboratory and field research on detection of specific man-made particles contaminating the stratosphere. Examples of such research are: a search for lead, the determination of the ammonium-ions/nitrates ratio among nitrogen compounds, full characterization of volcanic particles, determination of the sulfur-32/sulfur-34 ratio of sulfates, and so on.

Heat Released

The waste heat that is cast off during the processes of energy generation and consumption is as much a climatic contaminant as are the gases and particles that we put into the atmosphere. We know that a concentration of heat sources (for example, a city) affects local climate (Peterson, 1969). The hard question is, At what point will our waste heat become a global climatic factor? Our ability to answer this question is hampered by the necessarily crude estimates of future power generation amounts and locations and by our still-uncertain understanding of the dynamic process by which this heat would affect global climate.

It is possible to calculate the approximate magnitude of the heat released now and over the next 30 years. Greenfield (1970), using current energy utilization figures and the assumed rates of growth listed in Table 2.3, estimated that thermal waste power of the world will rise from 5.5×10^6 megawatts in 1970 to 9.6

Table 2.3 Thermal Waste Energy (in units of 10^6 MW)

Geographic Location	Assumed Yearly Increase (percent)	1970	1980	2000
World	5.7	5.5	9.6	31.8
North America	4	2.2	3.4	7.5
(United States)[a]	(4)	(2.0)	(3)	(6.5)
(Canada)	(7)	(0.183)	(0.36)	(1.0)
Central America	6	0.12	0.2	0.68
South America	6	0.09	0.16	0.5
Western Europe	4	1.08	1.6	3.5
Western Asia	10	0.05	0.13	0.81
Far East	10	0.44	1.1	7.2
Oceania	8	0.069	0.145	0.64
(Australia)	(8)	(0.06)	(0.13)	(0.58)
Africa	6	0.1	0.18	0.57
East Europe	7	1.37	2.7	10.4
(Russia)	(8)	(0.98)	(2.0)	(9)

Source: United Nations, *World Energy Supplies.*
[a] ( ) indicates subregion of main region.

$\times$ 10^6 megawatts in 1980, and will reach 31.8 $\times$ 10^6 megawatts by 2000 (see Table 2.3). It should be noted that as a general matter all rates of growth are extrapolated from past history and are questionable when used to project much beyond five years. This is due to the fact that we are not yet capable of taking into account the effect of factors such as changing national goals and environmental constraints on power usage.

Lees (1970) has examined the 4,000 square mile area of the Los Angeles basin and has calculated that, at the present time, this area generates thermal power equivalent to more than 5 percent of the solar energy absorbed at the ground. He estimates that this will rise to about 18 percent in 2000 (see Figure 2.3).

A similar calculation was made (Greenfield, 1970) for a larger area, in the northeast section of the United States, where 40 percent of the national energy utilization occurs. In this area (roughly 350,000 square miles) the thermal waste is currently equal to about 1 percent of the absorbed solar energy at the ground and is projected to reach 5 percent by the year 2000 (Figure 2.4).

Waste thermal power is available to the atmosphere as either sensible or latent heat. The latent heat is produced primarily in

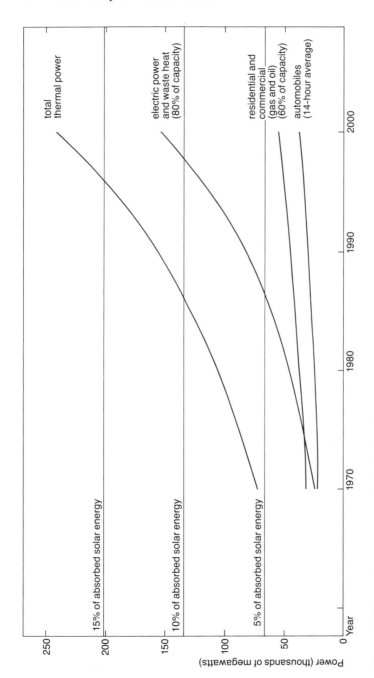

Figure 2.3 Thermal power generation in the Los Angeles Basin

Source: Lees, 1970.

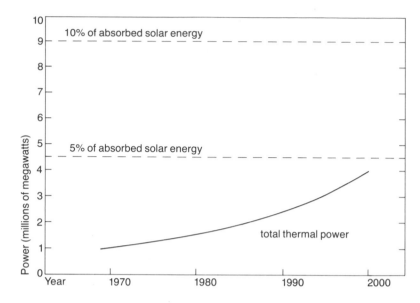

Figure 2.4 Thermal power generation in a climatically significant area in the United States

Source: Greenfield, 1970.
Area: East, North Central States (Mich., Ill., Ind., Wis., Ohio); Middle Atlantic States (N.Y., N.J., Pa.) 351,028 mi³
Assumptions:
1. Electric power + waste heat = 30 percent of all power
2. Electrical capacity increases by factor of 10, 1970–2000
3. All other energy doubles, 1970–2000

the generation of electrical power and is dissipated by deposition in water bodies or by circulation through cooling towers. It should be noted that the latent heat will generally be available to the atmosphere at higher altitudes than the sensible heat and probably at some distance from the source. Based on the fractionation between waste heat from electrical power generation and from other sources, we may estimate from Figure 2.3 that, in 1970, 15 to 20 percent of the total waste heat will enter the atmosphere through the evaporation-latent heat route in the Los Angeles area, and that this will increase to approximately 30 percent by the year 2000. There is no apparent reason for not applying this estimate to other areas of the country as well.

Jet Aircraft Contributions to the High Atmosphere
There are two aspects of jet aircraft operations that deserve attention as possible sources of pollution, and consequent weather and

climate modification: the condensation trails ("contrails") formed in the upper troposphere by subsonic commercial jets, and the contamination of the stratosphere by supersonic transports (SSTs) flying at about 20 km (65,000 feet).

CIRRUS FROM JET CONTRAILS

The relative humidity, or proximity to saturation of water vapor, tends to increase with altitude in the troposphere so that the upper troposphere is most vulnerable to cloud formation from the addition of moisture. Also, the average residence time of a water vapor molecule is longer in the upper troposphere than in the lower troposphere; therefore, on the average, more water vapor can accumulate before being removed by precipitation. For these reasons, contrail formation, so common near the world's air routes, is more likely to occur when jets fly in the upper troposphere than when they fly in the lower troposphere.

There are very few, if any, statistics that permit us to determine whether the advent of commercial jet aircraft in about 1958 has altered the frequency of occurrence, or other properties, of cirrus clouds. Appleman of the U.S. Air Force Air Weather Service (1965) analyzed the trend of cirrus cloudiness on days with five-tenths or less low and/or middle cloudiness at 30 U.S. and non-U.S. Air Force bases, from about 1949 to 1965. He found a general increase in cirrus clouds during that period, which may not be statistically significant. Machta and Carpenter (1970) have analyzed the trends of cirrus cloudiness at Salt Lake City and Denver between 1949 and 1969, again for days with no or few obscuring lower clouds. The results, still very tentative, indicate that cirrus cloudiness has increased. These results are statistically significant. They noted that the cirrus cloudiness increase paralleled the increase in jet aircraft activity (as measured by the amount of fuel used by domestic airlines). However, slight decreases in middle cloudiness at Denver suggest that observers may have relabeled some of the middle clouds as upper clouds after jet pilots reported their true heights. It should also be noted that all stations whose records have been examined lie near or under commercial air routes.

Estimates of future subsonic jet operations, while uncertain, indicate perhaps a three- to sixfold increase in the years 1985–1990. These estimates would probably have to be increased if the SST, which would cruise in the stratosphere, instead of the troposphere, does not come into operation.

Recommendation

We recommend that the magnitude and distribution of increased cirrus cloudiness from subsonic jet operations in the upper troposphere be determined. A study of the phenomenon should be conducted by examining cloud observations at many weather stations, both near and remote from air routes.

SST CONTAMINATION OF THE STRATOSPHERE

Contamination from the future operation of SST (supersonic transport) aircraft requires special attention because of the portion of the atmosphere, the stratosphere, in which the contamination will be injected. The stratosphere differs from the troposphere, for our purposes, in two important ways: (1) contaminants will remain airborne for longer periods there (1–3 years compared to several weeks), and (2) the stratosphere may be sensitive to contaminants because concentrations of parts per million or less may participate in its photochemistry.

Of all the SST combustion products, carbon dioxide is the most abundant. The average life of a carbon dioxide molecule in the air before passing into the sea or the biosphere is about 5 years (estimates vary from about 3 to 30 years). This is appreciably longer than the 2 years that we shall use for the stratospheric residence time of an inert gas inserted at about 20 km (60,000–70,000 feet) by the SST. There would therefore be time for mixing of CO_2 to occur between stratosphere and troposphere. Further, during the 5 years, 1985–1990, atmospheric carbon dioxide would increase by about 0.5 ppm (by volume) from SSTs, and by about 10 ppm from other fossil fuel combustion at the ground (see Table 2.2). For both reasons, the contamination of the stratosphere by SST carbon dioxide does not appear to be as important as the contamination of the whole atmosphere by other, much larger sources of carbon dioxide.

The second most abundant product of the combustion of jet fuel is water vapor. In contrast to the wet troposphere, the stratosphere is exceedingly dry, containing only about 3 ppm (by mass) of water vapor. It is therefore possible to alter stratospheric humidity by SST activities. Indeed the proposed SST flights for

1985–1990 (Federal Aviation Administration, 1967) would increase the global average stratospheric concentration of water vapor by 0.2 ppm (from 3.0 to 3.2 ppm). However, this increase will be non-uniform, ranging from high values directly behind the aircraft to small increases in the south polar regions. Pending more rigorous calculations of the likely distribution, it will be assumed that significant volumes of the stratosphere, in terms of climate change, may be as much as 10 times (that is, 2 ppm) higher than before SST operations.

Carbon monoxide and nitrogen dioxide (a likely daughter product of nitric oxide) both absorb infrared terrestrial radiation. However, their small concentrations (parts per hundred million) and small absorption rates make it highly unlikely that either can directly compete with ozone, carbon dioxide, or water vapor in altering the stratospheric radiation budget significantly.

Both carbon monoxide and nitrogen in its various oxide forms can also play a role in stratospheric photochemistry, but despite greater uncertainties in the reaction rates of CO and NO_x than for water vapor, these contaminants would be much less significant than the added water vapor and may be neglected.

The currently accepted theory accounting for the predominance of sulfate in the stratospheric particulate layer (the Junge sulfate layer, at approximately SST cruising altitudes) argues that SO_2 from the lower atmosphere enters the equatorial stratosphere as a gas, where it is converted photochemically to particulate sulfates. Thus, direct introduction of SO_2 into the stratosphere would increase the amount of sulfate produced. Furthermore, reactions that occur in the lower atmosphere between NO and hydrocarbons can also take place in the stratosphere, converting hydrocarbons to particles in the presence of sunlight (the "smog" reaction). Unfortunately, the efficiency of these reactions in the stratospheric environment is uncertain. For the present purposes, we shall assume that *all* SO_2 is converted to sulfate particles and that the entire mass of emitted hydrocarbons is converted to organic particles (see Table 2.6 in the Appendix to this section). Some of the NO_x could be converted to nitrate particles but we believe that this conversion will be very small due to the unavailability of suitable positive ions and the fact that nitric acid is not hygroscopic.

The natural background of particles was measured in the early 1960s by Junge (1963) and again in the late 1960s by Cadle et al. (1970), who derived concentrations about thirtyfold higher. This increase is probably due to added particles injected into the stratosphere by the eruption of Mount Agung in 1963 and to subsequent tropical eruptions, but the large factor may also be partly due to inefficient collection systems used by Junge et al. (that is, the earlier numbers may be too low). Thus, it is likely that the true, nonvolcanic sulfate background lies between the two values shown in Table 2.6.

Because one cannot verify the stratospheric radiation balance using gases alone in the calculation, it is believed that the Junge sulfate particle layer may play a role in the heat balance of the lower stratosphere (Junge, 1963). Newell (1970a), for instance, has shown that shortly after the injection of debris by the volcanic eruption in March 1963 of Mount Agung in Bali, the temperature of the lower equatorial stratosphere rose about 6° or 7°C, presumably because of the absorption of solar radiation by the injected particles at that altitude. The position will be taken in this discussion that weather and climate effects are possible when the increase in concentration due to added particles is of the same order as the natural post-Agung concentration of particles in the Junge sulfate layer.

Table 2.6 of the Appendix to this section shows that the global average values for sulfates, hydrocarbons, and soot are about equal to or greater than the Junge pre-Agung values; and that the peak north temperate latitude concentrations of these particles approach or (in total) exceed even the post-Agung values.

Although the foregoing calculation is conservative in the sense that we have tried to avoid any overestimate of the formation of sulfate and hydrocarbon particles from SSTs, it is possible that the artificial particulate concentration might be larger than our estimate if: (1) the NO_x, which is produced in rather appreciable quantity, were to form particles as well as the SO_2, and thus increase particle production by a large factor; (2) the GE-4 turns out to approach present-day large-engine emission rates of soot and hydrocarbons which could raise the particles originating as both soot and hydrocarbons by over a factor of 5. Furthermore, it should be pointed out that the particle formation rates observed during static ground-level tests of jet engines, or derived from theoretical calculations, may underestimate the particle formation rate which will occur when the exhaust products condense in the very cold ($<$ $-65°$C) stratospheric environment.

On the other hand, if it were economically feasible to produce and use very low sulfur content fuel, namely, 0.01 percent S, instead of continuing with that currently available (0.05 percent S by mass), then the sulfate contribution would be cut to one-fifth of that shown. However, the total particulate addition would still be at least half of that estimated in Table 2.6, since there would be no reduction in the amount of soot and converted hydrocarbons. The total particulate addition would also still be comparable to the apparent, pre-Agung, global particle background.

Appendix: Statistics on SST Engine and Calculations of Stratospheric Contamination

The emissions from a single GE-4 engine to be used on the B2707-300, flying at 65,000 feet (about 20 km) in a cruise mode, at Mach 2.7, on a day with standard weather conditions, have been calculated (Table 2.4), assuming chemical equilibrium, by General Electric Company engineers (Hession, 1970; Thompson, 1970). In the future, the calculations will be compared with actual engine tests, but it is estimated by SST officials (Thompson, 1970) that the fraction of products per unit amont of consumed fuel may be accurate within 10 percent (except as indicated later). It should be noted that the engine will be equipped with an afterburner; however, it will be in only partial operation in the stratosphere, as opposed to full operation during takeoff; that is, the temperature in the afterburner will be 2,800°R (degrees Réaumur) while cruising in the stratosphere, but will be 3,500°R during takeoff.

An estimate suggested by the Federal Aviation Administration (FAA) calls for 500 SST aircraft to be flying during the period 1985–1990. Each of these jets would fly in the stratosphere for almost 7 hours a day (2,500 hours per year). Of the 500 jets, 334 would be U.S. and be equipped with four engines, and 166 would be of non-U.S. fabrication and would have the equivalent of two engines.

Table 2.4 Statistics of Emissions from One GE-4 Engine, Cruise Mode

Constituent	Pounds per Hour
Ingested air and consumed fuel	
Air	1,380,000
Fuel	33,000
Unused air	
N_2	1,039,000
O_2	208,000
Ar	19,300
Combustion products	
CO_2	103,500
H_2O	41,400
CO	1,400
NO^a	1,400
$SO_2{}^b$	33
Soot (Particles)	5
Unused fuel	
Hydrocarbons[c]	16.5

[a] The General Electric Company advises (Thompson, 1970) that the true NO output is likely to be no more than one-half to one-third of the calculated value. A few past comparisons suggest that measured values will be values that are 10 to 15 percent of the calculated numbers (Thompson, 1970).

[b] Sulfur content of SST fuel will be specified as no more than 0.3 percent sulfur by weight. The 33 pounds per hour given in the table corresponds to 0.05 percent sulfur by weight, which is the average sulfur content of currently available jet fuel, as determined by the Department of the Interior. An appreciable amount of jet fuel with less than 0.016 percent S is available now, according to a Boeing Company survey (Swihart, 1970). A leading producer has told us that it is technologically feasible to produce jet fuel with only 0.01 percent S, but that present production facilities are not adequate to supply a large market.

[c] The General Electric Company expects the true hydrocarbon emission to be vanishingly small, due to the high temperature of the afterburner (Thompson, 1970).

Tables 2.5 and 2.6 contain our estimates of equilibrium or steady-state concentrations, after several years of operation, assuming a 2-year mean residence time for all products.

Conclusions

The following conclusions are based on the assumed operation of 500 commercial SSTs in the period 1985–1990, and on their emissions as provided by the federal SST office (except as noted).

1.   Because of the long atmospheric residence time of CO_2 and the relatively small CO_2 contribution from SSTs compared with

Table 2.5 Gaseous Concentrations

	10^{11} Grams/Yr[a]	World Average Concentration[b]	Possible Peak Concentration[c]
CO_2	1,960.	Not relevant	
H_2O	783.	0.20 ppm (0.3)	2.0 ppm (3.0)
CO	26.3	6.8 ppb (7.3)	0.068 ppm (.073)
NO	26.3	6.8 ppb (7.3)	0.068 ppm (.073)
SO_2	0.63	0.16 ppb (0.20)	1.6 ppb (2.0)
Hydrocarbon	0.31	0.081 ppb (0.087)	0.81 (0.87)

[a] This column is computed using data from Table 2.4 and assumptions for full-scale SST operations for 1985–1990 given in the text.

[b] This column is computed using the data from the first column and the assumption of a two-year mean residence time for all products in this portion of the stratosphere. The mass of the atmosphere is taken as 5.14×10^{21} grams, with the stratosphere containing 15 percent of that, or 0.77×10^{21} grams. The numbers in parentheses represent rounded values obtained by increasing the SST-produced concentration to accommodate subjective estimates of military injections of contaminants into the stratosphere. All concentrations are by mass.

[c] The last column is obtained by multiplying the world average by 10 and represents a tentative upper limit for certain temperature change considerations that would apply to a region of high SST activity.

Table 2.6 Particle Concentrations (10^{-4} ppm by mass)

	Formed from SST Exhaust		Measured at SST Altitude	
	Global Average[a]	Peak N. Hem.[b]	Pre-Agung[c]	Post-Agung[d]
SO_4^-	2.4 (4.8)[e]	24 (48)	1.2	36
Hydrocarbon	0.81 (1.6)	8.1 (16)	——	——
Soot	0.25 (0.5)	2.5 (5)	——	——
Total	3.46 (6.9)	34.6 (69)	1.2	36

[a] The SO_4^- figure was calculated using data in Table 2.5 and assuming all SO_2 is converted to SO_4^-. (Note that SO_4^- is heavier than SO_2.) The hydrocarbon figure is taken from Table 2.5 and assumes total conversion of gas to particles. The soot figure is computed using data from Table 2.4 and assumptions for full-scale SST operations for 1985–1990 given in the text.

[b] A factor of 10 is applied to the first column in order to estimate the peak concentration within the nonuniform distribution over the globe and would apply to the regions, mostly in the Northern Hemisphere, of large SST activity.

[c] Source: Junge, 1963. No data exist for carbon or hydrocarbon concentrations in the stratosphere.

[d] Source: Cadle et al., 1970.

[e] Numbers in ( ) are twice the calculated value, to account for particle settling. It is believed that sulfate particles, because of their slow settling speeds, remain concentrated in the sulfate layer rather than disperse throughout the vertical extent of the atmosphere. The natural sulfate layer happens to be at about the same altitude as the SST injections. The net effect is probably an increase by about a factor of 2 over the concentration calculated for a gas, which would diffuse upward as well as downward. This factor of 2 has been applied in Table 2.6 (the values in parentheses) to the calculated concentrations of all three particles—sulfates, hydrocarbons (HC), and soot—for the same reason.

that from other sources of fossil fuel combustion, there will be no special CO_2 problem due to SST operations.

2.   Stratospheric water vapor will increase, on a global average, by 0.2 ppm by mass (from 3.0 to 3.2 ppm). Since there will be more water vapor added to the north temperate latitudes, parts of this region may perhaps have a contribution to standing concentration as much as tenfold higher than the increase of the global average (that is, grow from 3 to 5 ppm of water vapor).

3.   Concentrations (by mass) of CO, NO, SO_2, HC, and soot will range from fractions of a ppb for soot to 68 ppb for CO and NO from 500 SSTs in the north temperate latitude, peak of SST activity. Emissions from present large engines are appreciably different from those predicted for the SST GE-4 engines. The current NO emission rates per gallon of fuel for certain jet engines are as much as twentyfold lower, but the HC can be fourfold greater and the particles more than tenfold greater. Fuel with a 0.3 percent sulfur content (the maximum allowable limit) would increase the emissions of SO_2 by more than sixfold over our SST estimates, and use of 0.01 percent sulfur fuel (technically feasible) would reduce the SO_2 emission to one-fifth of our estimates, since the calculations here are based on 0.05 percent sulfur content. Further, there are doubts concerning the applicability of particle emission information derived from both theoretical calculations and static field tests to the real atmosphere and operating conditions. It is felt that realistic operation may produce larger numbers of particles. Finally, we have no information on fuel additives.

Oxygen: Nonproblem

Atmospheric oxygen constitutes about 21 percent of the atmosphere by volume. There is virtual equilibrium between photosynthetic production of it and its utilization by animals and bacteria. The mean residence time may be of the order of 10,000 years (Johnson, 1969). From time to time there have been predictions of a disruption of this balance; most recently it has been speculated that there may be a reduction in photosynthetic organisms in the ocean by herbicides and pesticides. To obtain a current base-line point in order to detect any (unlikely) changes in atmospheric oxygen, Machta and Hughes (1970) measured oxygen concentrations in a large number of clean air locations. The mean value was found to be 20.946 percent by volume with virtually no discernible geographic variation. The accuracy of the comparison standard was ±0.006 percent by volume (one standard deviation).

Comparison with measurements taken since 1910 suggests no detectable change in concentration. The best earlier measurements also indicate 20.946 percent by volume. All others that are reliable lie within the uncertainty of the measurements. Past data show only a negligible lowering of oxygen concentration in the most polluted open areas.

Finally, calculations of the consumption of oxygen that would be caused by the burning of all known recoverable fossil fuels indicate that oxygen concentration would be reduced to 20.800 percent by volume. Such a reduction would have a negligible effect on human or animal respiration.

Conclusion
There has been no detectable change in atmospheric oxygen in this century. Future changes, if any will occur very slowly.

Recommendation
We recommend that monitoring be conducted only at very infrequent intervals, such as every 10 years, in light of the theoretical and observed stability of atmospheric oxygen.

Radioactive Nuclides
Estimates of the production of various radioactive waste products by the expanding use of nuclear power are made in the report of Work Group 7. There is no question that two radioactive products will find their way into the air and water system of the earth: tritium (as tritiated water), and krypton 85 (a chemically inert noble gas). Both have half-lives of about 12 years, so they will have a chance to diffuse throughout the globe regardless of the location of their sources.

The complex question of how much of these two substances we can tolerate in our environment is related entirely to their effects on living things, including man. We shall not deal with that here.

As far as possible effects of these radioactive substances on the physical processes in the atmosphere are concerned, we can identify none at the concentrations which have been projected to the year 2000. The atmosphere in this case will be far more resistant to these nuclides than the biosphere.

Changes of the Earth's Surface

The most important properties of the earth's surface that have a bearing on climate and are likely to be affected by human activity are the following:

1.   *Reflectivity.* When the reflectivity (albedo) of the ground for shortwave radiation is modified, the amount of sunlight returned to space will be changed along with the energy available to heat the ground and the atmosphere just above it. Thus, decreasing the surface albedo usually raises the ground temperature, and, therefore, the air temperature immediately above the ground as well. The result of that effect is that the overall lapse rate, that is, rate of decrease of atmospheric temperature with altitude (not limited to the region immediately above the ground) is increased. The atmosphere becomes more unstable, and thus experiences increased vertical mixing.

2.   *Heat Capacity and Conductivity.* The smaller the heat conductivity and capacity of the surface, the larger will be the surface temperature during the day and the greater will be the lapse rate and the vertical dispersion of contaminants and, consequently, atmospheric properties.

3.   *Availability of Water.* Increased availability of water on the surface will increase the portion of the sun's energy that is used in evaporation. Therfore, less of the sun's energy will be used in surface heating, and the surface temperature will be cooler. In addition, the water vapor produced by this increased evaporation will be available in the atmosphere to form clouds and to cause precipitation downwind from the water source.

4.   *Availability of Dust.* Human activity may change the mechanical properties of the surface so that more or less dust can be carried from it into the atmosphere. The effect of such dust is of potentially great importance and is dealt with later.

5.   *Aerodynamic Roughness.* The rougher the ground, the weaker will be the wind above it and the stronger will be the turbulence of the air (irregular fluctuations of the wind, both in the vertical and in the horizontal). Therefore, for a given lapse rate,

vertical mixing will be stronger over rough terrain than over smooth terrain—for example, vertical mixing will be stronger over a city than over a water body.

6. *Emissivity in the Infrared Band.* The emissivity of the surface in the infrared band is always nearly 100 percent, but the small change in emissivity produced by a change in the character of the surface may have some significance.

7. *Heat Released to the Ground.* Man's activities produce large amounts of heat is released at or near the earth's surface. This matter is dealt with elsewhere in this report.

Theory of Climate and Possibilities of Climatic Change

While the main thrust of the Study has been to determine where man may be influencing and changing his environment adversely on a global scale, we must not forget that the atmospheric environment has changed many times during the history of the earth. Man might now be forcing his way with or against the natural tide of climatic change. We do not know whether such a powerful tide can be stemmed, but man could quite conceivably abet it. In any case, we must take naturally occurring changes into account when we think of the future decades and centuries, since they have had a major impact on mankind in the past and promise to do so again—sometime in the future.

In the following sections, some of the lessons of the past are reviewed so that we may proceed, with a better perspective, to a discussion of man's present influence. It will be also clear that we need to know much more about the natural forces that are at work, at work more inexorably, perhaps, than we realize.

Mathematical Models of the Atmosphere

On both practical and conceptual grounds, the use of mathematical computer models of the atmosphere is indispensable in achieving a satisfactory understanding both of climatic changes, as we observe them, and of the potential role that man may play in causing them. The conceptual reason is that, without knowledge of the alternative behaviors with which the atmosphere could have consistently responded to the changing conditions of the planet, we shall be unable to ascribe observed effects unambiguously to possible causes. From a practical viewpoint, the dynamic behavior of the earth's fluid envelope presents us with so many interdependent phenomena and processes, governed by so many feedback cycles (some stabilizing and others destabilizing) that the only way that we now conceive of exploring the tangle of relations which describes all this is the numerical solution of specific examples.

Corresponding to this pair of focal reasons for the mathematical simulation of global atmospheric behavior are two characteristic uses of such computer models. First, they serve, one might say, as the apparatus for conducting laboratory-type experiments on the atmosphere-ocean system which are impossible to conduct on the actual system. Used in this mode, models contribute to our understanding of climate dynamics. Second, models are run in such a way as to produce results of possible operational usefulness, such as longer-term forecasts of global atmospheric conditions.

Global models of the earth's fluid envelope, to be of any value, are necessarily elaborate and therefore costly to design and operate. Nevertheless, our dependence on them for making progress in this field is such that the effort of developing them cannot be avoided.

MODEL DESIGN

Mathematical models of the atmosphere are derived from consideration of the atmosphere's behavior as an aerothermodynamical process. A minimum of seven quantities are required to describe its state at any point: pressure, temperature, density, the three components of the velocity vector, and moisture (usually the fraction, by weight, of water vapor). These quantities and their time rates of change are related by four basic laws of physics (the "primitive equations"): Newton's second law (stated in two relations—for the horizontal motion and the hydrostatic approximation in the vertical); continuity of mass (two relations—one for total mass and one for water mass, to allow for the latter's change of phase); the first law of thermodynamics; and the equations of state (gas law). To these equations must be added the relations that define boundary constraints and external influences, especially the incidence of solar radiation, the flux of outgoing infrared radiation, and turbulent transport of heat, water vapor, and momentum at the lower boundary in order to complete this, in every respect primitively minimal, model of the atmosphere.

Although current numerical models of the general circulation are satisfactory for quite a few purposes, they run almost immediately into a number of difficulties because they fail to specify the behavior of the atmosphere completely. For example, they make allowance for the condensation of water vapor at specified states of that gas, but they do not determine whether the condensed water remains suspended in the form of clouds or falls out as rain. Another example is found in the first law of thermodynamics, which contains a term measuring the heat added to the air. This term depends on the latent heat budget and on radiational effects. Even if the former is specified as a function of the state of the air, the latter can cause difficulty by its dependence on the composition of the air. Eliminating this difficulty, by assuming the compositions of the atmosphere to be constant, automatically builds a major element of what might be called "now-climate" into the models. A third deficiency is that the variable effect of cloudiness on the radiation budget is not treated explicitly in current models.

A similar set of variables and equations can be used to describe the ocean, using salinity instead of moisture as the main variable for characterizing composition. Comparable degrees of incompleteness affect present ocean models, which similarly limit the usefulness of the models in simulating climatic changes. The atmospheric and ocean models can in principle be coupled, and in present designs are coupled, by making them respond to each other across their interface (that is, by requiring continuity of heat, momentum, and mass flow at each point).

MODEL OPERATION

A model is run by numerically integrating its equations, starting with some initial atmospheric regime. For this purpose the size of the time steps must be properly related to the spacings between points in the grid of discrete points that replaces the continuous atmosphere. If the steps are not properly related, the propagating computational errors, and not the conditions of the problem, will determine the computed results. This constraint takes the form of a practical ceiling on the time step, which, in the case of our atmosphere, is about 5 minutes for a horizontal grid spacing of 300 km, and decreases proportionally with the latter. Clearly, the integration of even a simplified version of such a model over any length of time becomes a formidable undertaking. Advanced models require two or three hours (or more) for each model day, using the largest computers available.

It is becoming increasingly clear that the atmosphere is capable of motions of practically all scales, varying from millimeters to thousands of kilometers and from a few seconds to millennia. The operation of a single model that would simulate all of these processes is therefore completely beyond us. We must be satisfied with models which are reasonably faithful over the range of scales that are of interest for a particular purpose. This poses the problem of how to simulate what might be called the leakage of energy and momentum to and from the scale ranges that lie beyond our simulation capability. Up to a point, various forms of ad hoc lumping and aggregation have provided satisfactory parameterizations of the interaction of model quantities with processes that lie below the scale of the model. However, there is a vexing dearth of such techniques at larger scales. For example, the very important effects of cumulus convection, which account for a major part of the vertical exchange of heat, moisture, and momentum in the tropics, are still not satisfactorily parameterized in current models.

CHARACTERISTICS OF NUMERICAL GENERAL
CIRCULATION MODELS

Up to now we have been describing the kind of general circulation models that have been developed to simulate the atmospheric motion in some detail and that lend themselves to making forecasts for a few days when real initial conditions are applied.

In order to create a model that can simulate the climate and its long-term changes, it becomes essential to introduce some factors that do not have to be included in the shorter-term simulations, the ocean circulation being the most important. An attempt to run a combined ocean-atmosphere model has been made (Manabe and Bryan, 1969) but it had to be a highly simplified one so that it could run long enough for the ocean to approach some sort of equilibrium with the atmosphere—several centuries at least. The model did simply account for the variation of snow cover and sea ice but not variable cloud cover and its effects on the radiation budget. Variable cloud cover has not been satisfactorily handled in any numerical model so far.

The problem of estimating climate modifications that result from the rather puny activities of man is even more demanding. It requires models that are capable of simulating the natural climate and its variability in sufficient detail, and that must consider the effects of changing aerosols as well as carbon dioxide and energy inputs at the surface. Progress toward that end has been made, but the objective is far from attainment. It is not believed that models now operating can simulate the difference between the global climate of 100 years ago and today, because they do not take into account the long-term factors that we have mentioned.

Most of the inadvertent climate changes which man might cause represent variations that are comparable to the noise level of the natural variability of the atmosphere, as well as being within the uncertainty range of current modeling techniques. The real atmosphere is apparently capable of anomalies that last longer than a year—we hardly know how long—and some models appear to show the same characteristic. To obtain discriminating statistics will require long runs with more complete models, a rather costly proposition.

At a more fundamental level, however, the goal is to free progressively the models that we operate from their dependence on empirical relations derived from the climate as we now happen to find it. At the distant, probably unattainable, end is the replacement of those relations by an application of the basic laws of physics, resulting in models that would simulate with equal facility the climates of Earth, Venus, or Jupiter. It is more likely that we shall have to be satisfied with replacing empirical relations by increasingly general climatological laws, toward which we shall gradually grope our way as we amalgamate not only observations on the present climate but also the experience that we acquire from working with ingenious models, explicitly devised for this purpose.

Recommendations
We recommend that current models be improved by, first, including more realistic simulations of clouds and air-sea interaction

and, second, that attempts be made to include particles in the radiation balance calculation when their optical properties become better known.

Such models should be run on many time scales, but at least for periods of several simulated years. With ocean circulation included in the atmosphere-ocean system, it will take several simulated centuries even to approach a quasi-steady state. Possibilities should be investigated for simplifying existing models to provide a better understanding of climatic changes. Simultaneously, a search should be made for alternative types of models which are more suitable for handling problems of climatic change.

Effects of Changing CO₂

If we assume that the earth-atmosphere system is in radiative equilibrium with the sun (that is, that it emits as much radiation as it receives), we can define an equivalent temperature ("brightness") for the planet as

$$T_s = \left[\frac{(1 - A)S}{4\sigma} \right]^{1/4}$$

where A is the planetary albedo, S is the solar constant and σ is the Stefan-Boltzmann constant. The factor 4 appears because the solar constant is defined for a surface perpendicular to the sun's rays, but the earth-atmosphere system radiates energy outward in all directions. For an average planetary albedo of approximately 33 percent (London and Sasamori, 1970), the radiative-equilibrium temperature of the earth-atmosphere system is 253°K (−20°C), a value that is very close to recent estimates from satellite observations (Raschke and Bandeen, 1970).

The average surface temperature T_s is higher than the earth-atmosphere radiative equilibrium temperature, as just defined, because the atmosphere is semitransparent for solar radiation but is fairly opaque for terrestrial (infrared) radiation. For instance, in the clear atmosphere (containing an average amount of water vapor, carbon dioxide, ozone, and dust), about 65 percent of the incoming solar radiation reaches and is absorbed at the earth's surface. For the same atmospheric conditions, however, only about 10 percent of the total radiation leaving the surface is directly transmitted back to space (that is, does not get absorbed by the atmosphere). For an average cloudy atmosphere these numbers are 45 percent and 5 percent, respectively (London and Sasamori, 1970). The remaining portion of the surface radiation is absorbed by trace atmospheric gases (primarily water vapor and carbon dioxide) and by clouds and is then reemitted downward to the ground and upward to space. The returned (downward) infrared radiation helps to keep the surface temperature relatively high.

The mean surface temperature is about 286°K (13°C). The ratio T_s/T_e is sometimes referred to as the "greenhouse" coefficient. For the earth this is $(286/253) = 1.13$. For Venus, the "greenhouse" coefficient is approximately $(700/265) = 2.60$.

As is well known, water vapor, carbon dioxide, ozone, and clouds are the principal active radiating substances in the earth's atmosphere. It is important, therefore, for present purposes, to comment on the importance of carbon dioxide, as compared with these other substances, in affecting the atmospheric radiation budget. Calculations show that in the troposphere both the radiative heating (solar radiation absorption) and cooling (infrared flux divergence) by water vapor is about ten times larger than that for carbon dioxide. In the vicinity of the tropical tropopause, infrared flux due to CO_2 actually leads to heating. This radiative convergence results because the carbon dioxide in the cold air at the tropopause "sees" warm air from above and below.

In the stratosphere, radiative heating due to the absorption of solar radiation by ozone is dominant. The radiative cooling that does occur is carried on primarily by CO_2. Water vapor plays a minor role above about 20 km.

In general, radiative processes in the atmosphere act to cool the atmosphere. Since, in the troposphere, the *net* radiative cooling is larger at higher latitudes (where the temperature is low) than in equatorial latitudes (where the temperature is high), energy is available for atmospheric motions in the troposphere (see, for instance, London and Sasamori, 1970). In addition, net radiative processes in the troposphere lead to maximum cooling rates at levels of 5 to 10 km. A vertical column of air tends toward hydrostatic instability when the air gets colder aloft relative to the air below. Thus, radiation acts to produce less hydrostatic stability in the lower troposphere, and greater stability above.

The thermal radiative component due to carbon dioxide slightly reduces this generation of available horizontal and vertical potential energy (Newell and Dopplick, 1970).

Since carbon dioxide represents one of the components affecting the earth's radiation budget, changes in the carbon dioxide concentration (natural or man-made) will produce some changes in the radiation budget in general and in the surface temperature in particular.

It was first suggested by Tyndall in 1863 that the blanketing effect of increased carbon dioxide would cause climatic changes through variation of the surface temperature. Increased carbon dioxide, because of its strong absorption (and therefore emission) of infrared radiation at 12 to 18 μ, would reradiate energy downward to the earth's surface and further inhibit the radiative cooling at the ground. Water vapor, ozone, and clouds have qualitatively similar effects.

At the end of the nineteenth century, Arrhenius computed that a surface temperature rise of 9°C would result from a threefold increase of atmospheric carbon dioxide (Arrhenius, 1896). Renewed interest in this problem was sparked by the computations of Plass (1956), starting about 15 years ago, who used a fairly realistic model of the carbon dioxide 15-μ absorption band and high-speed computers to derive changes in surface temperature that would result from a change in carbon dioxide concentration from 300 ppm to 600 ppm or to 150 ppm. His calculations, assuming only the 15-μ CO_2 band, no overlap with water vapor absorption bands, and no surface evaporation or convection, gave a value for ΔT_s of +3.6°C or −3.8°C, depending on whether the CO_2 concentration was doubled or halved, respectively.

Subsequent calculations have included clouds, the overlap of water vapor and carbon dioxide absorption bands, and the impact of increased water vapor (resulting from an increased surface temperature) as a feedback mechanism (Möller, 1963). Other computations included, in addition to the preceding, the effect of ozone and of a "convective adjustment" so that the lapse rate in the troposphere would not become unstable (Manabe and Wetherald, 1967). The influence of the absorption of near infrared solar radiation by carbon dioxide has been studied by Gebhart (1967).

The various results indicate that, if the surface heating due to CO_2 is accompanied by evaporation (to maintain a constant relative humidity), the "greenhouse effect" is amplified slightly. But if average cloudiness is included in the radiative model and/or the overlap of water vapor and carbon dioxide infrared bands is accounted for, the surface radiative equilibrium heating due to an increase in atmospheric carbon dioxide concentration is damped to about one-half or less of what it otherwise would be. The average increase of surface temperature predicted by radiative equilibrium studies, which include a "convective adjustment" but no other dynamic or thermodynamic effects, is approximately 2°C for a doubling of the carbon dioxide concentration (Manabe and Wetherald, 1967).

In addition to the results of radiative equilibrium studies of the influence of carbon dioxide on the surface temperature, some computations have been made of the changes in the vertical radiative equilibrium temperature distribution and of the heating and cooling rates that would result from changes in the concentration of carbon dioxide (see, for instance, Gebhart, 1967; Newell and Dopplick, 1970). Figure 2.5, taken from Dopplick (1970), shows the contributions to the radiative heating and cooling produced by the principal trace gases involved: H_2O, CO_2, and O_3.

The results of these studies show that, if radiative processes alone are considered, the lower and middle troposphere would warm slightly and the stratosphere would cool if the carbon dioxide content were doubled. For example, stratospheric radiative equilibrium temperature would decrease by 2°C at 25 km and by about 12°C at 40 km. Radiative cooling at these levels would increase from about 1.0° to 1.1°C/day at 25 km and from 4.0° to 5.0°C/day at 40 km for an increase of carbon dioxide from 300 ppm to 600 ppm. The increased cooling (or lower radiative equilibrium temperature), if maintained when atmospheric motions are considered, would decrease the frost point required for water vapor saturation in the stratosphere and, conceivably, result in an increase in the occurrence and distribution of stratospheric clouds. It is probable, however, based on the computations of Manabe and Wetherald (1967), that these clouds would exert very minor influences (if any) on the temperature distribution in the troposphere. Both low-level warming and higher-level cooling would be

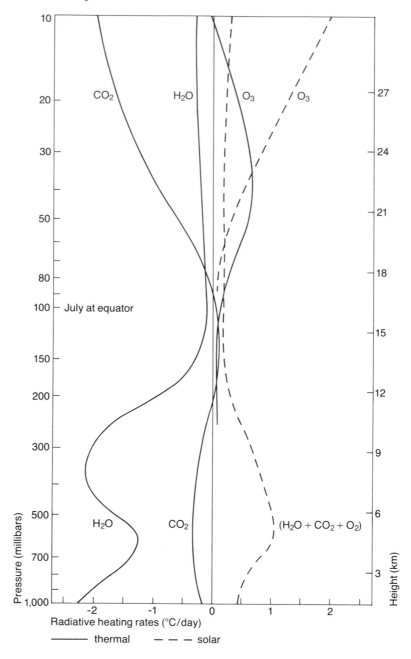

Figure 2.5 Radiative heating rates of three atmospheric constituents
Source: Dopplick, 1970.

reduced considerably at high latitudes during winter, as compared to summer.

The vertical and horizontal variation of radiative temperature changes due to carbon dioxide variations can affect the creation of atmospheric available potential energy, since radiative heating will occur in regions of the troposphere where it is relatively warm (lower troposphere at lower latitudes).

Thus, the radiative temperature changes should have an effect on the dynamics of atmospheric motions. However, it is important to note that the motions of the atmosphere interact in turn with the atmospheric temperature field generally in such a way as to reduce the temperature differences that exist.

In this discussion, and in all computations that have been made so far, three primary physical processes have been left out of the thermodynamic-hydrodynamic system that produces the aspect of climatic change being considered here. The three processes are the role of evaporation and condensation as energy transport mechanisms, the role of cloudiness (and the variation of cloudiness) in regulating the field of radiative flux in the earth-atmosphere system, and the role of the oceans in providing a heat reservoir to damp short-period temperature oscillations at the earth's surface. Although much physical insight can be (and has been) gained by the study of so-called radiative equilibrium models of the "greenhouse atmosphere," no definitive statement can be made regarding the influence of carbon dioxide variations on the temperature distribution of the atmosphere, or their effect on climate, until the radiative and dynamic effects are combined.

Conclusions

Radiative equilibrium computations, including a convective adjustment, suggest that the projected 18 percent increase of the carbon dioxide concentration by the year 2000 (to about 379 ppm) would result in an increase of the surface temperature of about one-half degree and a stratospheric cooling of 0.5° to 1°C at 20 to 25 km; a doubling of the carbon dioxide concentration over the present level would result in an increase of the surface temperature of about 2°C, and a 2° to 4°C decrease in the stratosphere at the same level. We would like to emphasize, however, that these computations neglect the important interacting dynamics and thermodynamics of the atmosphere, as well as the ocean-atmosphere interaction. This neglect makes the computed temperature changes very uncertain.

Recommendations

1.   We recommend that comprehensive global numerical-dynamical models, including ocean-atmosphere interaction and cloud variation, be developed and applied to the study of expected circulation, precipitation, and temperature patterns for the levels of CO_2 anticipated for the future.

2.   We recommend that the effects of stratospheric cooling be studied by developing models that permit cloud formation at these heights.

Effects of Changing Particles and Turbidity

All atmospheric particles in the size range 0.1- to 2.5-μ radius are of special interest to the earth's heat budget, since they are relatively abundant in the atmosphere and are well recognized as having the greatest effect in scattering, absorbing, and attenuating solar radiation. Particles in this size range also may have an effect on terrestrial long-wave radiation, which is usually ignored. Larger particles that might interfere more strongly with long-wave radiation are less abundant in most natural aerosols, but may become important locally around industrial pollution sources or in heavily dust-laden air (Bryson and Baerris, 1967).

It has commonly been assumed that the attenuation of direct solar radiation by particles is mainly attributable to scattering rather than to absorption (McCormick and Ludwig, 1967). For particles in the 0.1- to 2.5-μ radius range, scattering is predominantly in forward (downbeam) directions, with only a small fraction of the total (approximately 10 percent) in the backward (upbeam) directions. Any increase of backscatter tends to increase the albedo (reflectivity) of the atmospheric layer containing the particles. Whether increased backscatter increases or decreases the albedo of the combined atmosphere and underlying surface, however, depends upon the surface albedo.

Recently the neglect of *absorption* in atmospheric particulate layers has been criticized (Charlson and Pilat, 1969; Robinson, 1970). There are indications that in certain man-made aerosols, especially those of industrial origin containing carbon, iron oxides, and some other materials, the amount of absorption is of the same order as the amount of backscattering (Roach, 1961; Waldram, 1945). Backscattering of solar radiation by itself reduces the total heating within and below the scattering medium, since the fraction scattered back to space is not available for heating. Absorption, on the other hand, increases the heating within the absorbing medium. The net effect of the two processes on heating within and below an aerosol layer depends on their relative magnitudes. To determine this net effect in actual cases, it is considered necessary to measure the absorption and backscattering coefficients by the use of airborne instruments (Roach, 1961; Robinson, 1966).

Recent calculations by M. Atwater (1970) and by J. M. Mitchell (1970a) of the combined effects of absorption and scattering by aerosols and reflection from the surface indicate that a warming effect is attainable with relatively small (and nearly equal) absorption and backscattering.

It should be stressed that the situation with regard to stratospheric aerosols, or to high-tropospheric aerosol layers, is different from that of low- and middle-tropospheric aerosol layers in one very important respect. The heating caused by absorption within the former group of aerosol layers occurs well above the earth's surface. In such a case, surface air is likely to be cooled regardless of the magnitude of the absorption (and associated warming) in the aerosol layer itself. Direct observation of very small changes of atmospheric surface temperature following volcanic eruptions which hurled volcanic particles into the stratosphere (Mitchell, 1970b) tends to confirm this. (There is a further discussion of this in a later section, where the Agung eruption is compared to the potential SST particulate addition.)

Another quite different effect of aerosols, not related to their optical properties, is their role as condensation and freezing nuclei. Condensation nuclei, which are effective in forming water droplets at low supersaturation, are fairly well understood, but freezing nuclei are not. In particular, the wide fluctuations of freezing nuclei in space and time and their association with some human activities strongly suggest that precipitation from cold clouds may be inadvertently influenced by man. (Cloud seeding with silver iodide is a case of purposeful modification, of course.)

This could have an effect on a regional basis (Schaefer, 1969), but in view of the limited lifetimes of freezing nuclei in the atmosphere it seems highly unlikely that man will be able to influence the global freezing nuclei content.

Conclusions

1.   Tropospheric particles can produce changes in
a. albedo;
b. cloud reflectivity;
c. solar radiation reaching the surface and the upward flux of terrestrial infrared radiation;
d. cloud formation and precipitation.

While the specific effects on climate cannot be designated at this time, it appears clear that any such effects are enhanced as the number of atmospheric particles is increased. However, the nature of the impact on global climate by changes in the population of tropospheric particles remains undetermined in both sign and magnitude.

2.   Large variations in the stratospheric population of particles may produce correspondingly large fluctuations in stratospheric temperatures. For example, in the case of the Agung eruption, an increase of 6° to 7°C was found in the equatorial stratosphere. However, only a small decrease is apparent in the global surface temperature following volcanic eruptions.

3.   The area of greatest uncertainty in connection with the effects of aerosols on the heat balance of the atmosphere lies in our current lack of knowledge of their optical properties, in both the solar and terrestrial parts of the spectrum (approximately 0.3 to 30μ).

Recommendations

1.   We recommend that *in situ* scattering and absorption measurements be made under a variety of locales and conditions and that the radiative effects of particles be investigated.

2.   We recommend that laboratory and field research be conducted in many areas, including
a. sources, transport and dispersal, and purging mechanisms for contaminant particles;

b. nature and statistics of particle populations in the troposphere and the stratosphere;

c. particle production and transformation in the atmosphere;

d. particle effects on the physics of clouds.

The various tropospheric contamination patterns of particles should be identified and characterized by kind of contaminant population, geographic extent, and typical time histories downwind (in urban areas, coastal zones, deserts, and so on).

3.   We recommend that the effects of particles on radiative transfer be studied, and that the results be incorporated in mathematical models to determine the influence of particles on planetary circulation patterns.

Atmospheric Clouds
CLOUD DISTRIBUTION

The distribution of cloudiness is an important atmospheric parameter because of its central role in affecting the radiation budget of the atmosphere and, therefore, the evolution of climate.

Clouds reflect solar radiation (the reflectivity differs according to cloud type) and in general act as efficient radiators for terrestrial radiation. Also, because cloud transmissivity of solar radiation is higher than cloud transmissivity of terrestrial (infrared) radiation, clouds are important contributors to the so-called "greenhouse" effect.

As just indicated, clouds produce two opposite effects. Increased cloudiness results in an increased albedo which reduces the solar energy available to heat the surface but, at the same time, increases the back (downward) radiation to raise the surface temperature. It has been shown (Ohring and Mariano, 1964; Manabe and Wetherald, 1967) that the *net* result of increased low or middle cloudiness is to lower the surface temperature. Increased cirrus clouds, on the other hand, would warm the surface if the clouds were efficient (blackbody) radiators, or have practically no effect if the clouds transmitted infrared radiation. Some observations indicate that thin cirrus has an infrared transmissivity of about 95 percent, but thick cirrus (of the order of 5 km) may have a transmissivity of as little as 50 percent (see, for instance, Kuhn and Weickmann, 1969).

Before 1960 our knowledge of the cloudiness distribution was derived almost entirely from land and ship observations. Most of the observations were made over populated land areas and over narrow, but frequently used, shipping lanes. Relatively few observations were available in the tropics and in the Southern Hemisphere in general.

The ground-based observations indicate that the earth is covered about 52 percent by clouds with possible global variations from about 50 to 54 percent. These observations also suggest large seasonal variations in the tropics and subpolar latitudes and marked longitude variations at all latitudes. There are, however, no reliable data of the year-to-year variations of total global cloudiness, although the results of some studies have suggested a recent (the last ten years) increase in cirrus clouds.

Since 1960, satellite observations have been used to give some indications of the global cloudiness distribution. Because of the resolution and video systems used (there is a need in this area to set a proper threshold brightness for cloud detection), individual, small cloud groups are generally smeared out, and often thin cirrus clouds are not detected (although cirrus clouds could be inferred from infrared sensors). Horizontal spatial resolution of the order of 500 meters is needed therefore in order to get "correct" cloud population counts. We understand that this is within the capabilities of satellite systems now in operation and of those planned for the near future. If the brightness contrast is low, the threshold will be set at a low value and satellite-observed clouds may be reported in higher amounts than actually present. Some simultaneous observations by different satellite systems indicate that the problem of threshold brightness cannot be neglected if correct cloud populations are to be reported. Also, it is sometimes difficult to distinguish clouds from highly reflecting surface systems such as ice fields or desert areas. Despite these limitations, satellite observations have already shown the existence of unique cloud cluster formations and large cloudiness variability in the tropics and have provided considerable information on relative cloudiness in the Southern Hemisphere, where there was almost no prior data.

Since 1966, daily (and with the launch of ITOS-1 in 1970, twice daily) global cloud cover maps have been available *on a routine basis.* Detailed ground-based cloud cover data should be used to calibrate the satellite observations so the inferences made from satellite information in sparse data areas will be correct. This coordination will help in standardization of the satellite observations so that the absolute value of the total cloud cover and its variations can be determined.

Clouds appear in the atmosphere in different forms and at different levels. This information is, of course, exceedingly important in heat budget and climatic change studies. However, there is no practical method known, at present, for determining precisely the distribution of cloud amounts by type and height. Models of these distributions have, therefore, been developed, based on the regular synoptic, and aircraft, meteorological reports. It would be highly desirable if surface and satellite systems (using visible, infrared, and microwave sensors) could be coordinated to derive the necessary picture of the three-dimensional cloud patterns.

A global system of standardized cloud observations would provide the information for studies of the month-to-month, year-to-year and trend variations, as well as latitudinal and hemispheric differences, in cloudiness and cloudiness patterns (whether natural or man-made). It would be helpful if all this information were processed in a form so that it would be readily available for improved radiation studies and for the analysis of climatic trends, both "natural" and inadvertent.

OPTICAL PROPERTIES OF CLOUDS

Clouds reflect, absorb, and transmit electromagnetic radiation, but these properties depend very sensitively on the spectral range considered; on the solar zenith angle (for visible radiation); and on the cloud parameters such as the cloud's type, height, and thickness, droplet size and water content, and particulate concentration (that is, so-called clean or dirty clouds), and the nature of the surface below the cloud. There are many observations of optical characteristics and some theoretical studies of (ideal) clouds. The average cloud albedo varies from about 10 to 25 percent for cirrus to about 70 to 85 percent for thick cumulus clouds (Sellers, 1965; Kondratiev, 1969). The absorptivity of these clouds, although generally small (2 to 3 percent), can apparently be as high as about 30 percent (Robinson, 1970). The large measured variations in the transmission of solar radiation by clouds (differences of a factor of 2, in some cases for the same cloud type) cannot at prseent be explained.

Some specific optical properties that need to be studied, and their importance in environmental studies, are outlined below:

VISIBLE SPECTRUM

1.   Reflectively as a function of solar zenith angle, cloud parameters and, in the case of nonstratoform clouds, of the cloud array. Observations show that the reflectivity could vary from cloud to cloud within the same cloud type. Some of this variation might be due to the difference between a single cloud and the combined reflection of a group of clouds (for example, cumuluform) of the same type. This variation could be as high as 20 percent reflectivity. In view of the possible increase of cirrus clouds in the atmosphere, its reflective characteristic must be determined more accurately than is known at present.

2.   Absorption. The existing data (both theoretical and observational) give, for instance, absorptivities in cumulus clouds varying from 6 to 25 percent. This information needs to be verified and to be defined better as a function of the cloud parameters. The absorption of solar radiation in clouds can have a pronounced effect on the heating of the atmosphere.

3.   The contamination of clouds by nonwater substances could change the reflective and absorptive properties of the cloud. It is likely that "dirty" clouds have lower albedos and higher absorptivities than "clean" clouds. So-called dirty clouds are found in areas of high particulate concentrations.

INFRARED SPECTRUM

The radiative characteristics of clouds are known somewhat better for infrared radiation than for visible. The most important problem here is determining the emissivity ("blackness") of the various clouds as a function of thickness and composition (water or ice). In particular, it is necessary to determine the degree of blackness of cirrus clouds, both natural and those produced, for instance, by jet aircraft. The radiative properties of these clouds affect their detection by satellites but also could have some influence on temperature variation in the upper atmosphere. Although preliminary studies do not indicate that cirrus clouds play a large role in determining the surface temperature, the possible consequence of cloudiness variations for the earth's climate is very profound and requires further careful study.

Recommendations

1.   We recommend that there be global observations of cloud distribution and its temporal variations. The importance of cloudiness in the atmosphere stems from its relatively high albedo for solar radiation and its central role in the various processes involved in the heat budget of the earth-atmosphere system. High spatial resolution satellite observations (of the order of 500 m) are required to give "correct" cloud population counts and to establish the existence of long-term trends in cloudiness (if there are any).

2.   We recommend as being of fundamental importance studies of the optical (visible and infrared) properties of clouds as functions of the various relevant cloud and impinging radiation parameters. These studies, particularly for cirrus, should include the effect of particles on the reflectivity of clouds and a determination of the infrared "blackness" of clouds.

Effects of Heat Released

By the year 2000, the thermal power output of the world is estimated (Greenfield, 1970) to be of the order of 3×10^{13} watts, which is about 6 times the present output. Distributed over the globe, this is an insignificantly small amount. However, over cities the heating will be of the order of 1 percent of the solar constant, or about 10 percent of the solar heat absorbed at the ground. If this pattern continues, this will produce heat islands (cities or larger areas that are warmer than their surroundings) not much different in temperature elevation from those existing now, though more extensive in area (Peterson, 1969).

Recommendation

We recommend that the possibility of wider implications of the heat released be tested by numerical modeling. Specifically, the potential regional effects of aggregated heat islands should be examined.

Effects of Changing the Earth's Surface

INTRODUCTION

Some of the most important changes to the earth's surface are being produced by increasing urbanization, increasing land cultivation, and new water bodies. In addition, human activity may produce climatic changes that will affect the snow cover, thus producing a feedback.

URBANIZATION

Urbanization increases the area of cities or of city conglomerations. The heat island effect (the excess of temperature over the surroundings) is due largely to man-made heat sources in urban areas (Peterson, 1969). The temperature excess should remain at about the same magnitude as it is now; but the elevated temperatures should cover larger areas. Probably, the effects of such increased heat islands will be at least regional. For example, if the city borders an ocean or lake, the usual sea breeze would be suppressed by the diurnal variation of the heat island (that is, the city is relatively warm at night). The heat island effect causes air to rise over a city. This leads to a slightly larger precipitation probability over a city than over its surroundings.

Cities are rougher than most other regions. Hence, there is more turbulence and stronger vertical mixing over cities than elsewhere. On the other hand, the winds tend to be somewhat weaker.

Air pollution in cities, which is created by industry, incineration, automobiles, and space heating, causes snow to become dirty and to melt faster than it does in the cities' surroundings. The melting is further aided by the "heat island." Particles created over the city can travel large distances, dirtying snow in large areas and causing rapid melting. In this way, urbanization may affect the wintertime albedo in a large area. Normally, however, the albedo of cities is not substantially different from that of their surroundings.

EFFECTS OF WATER BODIES

Artificial lakes, produced by dams, have had negligible effects on the general climate and even on the shower activity downwind from them. This has occurred because the lakes have been created in otherwise quite dry areas. It is difficult to see how conventional damming operations now contemplated could affect the global climate.

However, certain more ambitious and speculative operations have been proposed. For example, if the Bering Straits were dammed and the Arctic Ocean were ice-free, it is possible that land glaciers would form in Canada and spread into the United States

(if, as Ewing and Donn suggested in 1956, the principal requirement for another ice age is the supply of water made available by
an ice-free Arctic Ocean). Similarly, damming the Congo would
affect the climate at least over a large portion of central Africa.
The anticipated area is large enough to suggest that global changes
of climate as a result of such projects cannot be completely ruled
out.

EFFECT OF CHANGES OF AGRICULTURAL PATTERNS

On a global scale, changes of agricultural patterns will affect only
a few percent of the earth's surface. New cropland will be produced, roughly half from grassland and half from forests. The
change from grass to crops has very little effect on albedo and
other surface features. Change from forest to crops increases the
albedo (particularly when snow covered), decreases the water available to the atmosphere (unless irrigated), and decreases the roughness. It is difficult to foresee any changes in global climate as a
result of these factors.

EFFECT OF OIL SPILLS

It has been suggested that oil spilled on the oceans may form extensive monomolecular layers which will prevent evaporation.
Observations with the kind of oil spilled at sea have shown, however, that the molecular structure of this oil is sufficiently irregular to permit evaporation (LaMer, 1962). In fact, evaporation is
quite vigorous over the Mediterranean, into which oil is being
spilled extensively. Also, the oil layer is usually perturbed by the
various motions at the sea surface. As a result, evaporation is reduced only in the immediate region of the original spills, and
such regions will not cover a significant area of the sea by the year
2000.

Also, oil spills appear cooler than their surroundings in the
infrared spectrum and warmer in the microwave region (see, for
example, Hanst, Lehmann, and Reichle, 1970). Again, however,
these temperature effects are small, even on a local basis.

ICE AND SNOW

If climatic change, man-made or natural, lowers the temperatures
in arctic regions, ice sheets will spread. Since snow and ice have
higher albedos than any other ground cover, the result would be
a further reduction of temperature, resulting in a further increase
in the ice. This behavior suggests an instability which, unless

checked by other factors, could lead to an equator-to-pole ice cover with an initial mean annual temperature drop of only about 2°C (Budyko, 1968, 1969, 1970; Sellers, 1969, 1970).

We know that even at the peak of the Pleistocene ice ages ice did not penetrate beyond middle latitudes. Hence factors not considered, such as cloudiness, the limited moisture supply, the oceans, and atmospheric motions counteract this "albedo instability." Nevertheless, it is an important process that must be included in theories of climatic change of any kind.

Conclusions

Surface changes which contribute to the production of CO_2 or particles are discussed elsewhere in this report and in the report of the Work Group on Monitoring.

Important among the remaining surface changes are, first, man's activities that modulate snow and ice cover, particularly in polar regions; and, second, some possible projects involving the production of new, very large water bodies.

Over and above these projects, only increased urbanization is of possible global importance, as it produces extended areas of contiguous cities. Still, it is not certain whether effects of urbanization extend far beyond the general region occupied by the cities.

Recommendation

We recommend that, before actions are taken which result in some of the very extensive surface changes described, mathematical models be constructed which simulate their effects on the climate of a region or, possibly, of the earth.

Effects of Jet Aircraft

There are two quite distinct subjects that we must deal with here: (1) the question of the effects of condensation trails (contrails) formed by current, commercial, subsonic, jet aircraft flying in the upper troposphere, and occasionally in the lowest part of the stratosphere (below about 12 km, or 40,000 feet); and (2) the question of the possible future effects of the contamination by SSTs of the higher stratosphere (at about 20 km, or 65,000 feet).

EFFECTS OF CIRRUS FROM JET CONTRAILS

Added contrails or cirrus clouds can affect the climate through both the heat balance and the nucleating role of falling ice crystals.

If an increase in cirrus cloudiness were to become significant as a result of jet activity in the upper atmosphere, the most important global effect would be an increase in the earth-atmosphere albedo. Less solar energy would be available to drive the atmosphere; however, as in all such suggestions, we must caution that until a realistic simulation of the process is available we cannot be sure that the atmosphere might not compensate for the artificial increase in cloudiness. There have been measurements and theoretical estimates of the immediate, static effects of added cirrus clouds on the temperature near the ground. The answers depend sensitively on the reflectivity, emissivity, and temperature of the clouds. In fact, one may reverse the sign of the temperature change by changes of the emissivity, for example.

It is known that falling ice crystals can nucleate supercooled water clouds into which they fall. The result would be a quicker initiation of the precipitation process than would otherwise occur. Much more work in this field would be needed before the importance of such artificial seeding effects can be evaluated.

Conclusion
Two weather effects from enhanced cirrus cloudiness are possible. First, the radiation balance may be slightly upset and, second, cloud seeding by falling ice crystals might initiate precipitation sooner than it would otherwise occur.

Recommendations
1.   We recommend that changes in cirrus cloud population be monitored directly through the program recommended earlier in the section on Atmospheric Clouds.
2.   We recommend that the radiative properties of representative contrails and contrail-produced cirrus clouds be determined.
3.   We recommend that the significance, if any, of ice crystals falling from contrail clouds as a source of freezing nuclei for lower clouds be determined.

EFFECTS OF SST CONTAMINANTS IN THE STRATOSPHERE
Two products of the SST are of concern: water vapor and particles. The increased water vapor in the stratosphere may have three consequences: (1) a "greenhouse" radiation effect, (2) ozone depletion as suggested by Hampson (1964, 1966), or (3) cloud formation.

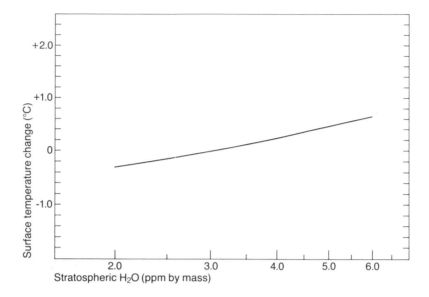

Figure 2.6 Surface air temperature change due to increased stratospheric water vapor, with average cloudiness and corrective-radiative equilibrium

Source: Manabe, 1970, after Manabe and Wetherald, 1967.

1.   The greenhouse radiation effect has been calculated by Manabe and Wetherald (1967) for a quasi-static atmosphere and is reproduced as Figure 2.6. The increase in ground level temperature, derived from Figure 2.6, is less than 0.10°C for the world average increase of 0.2 ppm of water vapor by mass.

2.   The inclusion of hydrogen (water vapor) in photochemical reactions will produce a smaller equilibrium ozone concentration than will a dry atmosphere. The reduced ozone would then admit more ultraviolet radiation to the lower atmosphere, a potentially undesirable condition for man and the environment. The quantitative aspects of the inclusion of moisture have been treated by London and Park (1970) and are illustrated by Figure 2.7. With the best present-day reaction rates and solar insolation amounts, equilibrium ozone distributions were calculated and the results are shown in Figure 2.7; one with no water vapor, curve 3; a second with 5 ppm (5×10^{-6}), curve 2; and a third with 20 ppm (2×10^{-5}), curve 1. It will be noted that water vapor alters the ozone concentration mainly above 50 km. Since the amount of

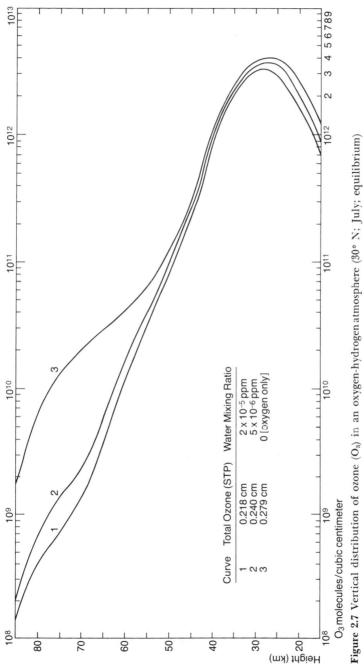

Figure 2.7 Vertical distribution of ozone (O_3) in an oxygen-hydrogen atmosphere (30° N; July; equilibrium)

Source: London and Park, 1970.

ozone above 50 km is very small compared with that below, major changes at the higher altitudes are of little significance in changing the total ozone in the entire column which controls the ground receipt of ultraviolet radiation. It must be noted that water vapor increases—either from 3 to 5 ppm (the likely maximum) or from 3 to 3.2 ppm (the world average)—fall between the curves, thus producing very small changes in total ozone. Interpolating, one finds a decrease of about 0.02 cm of ozone (standard temperature and pressure [STP]) for an increase in stratospheric moisture of 2 ppm, and 0.002 cm for the average global increase in stratospheric moisture of 0.2 ppm. Both of these changes in total ozone lie well within the routine, day-to-day, and geographical, variability.

These calculations must, however, be considered provisional. The photochemical reaction rates are, in some instances, imperfectly known; and in these calculations there is no adjustment allowed for transport or for the depletion of water vapor by photochemistry (which may reduce the new equilibrium water vapor concentration and consequently also reduce its influence).

3.   Nacreous, or mother-of-pearl, clouds occur in the height range 20 to 30 km and are most often seen over Norway in winter when the sun is below the horizon by several degrees. A condition for their occurrence seems to be a temperature near $-80°C$ or lower. Their lifetimes are not known with certainty, as the conditions for good visibility do not last long (World Meteorological Organization, 1956; Hesstvedt, 1959).

In the Southern Hemisphere over the Antarctic, a thin cloud veil in the lower stratosphere is often observed from midwinter to the beginning of October. The cloud frequently appears to cover the whole sky (Liljequist, 1956). Clouds are also observed near 80 km in summer at high latitudes (Ludlam, 1957).

The low temperature associated with nacreous clouds above the tropopause, coupled with their seasonal and geographical variations, lends support to the hypothesis that they are produced when local saturation occurs. At 24 km (30 mb), for example, when a mixing ratio of 3×10^{-6} g/g is assumed, the frost point is $-87°C$. It is quite conceivable that this temperature can be reached on individual days, particularly if the local vertical velocity is reinforced by mountain waves, as many people have suggested.

Suggestions concerning modification of high-level clouds from the injection of water vapor by the SST are based on the idea that a higher mixing ratio would permit clouds to form because the frost point could be reached at a higher temperature and, therefore, over a wider geographical and altitude range.

Recently Mastenbrook (1970) showed that the humidity mixing ratio over the Naval Research Laboratory's Chesapeake Bay Station east of Washington, D.C., had increased from about 2 ppm (by mass) in 1964 to 3 ppm in 1970. The upward trend is continuing and has no evident explanation. There are no other measurements adequate to confirm or reject Mastenbrook's finding or to indicate its geographic extent. A continued upward trend would clearly increase the likelihood of both natural cloudiness and artificial clouds from SST operations.

A further argument favoring cloudiness formation lies in the predictions of stratospheric cooling by the increase of atmospheric CO_2 (irrespective of the source of the CO_2). Manabe and Wetherald's calculation (1967) suggests that a tenfold greater cooling in the stratosphere than heating of the lower atmosphere would be caused by an increase in the CO_2 "greenhouse" effect. By 1985–1990 this could amount to several degrees C cooling of the stratosphere from the projected CO_2 increase due to the burning of fossil fuels by all sources (not only the SST).

The effects of SST particles (soot, converted hydrocarbons, and converted SO_2) can be assessed by analogy with the Agung dust cloud, in which nature may be said to have performed a most helpful experiment for us. The temperature in the equatorial stratosphere following the Agung eruption rose 6° to 7°C, and remained at 2° to 3°C above its pre-Agung level for several years, perhaps due to other volcanic activity in the tropics that has maintained a relatively high particle content (see Figure 2.8). Thus, something like a thirtyfold increase in stratospheric particles made itself felt in ordinary radiosonde temperature observations. The effect of Agung was observed for about 15 degrees of latitude, north and south of the source, at stations around the world. Such data have been analyzed, for example, from Australian stations 027 (Lae 6.5°S, 147°E), 120 (Darwin 12°S, 131°E), 461 (Giles 25°S, 128°E), 865 (Laverton 38°S, 145°E), and New Zealand station 780 (Christchurch 43°S, 173°E) (Newell, 1970a).

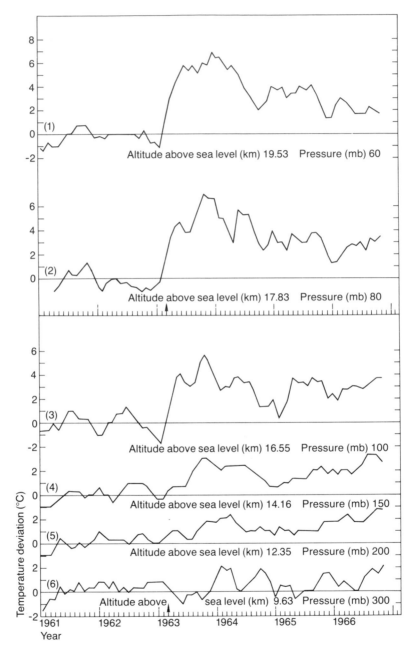

Figure 2.8 Temperatures at different pressures above Port Hedland, Australia
Source: Newell, 1970a.
Note: Arrow at March 1963 notes the Mount Agung eruption.

Although the analogy between the insertions of particles into the stratosphere by Agung and by the SSTs is tempting, there are several rather basic differences between the two situations. For example, the particles from Agung must have spread through a vertically much thicker layer, whereas the SST exhaust will all be injected in the limited range of its flight altitudes. It should also be noted that Agung is located near the equator. There, the temperature is lower than to the south or north; and it also increases upward from 17 km to 40 km. The SSTs, on the other hand, are likely to produce most of their particles at middle and high latitudes. In winter the temperature at 55°N at 20 km is actually larger than either to the south or to the north, and varies little in the vertical from 17 to 30 km. In summer the pole is warmer than other latitudes, but horizontal temperature gradients are weak.

For these and other reasons, the effects produced by Agung and by the SSTs may be quite different, even if the number of particles introduced is the same.

Our calculations show that the peak particulate loadings from the SST operations may be the same as the post-Agung observed concentrations for sulfates, and if one accepts the smaller pre-Agung values (which are less certain), then almost all the present estimates of particle concentrations from SST activities exceed the natural background concentrations. These increased particulate loadings may raise the temperature and play an important role in the stratospheric heat balance, as did the Agung sulfate cloud.

Conclusions

1.   The added SST water vapor in the stratosphere may introduce the following three effects (in order of likely importance):

a. Stratospheric clouds, already observed in the polar night, may increase in frequency, thickness, and extent. The effect of SST water vapor will be heightened by the increasing trends in CO_2 and natural stratospheric water vapor (observed only over Washington, D.C.).

b. Direct radiation effects will, according to Manabe and Wetherald's quasi-static radiation calculations, result in warming of air at ground level in regions of peak moisture concentration by less than a tenth of a degree C on a worldwide basis and cooling in

the stratosphere by a few degrees C. (The actual global effect would be smaller than the expected changes due to CO_2 increases.)
c. The reduction of ozone due to water vapor interaction (in a static photochemical model) has been estimated to lie well within the present day-to-day and geographical variability of total ozone.
2.   The direct role of quantities of CO, CO_2, NO, NO_2, SO_2, and hydrocarbons in altering the heat budget is small. It is also unlikely that their involvement in ozone photochemistry is as significant as water vapor.
3.   The SO_2, NO_x, and hydrocarbons can undergo complex reactions that produce particles. Increased particulate loadings may raise the temperature and play an important role in the stratospheric heat balance.

Recommendations

A feeling of genuine concern has emerged as a result of these facts and conclusions concerning the SST operations; we perceive that man's activities as he flies the projected 500 SSTs can have a clearly measurable effect in large regions of the world where they will fly, and quite possibly on a global scale. The effects will be most pronounced in the stratosphere, but we cannot exclude the possibility of significant effects at the surface. We must emphasize that, due to the uncertainties in the available information and its interpretation, we cannot be certain about the magnitude of the various effects.

Therefore, we recommend that uncertainties about SST contamination and its effects be resolved before large-scale operation of SSTs is implemented.

Specifically, we recommend the following program of action:
1.   Begin now to monitor the lower stratosphere for water vapor, cloudiness, oxides of nitrogen and sulfur, hydrocarbons, and particles (including the latter's composition and size distribution).
2.   Determine whether additional cloudiness or persistent contrails will occur in the stratosphere as a result of SST operations particularly in cold areas, and the consequences of such changes.
3.   Obtain better estimates of contaminant emissions, especially those leading to particles, under simulated flight conditions and under real flight conditions, at the earliest opportunity.
4.   Using the data obtained in carrying out these recommenda-

tions, estimate the change in particle concentration in the stratosphere attributable to future SSTs and its impact on weather and climate.

References

American Chemical Society (ACS), 1969. *Cleaning Our Environment: The Chemical Basis for Action* (Washington, D.C.: ACS).

Appleman, H., 1965. Investigation of the effect of contrails on cirrus frequency and coverage (unpublished).

Arrhenius, S., 1896. The influence of the carbonic acid in the air upon the temperature of the ground, *Philosophical Magazine, 41*.

Arrhenius, S., 1903. *Lehrbuch der kosmichen Physik 2* (Leipzig: Hirzel).

Atwater, M. A., 1970. Investigation of the radiation balance for polluted layers of the urban environment (unpublished).

Bainbridge, A. E., 1970. Data on the secular increase of atmospheric carbon dioxide at Mauna Loa, 1963–1969 (private communication to L. Machta).

Bhandari, N., Lal, D., and Rama, 1966. Stratospheric circulation studies based on natural and artificial radioactive tracer elements, *Tellus, 18*.

Bolin, B., and Bischof, W., 1969. Variations of the carbon dioxide content of the atmosphere (Stockholm: University of Stockholm, Institute of Meteorology), Report AC-2.

Brookhaven National Laboratory (BNL), 1969. *The Atmospheric Diagnostic Program at Brookhaven National Laboratory: Second Status Report, November 1969* (Long Island, New York: BNL). Report #BNL 50206 (T-553).

Brown, C. W., and Keeling, C. D., 1965. The concentration of atmospheric carbon dioxide in Antarctica, *Journal of Geophysical Research, 70*.

Bryson, R. A., 1968. All other factors being constant, *Weatherwise, 21*.

Bryson, R. A., and Baerris, D. A., 1967. Possibilities of major climate modification and their implications: Northwest India, a case study, *Bulletin of the American Meteorological Society, 48*.

Budyko, M. I., 1968. On the origin of glacial epochs, *Meteorologiia i Gidrologiia, 11*.

Budyko, M. I., 1969. The effect of solar radiation variations on the climate of the Earth, *Tellus, 21*.

Budyko, M. I., 1970. Comments on A global climatic model based on the energy balance of the earth-atmosphere system, *Journal of Applied Meteorology, 9*.

Cadle, R. D., Bleck, R., Shedlovsky, J. P., Blifford, I. A., Rosinski, J., and Lazrus, A. L., 1969. Trace constituents in the vicinity of jet streams, *Journal of Applied Meteorology, 8*.

Cadle, R. D., Lazrus, A. L., Pollock, W. H., and Shedlovsky, J. P., 1970. The chemical composition of aerosol particles in the tropical stratosphere, *Proceedings of the American Meteorological Society Symposium on Tropical Meteorology* (unpublished).

Chamberlain, T. C., 1899. An attempt to frame a working hypothesis of the cause of glacial periods on an atmospheric basis, *Journal of Geology, 7*.

Charlson, R. J., and Pilat, M. J., 1969. Climate: the influence of aerosols, *Journal of Applied Meteorology, 8*.

Cobb, W. E., and Wells, H. J., 1970. The electrical conductivity of oceanic air and its correlation to global atmospheric pollution, *Journal of Atmospheric Science*, forthcoming.

Cotten, J., 1970. Analysis and third degree polynomial fit of carbon dioxide concentration data from Mauna Loa (private communication to L. Machta).

Dopplick, T. G., 1972. Radiative Heating of the Tropical Atmosphere, in *The General Circulation of the Tropical Atmosphere* (Cambridge: The M.I.T. Press), forthcoming.

Ewing, M., and Donn, W. L., 1956. A theory of ice ages, *Science, 123*.

Federal Aviation Administration (FAA), 1967. *The U.S. SST, a Report on Economic Feasibility* (Washington, D.C.: FAA).

Fletcher, J. O., 1969. Controlling the planet's climate, in *Impact of Science on Society, 19*.

Flohn, H., 1969. *Climate and Weather*, B. V. deG. Walden, trans. (New York: McGraw-Hill).

Gebhart, R., 1967. On the significance of the shortwave CO_2 absorption in investigations concerning the CO_2 theory of climatic change, *Archiv für Meteorologie, Geophysik und Bioklimatologie, B 15*.

Greenfield, S. M., 1970. Projection and distribution of waste thermal energy, background paper prepared for SCEP (unpublished).

Hampson, J., 1964. Photochemical behaviour of the ozone layer, *Canadian Armament Research Division Establishment Technical Note #1627* (Val Cartier, Quebec: C.A.R.D.E.).

Hampson, J., 1966. Chemiluminescent emissions observed in the stratosphere and mesophere, *Les Problèmes Météorologiques de la Stratosphère et de la Mésosphère* edited by Centre National d'Etudes Spatiales (Paris: Presses Universitaires de France).

Hanst, P., Lehmann, J., and Reichle, H., 1970. Remote detection of air and water pollution from satellites and/or airplanes, background paper prepared for SCEP (unpublished).

Hess, S. L., 1959. *Introduction to Theoretical Meteorology* (New York: Henry Holt and Company).

Hession, J. P., 1970 (private communication to A. K. Forney).

Hesstvedt, E., 1959. Mother of pearl clouds in Norway, *Geophysica Norvegica, 20*.

Johnson, F. S., 1969. Origin of planetary atmospheres, *Space Science Reviews, 9*.

Junge, C. E., 1963. *Air Chemistry and Radioactivity* (New York: Academic Press).

Keeling, C. D., 1970. Atmospheric carbon dioxide: long-term trends, background paper prepared for SCEP (unpublished).

Kelley, J. J., Jr., 1969. An analysis of carbon dioxide in the arctic atmosphere near Barrow, Alaska, 1961–1967 (University of Washington), Report #NR 307-252.

Kondratiev, K. Y., 1969. *Radiation in the Atmosphere* (New York and London: Academic Press).

Kuhn, P. M., and Weickmann, H. K., 1969. High altitude radiometric measurements of cirrus, *Journal of Applied Meteorology, 8*.

Lamb, H. H., 1966. *The Changing Climate* (London: Methuen Books).

LaMer, V. K. (editor), 1962. *Retardation of Evaporation by Mono-Layers: Transport Processes* (New York: Academic Press).

Lees, L., 1970. Waste heat in the Los Angeles Basin (private communication to S. M. Greenfield).

Leipunskii, O. I., Konstantinov, J. E., Fedorov, G. A., and Scotnikova, O. G., 1970. Mean residence time of radioactive aerosols in the upper layers of the atmosphere based on fallout of high altitude tracers, *Journal of Geophysical Research, 75.*

Liljequist, G. H., 1956. *Norwegian-British-Swedish Antarctic Expedition, 1949–1952, Scientific Results,* Vol. II (Oslo: Norsk Polarinstitutt).

London, J., and Park, J., 1970. Ozone photochemistry in a wet atmosphere, forthcoming.

London, J., and Sasamori, T., 1970. Radiative energy budget of the atmosphere, *Space Research,* forthcoming.

Lorenz, E. N., 1970. Climatic change as a mathematical problem, *Journal of Applied Meteorology, 9.*

Ludlam, F. H., 1957. Noctilucent clouds, *Tellus, 9.*

Ludwig, J. H., Morgan, G. B., and McMullen, T. B., 1969. Trends in urban air quality (paper presented at the American Geophysical Union meeting, San Francisco).

McCormick, R. A., and Ludwig, J. H., 1967. Climate modification by atmo- ✓
spheric aerosols, *Science, 156.*

Machta, L., and Carpenter, T., 1970. Secular growth of cirrus cloudiness, forthcoming.

Machta, L., and Hughes, E., 1970. Atmospheric oxygen in 1967 to 1970, *Science, 168.*

Manabe, S., 1970. Surface air temperature change due to increased stratospheric water vapor, with average cloudiness and convection—radiative equilibrium (private communication to L. Machta).

Manabe, S., and Bryan, K., 1969. Climate calculations with a combined ocean-atmosphere model, *Journal of Atmospheric Science, 26.*

Manabe, S., and Wetherald, R. T., 1967. Thermal equilibrium of the atmosphere with a given distribution of relative humidity, *Journal of Atmospheric Science, 24.*

Martell, E. A., 1970. *Transport Patterns and Residence Times for Atmospheric Trace Constituents vs. Altitude,* Advances in Chemistry Series, 93 (Washington, D.C.: American Chemical Society).

Mastenbrook, H. J., 1970. Concurrent measurements of water vapor and ozone over Washington, D.C., during 1969 and 1970 (private communication to L. Machta).

Mitchell, J. M., Jr., 1965. Theoretical paleoclimatology, *The Quaternary of the United States,* edited by H. E. Wright, Jr. and D. G. Frey (Princeton: Princeton University Press).

Mitchell, J. M., Jr., 1968. Causes of climatic change, *Meteorological Monographs, 8,* No. 30 (Lancaster: Lancaster Press, Inc.).

Mitchell, J. M., Jr., 1970a. The effect of atmospheric particulates on radiation and temperature, background paper prepared for SCEP (unpublished).

Mitchell, J. M., Jr., 1970b. A preliminary evaluation of atmospheric pollution as a cause of the global temperature fluctuation of the past century, *Global Effects of Environmental Pollution*, edited by J. F. Singer (Dordrecht: Reidel Publishing Company), forthcoming.

Möller, F., 1963. On the influence of changes in the CO_2 concentration in air on the radiation balance at the Earth's surface and on the climate, *Journal of Geophysical Research, 68*.

Newell, R. E., 1970a. Stratospheric temperature change from the Mount Agung volcanic eruption of 1963, *Journal of Atmospheric Science*, forthcoming.

Newell, R. E., 1970b. Modification of stratospheric properties by trace constituent changes, *Nature, 227*.

Newell, R. E., and Dopplick, T. G., 1970. The effect of changing CO_2 concentration on radiative heating rates (unpublished).

Ohring, G., and Mariano, J., 1964. Changes in the amount of cloudiness and the average surface temperature of the Earth, *Journal of Atmospheric Science, 21*.

Pales, J. C., and Keeling, C. D., 1965. The concentration of atmospheric carbon dioxide in Hawaii, *Journal of Geophysical Research, 70*.

Peterson, J. T., 1969. *The Climate of Cities: A Survey of Recent Literature* (Washington, D.C.: National Air Pollution Control Administration, U.S. Government Printing Office).

Plass, G., 1956. The carbon dioxide theory of climatic change, *Tellus, 8*.

President's Science Advisory Committee (PSAC), 1965. *Restoring the Quality of Our Environment* (Washington, D.C.: U.S. Government Printing Office).

Raschke, E., and Bandeen, W. R., 1970. The radiation balance of the planet Earth from radiation measurements of the satellite Nimbus II, *Journal of Applied Meteorology, 9*.

Roach, W. T., 1961. Some aircraft observations of fluxes of solar radiation in the atmosphere, *Quarterly Journal of the Royal Meteorological Society, 87*.

Robinson, G. D., 1966. Some determinations of atmospheric absorption by measurement of solar radiation from aircraft and at the surface, *Quarterly Journal of the Royal Meteorological Society, 92*.

Robinson, G. D., 1970. Some meteorological aspects of radiation and radiation measurement, *Advances in Geophysics*, edited by H. E. Landsberg and J. Van Mieghem (London: Academic Press).

Schaefer, V. J., 1966. Ice nuclei from automobile exhaust and iodine vapor, *Science, 154*.

Schaefer, V. J., 1969. The inadvertent modification of the atmosphere by air pollution, *Bulletin of the American Meteorological Society, 50*.

Sellers, E. D., 1965. *Physical Climatology* (Chicago: University of Chicago Press).

Sellers, E. D., 1969. A global climatic model based on the energy balance of the earth-atmosphere system, *Journal of Applied Meteorology, 8*.

Sellers, E. D., 1970. Reply to comments of M. I. Budyko, *Journal of Applied Meteorology, 9*.

Spirtas, R., and Levin, H. J., 1970. *Characteristics of Particulate Patterns* (Washington, D.C.: National Air Pollution Control Administration, U.S. Government Printing Office).

Suess, H. E., 1965. Secular variations of the cosmic ray produced carbon 14 in the atmosphere and their interpretations, *Journal of Geophysical Research, 70.*

Swihart, J., 1970. Results of survey by the Boeing Company of sulfur content of present jet fuels (private communication to W. W. Kellogg).

Telegadas, K., and List, R. J., 1964. Global history of the 1958 nuclear debris and its meteorological implications, *Journal of Geophysical Research, 69.*

Thompson, J., 1970. Information on the calculated emissions of the GE-4 engine (private communication to L. Machta).

Tyndall, J., 1863. On Radiation Through the Earth's Atmosphere, *Philosophical Magazine* (4th series), *25.*

United Nations, 1956. World Energy Requirements, in *Proceedings of the International Conference on the Peaceful Uses of Atomic Energy*, Vol. I (New York: United Nations Department of Economic and Social Affairs, United Nations).

United Nations. *World Energy Supplies*, Statistical Papers, Series J (New York: Statistical Office, Department of Economic and Social Affairs, United Nations).

Waldram, J. M., 1945. Measurements of the photometric properties of the atmosphere, *Quarterly Journal of the Royal Meteorological Society, 71.*

World Meteorological Organization, 1956. *International Cloud Atlas* (Geneva: Atar S. A.).

3

The Concept of Monitoring

When the miner's canary died, it was time to get out of the mine. The canary "monitored" the mine air and gave an indication of potential disaster due to odorless, invisible methane. The immediate action necessary was clear; long-term solutions could be considered later.

But when we are concerned with a global environmental problem, this type of monitoring is insufficient. Because we cannot escape from the earth, we must have more than a sentinel to sound an alarm if a critical threshold is passed; we must know what it is that kills our "canary," where it comes from, and how to turn it off at the source.

Accordingly, we think "monitoring" is best conceived of as systematic observations of parameters related to a specific problem, designed to provide information on the characteristics of the problem and their changes with time. The parameters and problems with which we have been concerned are those of the global environment. And though any monitoring program will provide information useful to dealing with local and regional problems, our concern has been with identifying existing and potential monitoring systems capable of securing the information necessary to deal with the critical global problems identified by the Study of Critical Environmental Problems (SCEP).

For every one of the global problems that have been identified, we find we have insufficient knowledge of either the workings or the present state of the environmental system. This hinders us as we attempt to design monitoring that will not only warn us of change but also provide information upon which we can base rational and efficient remedial action. In most instances we can suggest a likely analogue of the canary, but we do not know what action would be best once our bird shows up sick. Further, we are persuaded by our colleagues that global systems both physical and biological are so complex that the ultimate consequences of any disturbance cannot at present be predicted with confidence.

Reprinted from Study of Critical Environmental Problems (SCEP), 1970. *Man's Impact on the Global Environment* (Cambridge, Massachusetts: The M.I.T. Press), pp. 168–222.

For these reasons our report is concerned not only with monitoring in its sense of providing warning of critical changes but also with measurements of the present state of the system (the "base line") and with measurements in support of research into the workings of the system. We mention the need for this research where it is apparent to us; we have not attempted to provide a complete assessment of research needs. In general, however, we have agreed that research is most needed in providing a closer specification of the present state of the planet and in developing a more complete understanding of the mechanisms of interaction between atmosphere, ocean, and ecosystem.

Monitoring Techniques and Systems
Before we turn to an analysis of the monitoring aspects of the critical global environmental problems identified by SCEP, we feel it necessary to look more generally at the current state of monitoring techniques and systems, both to provide a framework within which the later recommendations can be seen and to point out areas on which emphasis should be placed as the effort to obtain information for environmental management continues.

Economic and Statistical Monitoring
If we are concerned with predicting the accumulation of a pollutant in the environment, its rate of input must be known. For example, to evaluate the global CO_2 problem, we require a long-term prediction of the total atmospheric content, which, in turn, depends in part on statistical projections of fuel production and consumption. To evaluate the contribution of SO_2 to the global particulate problem, we require projections of natural, industrial, and energy-conversion emissions of SO_2 and SO_2-precursors which take into account possible control and abatement measures.

If we are concerned with evaluating the effects of alternative control technologies on pollutant levels, we need quantitative information about the flow of materials which will be altered by control technology to include inputs, wastes, and end products at each stage of the process (Ayres and Kneese, 1968).

Both these kinds of activities seem to us to be essential forms of monitoring. To some degree they are already being carried out, for example, in industry as a part of the management process, and in government as a part of economic policy making and of already existing regulatory activities.

Yet, the focus of the gathering and synthesis of data concerning industrial, agricultural, domestic, and energy-producing activities has not commonly included an "environmental effects" component. Moreover, research into the natural pathways and degradation of pollutants once they are deposited in the environment has as yet yielded little quantitative data. Nor is our knowledge of the functioning of particular ecosystems sufficient to allow us to quantify effects of a particular pollutant when it, for example, eradicates one or a group of species within the ecosystem. This has meant that we lack information on which to base projections and models for decision making. It has also meant that the organization of data on which to base an analysis of the global environmental problems considered by this conference has been a tedious, often approximate, and sometimes impossible task.

New methods for gathering and organizing economic and statistical data must be developed if we are to have the "handle" we need to deal with environmental problems. New centralized collection and collation points are needed. Federal regulatory bodies in the United States, such as the National Air Pollution Control Administration (NAPCA), have begun this task for their own areas of responsibility. But there exists no effective organization summing data across traditional areas of environmental responsibility, such as air and water pollution. Nor do we have any comparable international organization or, indeed, any effective standards to ensure, for example, that the industrial production data collection going on across the world will be of comparable precision and focus. All these tasks must be accomplished if we are to use effectively our economic and statistical monitoring potential.

Recommendations
1. We recommend the development of new methods for gathering and compiling global economic and statistical information, which organize data across traditional areas of environmental responsibility, such as air and water pollution.
2. We recommend the propagation of uniform data-collection standards to ensure, for example, that industrial production data

collection being carried out across the world will be of comparable precision and focus.

Physical and Chemical Monitoring

Physical and chemical monitoring methods are used to determine the amount of a contaminant in a sample of soil, water, air, or organism. Physical methods are also used to determine a property of an environmental system as a whole, such as the refractive index or the albedo of the atmosphere.

There are numerous examples of this type of monitoring. NAPCA's network of sampling stations monitors the quality of urban air. Weather satellites monitor the formation of hurricanes. The essence of good monitoring of this type is to measure what is needed, and no more, with the precision that is needed, and no more, and to maintain standards indefinitely.

Traditionally, monitoring of this type is carried out in networks of fixed stations. The entire operation may be completed at these stations or a sample may be taken to a central laboratory for examination or analysis. In either case, central coordination of methods and central standardization is necessary. Monitoring is now extended to measurements on ships, aircraft, and satellites. Moreover, it has become an international activity, and international coordination of standards is necessary.

Measurements of solar radiation at the ground form a good if little-known example. They are made in numerous countries by national meteorological services, universities, agricultural research stations, and so forth. They are collected, edited, and published at a Soviet observatory under arrangements guided by the World Meteorological Organization. These measurements are not of uniform quality, but the best, which can be identified, have been standardized by instruments compared internationally on an ad hoc basis. The last major comparison was arranged bilaterally between an American manufacturer and a Soviet university. The necessary nominal institutional anchor for these comparisons is the Radiation Commission of the International Association of Meteorology and Atmospheric Physics within the framework of the International Council of Scientific Unions. To give another example, our knowledge of the CO_2 content of the atmosphere is due to the interest, skill, and cooperation of small groups of scientists in the United States and Sweden (Pales and Keeling, 1965; Keeling, 1970).

Recommendation

We recommend the development and expansion of current physical and chemical monitoring systems and techniques. Specific recommendations are included later in this chapter.

Modern Technology and Monitoring

A major problem in those monitoring systems that require precise measurement and rigorous attention to detail is to combine the continuity and security of institutional control with the continued devotion of the scientists and skilled technicians who are often required to work in isolated regions. Further, the extreme care that is necessary in sample handling makes the standardization and interpretation problem extremely difficult when measurements must be made independently at each one of many stations within a global network (*in situ* monitoring) (Pales and Keeling, 1965).

For these and other reasons we must continue to investigate the possibility of monitoring remotely, by automatically reporting or man-operated instruments carried in earth satellites or airplanes.

The unmanned earth satellite, especially, has many virtues as a vehicle for monitoring equipment. It can provide global coverage in very short periods of time. It can carry equipment without subjecting it to environmental stress. If large amounts of information from widely distributed points are required or if a particular monitoring system can be "piggybacked" on a satellite with other functions, there may be cost advantages. *In situ* techniques in general yield data whose precision has not yet been equaled by remote techniques for measuring the same parameter. The techniques now available for use in satellites must be improved if they are to provide the high-precision data that are required.

It must be pointed out that interesting and useful data have already been obtained from satellites for use in the atmospheric and oceanographic sciences. This, combined with preliminary information about satellite monitoring experiments planned in the next few years, indicates that satellite solutions to many environmental data-gathering problems could be available in the not-too-distant future.

What is clearly needed is a series of evaluations of appropriate satellite techniques, including both scientific feasibility studies and cost-benefit analyses. Once scientific feasibility is established, the aim of such an evaluation would be to determine the optimum system for providing the required information, be it a satellite, ground, or mixed system.

In addition to this comparative evaluation of remote and *in situ* monitoring, there is need for continuing research into monitoring per se. In recent years new and powerful techniques have been developed in radiometry, radar, and spectroscopy which are operated in the microwave, infrared, and visible regions of the electromagnetic spectrum. Furthermore, digital data-processing systems have been developed which permit rapid handling and analysis of large data flows. There is a need for continuing research into these systems, as well as into the integration of diverse requirements to ensure optimum use of resources, and into scientific and administrative coordination.

Recommendations

1.   We recommend generally a series of evaluations of appropriate satellite measurement and monitoring techniques, including both scientific feasibility studies and cost-benefit analyses, aimed at determining the role of satellites in an optimum monitoring system for the problems dealt with by SCEP.

2.   We recommend continuing research into the application to environmental monitoring of radiometry, radar, spectroscopy, digital data processing and other newly available techniques.

Monitoring Critical Global Problems

Our conception of monitoring associates the activity with identified problems. The deliberations of SCEP appear to us to have isolated the following as currently critical global problems:

1.   Ecological effects of DDT and other toxic persistent chlorinated hydrocarbons.

2.   Ecological effects of mercury and other toxic heavy metals.

3.   Climatic effects of increasing carbon dioxide content of the atmosphere.

4.   Climatic effects of the particle load of the atmosphere.

5. Climatic effects of contamination of the troposphere and stratosphere by subsonic and supersonic transport aircraft.
6. Ecological effects of petroleum oil in the oceans.
7. Ecological and climatic effects of physical and thermal changes at the earth's surface, including changes in land use.
8. Ecological effects of nutrients in estuaries, lakes, and rivers.

In this section we consider in turn the nature of the monitoring effort associated with each of these problems.

Where we feel competent, we have added estimates of the cost of some parts of the effort. We add the caveat that these estimations are based on our visualization of how the measurements might be made. Furthermore, time did not permit the rationalization of estimates made by different members of the group. If monitoring schemes are activated, others will doubtless activate them, and they will differ in some degree, perhaps considerably, from our visualization. The basis of costing will therefore also differ.

We add a second caveat that is even more weighty. It would be far better to monitor global problems by using an integrated system than to tackle them piecemeal. To some degree we have assumed integrated systems in making our cost estimates and in estimating the number of sampling stations and the kind of data processing required. Yet because of the lack of base-line data and the nature of the SCEP exercise, it has been impossible for us to do much more than state the need for a study of the components of an optimum integrated monitoring system for global problems.

Atmospheric Composition and Possible Climatic Changes
The possibility has been widely publicized that man, by changing the chemical composition of the atmosphere by burning fossil fuel, might also inadvertently change the physical state and motions of the atmosphere and thus the climate (Shapley, 1953; Callender, 1961; Mitchell, 1969). Elsewhere in this report it is emphasized that the present state of atmospheric science is such that no certain answer can be given to the question, "Can man's activities produce catastrophic changes of climate?"

What is clear to us is that the stake in terms of human welfare is so high that the relatively small investment in monitoring and research should be made. In relation to climate two atmospheric constituents appear to be of importance, carbon dioxide and particles. They present monitoring problems of diverse kinds, but so far as measurements in the atmosphere are concerned they have some things in common. We shall examine these problems after we have looked more generally at atmospheric monitoring.

The physical state of the atmosphere is continuously observed in great detail for general meteorological purposes, so that if we were interested in the distribution of a contaminant and knew where it entered the atmosphere, we could trace its path with useful precision, and estimate the extent to which it spreads by diffusion. If we wished to monitor any pollutant, we could locate sampling stations by reference to our knowledge of atmospheric motions. However, we have now very little knowledge of the distribution within the atmosphere of some pollutants.

Reasoning of this kind leads us to suggest that global monitoring of a trace contaminant or constituent with a long residence time, for example, CO_2, can be accomplished by about 10 stations. Constituents with short residence times with a number of localized sources, for example, some particles, we judge to require about 100 stations (Commission for Air Chemistry and Radioactivity, 1970). Experience for 1 or 2 years with measurements at about 100 stations, including aircraft collection, would allow us to reassess the need, decreasing or increasing the number of stations or relocating some if necessary. In this sense we would acquire an atmospheric "base line." Development of a comprehensive, permanent monitoring system requires these initial measurements.

PREDICTION OF THE CARBON DIOXIDE CONTENT
OF THE ATMOSPHERE

The earth's accessible reservoirs of CO_2 are in the atmosphere, the oceans, the living and decaying biomass. There is continuous exchange between these reservoirs, and man is continuously releasing the CO_2 that is stored in fossil fuels. To allow study of the possibilities of climatic change, we must know what proportion of the released gas remains in the atmosphere, we must be able to extrapolate the trend for many years ahead, and we must be able to update the extrapolation continually. If the existence of a climatic trend were to be firmly established, the obvious corrective actions—curtailment of combustion and initiation of major population movements—would imply change in the whole pattern of human life. We should need a long lead time in order to make the change; monitoring the CO_2 problem therefore calls for the longest possible extrapolation of the average CO_2 concentration in the atmosphere. This estimate must be based on

1.   Accurate and reliable data obtained from a monitoring system that senses mean global CO_2 changes, free from perturbations from local sources and sinks.

2.   A measurement program that investigates the earth's CO_2 system, including the three interacting components of source, route, and reservoir.

THE PRINCIPAL SOURCES OF CO_2

Possible sources of CO_2 are changes in the biomass (mostly forests), the combustion of fossil fuels by man, and net release or take-up of CO_2 by the oceans. Monitoring, which yields a global map of source position (and intensity at each position), is done by obtaining and analyzing the available data on man's activities such as the burning of fossil fuels, lumbering, or forest clearing for agricultural purposes.

These data need be updated only every 5 years since these sources do not vary significantly in shorter periods of time. Use should be made of the high resolution data available in space photography (Minzner and Oberholtzer, 1970). Use of such data could help determine the characteristics of sources that have not been documented in available records.

There appear to be very few data useful in assessing the effect of man on the carbon cycle in forests. A research program should be instituted to obtain and interpret these data, which would note in a direct manner the net rate of change of carbon that is fixed in the forests.

Data-Gathering Requirements

1.   Spatial resolution. The production and usage records should be available for each large industrial region and forest .area in each country.

2.   Temporal resolution. The needed resolution for industrial, domestic, commercial, and transportation combustion is seasonal, yielding annual averages on a global basis. These 12-month averages should be upgraded every 5 years.

3.   Accuracy. For monitoring, estimates of source changes should be made with the highest possible accuracy—±20 percent should certainly be attainable. For research, carbon changes in living

trees and woody debris should probably be measured to better than ± 1 ppm.

Data-Gathering Techniques

1.   Monitoring. Empirical source models would be developed from available records, augmented by photographs taken in the visible and infrared by satellite systems already in operation and those planned.

2.   Research. The carbon cycle data would be obtained by one or more of the following proved techniques: (a) Harvest techniques, which measure weight increase (and caloric equivalent and chemical composition) of net production, (b) enclosure studies, involving measurements of CO_2 exchange in plastic enclosures of parts of ecosystems and (c) flux techniques based on measurements of CO_2 levels in the forest environment (Woodwell and Whittaker, 1968). The last method, which may yield extremely high quality data, requires the development of new techniques.

Cost estimates. The costs for source monitoring and research are given here with the annual operating costs estimated to be the number of scientific and technical staff multiplied by an assumed total cost of $60,000 per staff member. The capital costs are for data-processing equipment and laboratory test equipment (see Table 3.1). The total cost would be approximately $300,000 for the first year.

THE PRINCIPAL ROUTES OF CO_2

The major route by which CO_2 is transported from one region of the earth to another is the atmosphere. Oceanic transport is much less effective.

The task of extrapolating global atmospheric CO_2 content calls for the institution of monitoring and research programs. The aim of the research would be to develop a quantitative understanding of CO_2 transport in the atmosphere as a basis for the extrapolation. The inputs to this transport model are the source data, whose characteristics in the future would be obtained by estimating future population increases and the use by this population of fossil fuel and wood.

The monitoring stations should be on land and in places where the measurements will be free of perturbations from local

Table 3.1 Cost Estimates for Monitoring and Research

	Capital Investment	Annual Operating Costs
	(thousands of dollars)	
Monitoring	40	120 (2)
Research	40	120 (2)
Total	80	240 (4)

The ( ) indicate the number of technical staff.

sources, wherever possible. If the research effort indicates a multi-source accumulation which may lead to regional climatic effects having possible widespread implications, then monitoring will also be necessary in these regions.

The research stations could be on land or in airplanes. Satellites seem to have potential here because they can, in principle, give continuous globally distributed data, including source effects, which the other monitoring systems cannot do. The satellite method is listed under the research effort because increases in precision and resolution are needed in the remote instrumentation which would have to be used; instrumentation for ground stations (Pales and Keeling, 1965) and aircraft is already developed.

Data-Gathering Requirements

1.   Spatial resolution. For near term monitoring, we need to measure the total CO_2 content in a thoroughly mixed atmosphere, free of sources. Since the effects of the sources are not well understood, prudence suggests more than one station. Four widely separated stations are suggested, two in the Northern Hemisphere (Alaska and Hawaii) and two in the Southern Hemisphere (South Indian Ocean island and Antarctica). Such diversity in position tends to minimize perturbations due to some special regional effect.

From a costing view, this arrangement may be less satisfactory than a single land station plus one airplane. The latter could be used to minimize perturbations in "clean air" data due to vertically distributed sources. In fact, this is already being done by Swedish experimenters who use commercial flights mainly over the Arctic Sea (Bolin and Bischof, 1970). They have collected about 700 good samples from 50 flights. However, the data quality obtained by using airplanes seems to be somewhat inferior at present to that obtained from ground stations. In the short term, the latter approach is preferred, although the feasibility of using airplanes should be investigated, along with the use of statistical techniques to optimize data analysis.

For research, 12 stations (which include the 4 monitoring stations) have been suggested, of which 3 would be separated widely in longitude (Indian Ocean, North Atlantic, and South Atlantic) while the other 9 would stretch in the Pacific from the North to the South Pole with a spacing of 20° or 2,400 km. These stations would record effects due to atmospheric and ocean currents and possible source effects. This recommendation is similar to one made by the Commission for Air Chemistry and Radioactivity: "To monitor seasonal and regional variations adequately, the minimum number of stations for CO_2 and constituents of similar character is about 10." (Commission for Air Chemistry and Radioactivity, 1970.) The altitude dependence, to be obtained by using an airplane with improved instrumentation, is a research effort; the suggested vertical spatial resolution is about 2 km with several horizontal passes at different altitudes taken during the flight. This kind of research data coverage would be quite expensive.

2.   Temporal resolution. Data have been taken almost continuously for some years at a station on Mauna Loa (Pales and Keeling, 1965). In further monitoring it might be possible to collect data at greater intervals, leading to decreased data-processing costs, by using appropriate statistical techniques.

3.   Accuracy. Data taken to date at Mauna Loa show diurnal variations of ±0.5 ppm, seasonal variations of ±3.5 ppm, and a mean annual increase of 0.7 ppm (Pales and Keeling, 1965). Present indications are that changes of order 0.1 ppm in the annual rate of increase are significant in the extrapolation problem. The essential requirement is that there should be no systematic drift in reference standards, or systematically changing errors in operation of the instrument. An accuracy in individual reading of ±0.1 ppm is sought at the Mauna Loa station. It seems advisable to continue this practice at all monitoring stations for the present. In time, some relaxation, perhaps to ±1 ppm, might be permissible.

Data-Gathering Techniques

1. Monitoring. The *in situ* techniques are ideal for monitoring in the immediate future. They are in general extremely accurate and very economical in a ground station. Occasionally, impurities in the intake pipes will degrade the data. Also "flask" data are about 1 ppm higher than continuous intake data for reasons which are not understood. The air sample is cooled to separate water vapor and the CO_2 concentration is measured by infrared techniques.

2. Research. When global coverage is desired, then *in situ* measurements become impractical and remote measurements must be made. Such measurements become very difficult when an accuracy of ± 0.1 ppm or even ± 1 ppm is desired. This requires that research be done on the most appropriate techniques for making the measurement: passive (no radiation source) or active (use of a radiation source); use of visible, near infrared, or far infrared wavelength bands; vertical viewing or horizontal viewing; and so forth. These, and other appropriate techniques should be placed in the study phase category. As noted earlier, this would involve feasibility studies which would include some instrumentation testing on the ground and in an airplane. These techniques will most likely be useful in obtaining data on other gaseous pollutants and particles.

Cost Estimates

A suggested cost schedule for the atmospheric route effort in the first year is shown in Table 3.2. The first is a site selection study

Table 3.2 Cost Schedule for the Atmospherie Route Effort

	Capital Investment	Annual Operating Costs
	(thousands of dollars)	
Monitoring*	2,000	960 (16)
Research†	600	480 (8)
Total	2,600	1,440 (24)

* Indicates that the Mauna Loa station (upgraded at a cost of about $150,000) and 3 new stations (each with an initial capital investment of about $600,000) would be prepared for monitoring during the first year.
† Indicates that the first year's research effort is divided into two parts.

for the other 8 sites. This effort would use about 3 staff and
$100,000 (for site selection experimental equipment) in capital
investment the first year. The second effort is the beginning of an
experimental program to use advanced technology in instrumentation and station platforms, which would involve approximately
$600,000 for instrumentation and 5 staff; ( ) indicates the number
of technical staff. The total cost of the CO_2 "atmospheric route"
program would be approximately $4 million in the first year.

THE PRINCIPAL RESERVOIRS OF CO_2

Data-Gathering Requirements and Techniques

Since the ocean is the principal reservoir, high priority is placed
in gathering data pertaining to its characteristics. The ocean's
ability to store CO_2 varies very significantly with surface layer
temperature. Furthermore, an exchange of surface water with
deep ocean water, which is rich in CO_2, will cause the ocean to
release CO_2 to the atmosphere.

This upwelling phenomenon occurs in three areas: south-
west of Greenland, the Norwegian Sea, and the Weddell Sea.
These areas should be surveyed annually by oceanographic ships
that would look for intermittent large-scale upwelling of the
deep waters: the required measurements would be of C^{14}, the
partial pressure of CO_2, temperature, and salinity.

The average temperature of the top 400 meters of the ocean
should be monitored with sufficient accuracy to observe $0.5°C$
changes from year to year. Sea surface temperature is intensively
measured for meteorological purposes (8,000 observations per day
in the Northern Hemisphere shipping lanes) and there are plans
in the World Weather Watch program for use of satellite methods
and automatic data buoys. Meteorological requirements for surface water temperature measurements supply data sufficient for
the needs of CO_2 monitoring.

Available data on subsurface temperatures are less satisfactory, but there are several deepwater stations already in use. These
are ships that use submerged thermographs (thermistors) to measure temperature to $±0.3°C$ down to levels of 300 meters, with 0.3-
meter vertical spatial resolution. A study of the configuration of
observing methods best suited to meet all requirements for ocean
water temperatures may soon be needed.

The major uncertainty in the process of extrapolating atmospheric CO_2 content is allocation between the ocean and the biosphere of that proportion of the source which does not remain in the atmosphere. This proportion will almost certainly vary over time as fossil fuel is burned. At present, it is uncertain within a factor of 2. This uncertainty could be decreased by measuring the rate of downward mixing of ocean waters from the surface to depths of about 600 meters or more. This might be obtained by following the C^{14} released by the past nuclear tests. (The present C^{14} levels are 150 to 200 percent of the pretest values.) The necessary data could be obtained by 3 meridional transects, one in each of the three major oceans, in which C^{14} could be measured at different latitudes to depths of about 1,000 meters. The data should be taken with a 3-year spacing at first, then 5 years, then every 10 years. The first year would involve only the transect in the Pacific.

Cost Estimates

A suggested approximate cost schedule for the ocean reservoir effort in the first year follows (see Table 3.3).

In this schedule the temperature measurement effort is a study only and involves no capital investment. The upwelling and C^{14} measurement efforts involve ship costs that were estimated at the normal rate of $5,000/day. The capital investment for these two efforts includes the costs of instrumentation and data processing. The cost for the ocean program in the first year is approximately $1.2 million.

The total approximate cost of the CO_2 monitoring effort, based on these figures, is $5.5 million for the first year. It must be emphasized that this figure is to be used as a guide only and is not based on a thorough cost analysis. This cost does not involve

Table 3.3 Cost Schedule for Ocean Reservoir Effort

Required Information	Capital Investment	Annual Operating Costs
	(thousands of dollars)	
Upwelling	150	220 (6)
Temperature	0	120 (2)
C^{14}	100	620 (2)
Total	250	960 (6)

coupling to the other efforts in monitoring (which, of course, must be done) and is not realistic from that point of view.

Recommendations

1.   We recommend continuing attention to improving our estimates of present and future consumption of fossil fuel.

2.   We recommend similar attention to changes in the mass of living matter and its decay products.

3.   We recommend continuous measurement and study of the carbon dioxide content of the atmosphere in a few stations remote from sources—specifically 4 stations—and on some airplane flights. We particularly recommend that the existing record at Mauna Loa Observatory be continued indefinitely.

4.   We recommend continuing systematic study of the partition of carbon dioxide between the atmosphere, the oceans, and the biomass. Such research might require up to 12 stations.

THE PARTICULATE LOAD OF THE ATMOSPHERE

The emission of solid and liquid particles into urban atmospheres and the formation of particles there by chemical reactions between emitted gases yield what is to many people the most obvious and obnoxious sign of "atmospheric pollution"—smog (National Air Pollution Control Administration, 1969; Cadle, 1970). Emissions are extensively monitored on a local basis in many parts of the world. We are concerned here with the truly global problem: "What changes in the particulate loading of the atmosphere affect the world's climate?" (Mitchell, 1969; Shapley, 1953.) The question as it affects monitoring becomes, "Is the particle content of the atmosphere changing in a systematic way as a result of man's activities?" Monitoring can, at least in this instance, also help considerably with the further questions: "How serious is the climatic threat? How can we best correct the trend?"

Globally the particle load affects climate by changing the radiation balance of the earth. It acts in two ways: by changing the amount of solar energy available to the earth, that is, by changing the earth's albedo (reflectivity); and by changing the internal redistribution of the solar energy (CO_2 acts only in the latter way). Atmospheric particles have readily observed optical effects, which we monitor by optical devices, but these methods do not give us the opportunity to identify the sources of the particles.

Chemical methods do permit identification and thus provide the data for rational discussion of abatement. A very large proportion of atmospheric particles, both natural and manmade, are not emitted as particles but are formed within the atmosphere by chemical reaction between trace gases. Hence, monitoring the particles implies monitoring these trace gases (Cadle, 1966, 1970).

MONITORING PARTICLES BY OPTICAL METHODS

We are concerned with measurements of solar radiation or of similar radiation from artificial sources in the atmosphere, and we wish to isolate the effect of particles on this radiation. Particles are transported by the atmosphere, and their reservoir is the atmosphere. The residence times of particles are hours to days in the troposphere, weeks to years in the stratosphere. Because of the short tropospheric residence times, spatial variation of the total particulate load of the atmosphere is very high. Cities and windy deserts are major sources, and the load is high in their vicinity. The global monitoring problem can be attacked by choosing a few sites remote from surface sources, by making simple measurements at many sites and examining the results statistically, or by standing off in space and looking at the earth as a whole.

The last of the three methods is the most fundamentally sound attack. The potential climatic effect of warming or cooling would be the result of a change in the earth's albedo (McCormick and Ludwig, 1967). The proposal is to measure this change to the required accuracy by satellite. The difficulty is that estimates of the change suggest it is likely to be small and to build up slowly. So, although we believe that satellite measurement of an annual average of the whole earth albedo with a precision of a fraction of 1 percent may just be feasible, we suspect that an attempt to establish a trend might meet the problem of a high noise-to-signal ratio solvable only by integration over decades.

The first of the three methods, the few remote sites, involves the measurement of solar radiation at the earth's surface, to include both the radiation received directly from the sun and the diffuse radiation scattered by the atmosphere. Analysis of records on cloudless days is required. It is possible to deduce approximately the overall scattering and absorptive properties of the total atmospheric load of particles (Robinson, 1962). Comprehensive solar radiation monitoring equipment, such as that which would be required, has been obtained for the Mauna Loa Observatory at a cost of $50,000. We recommend use of about 10 units of this type.

Trials have shown the potential usefulness of the lidar (optical radar) technique, which displays the intensity of the scattered light returned from the atmosphere against the height of reflecting particles. Experiments are required to establish the long-term stability and intersite comparability of these methods. In principle, they are a valuable supplement to solar radiation records at remote sites (Collis, 1966). Equipment costs for one site would be on the order of $100,000.

The method of the widespread collection of observations of low precision relies on the absence of *systematic* error. Continued attention to centralized standardization of the measurements is required to avoid such error. A monitoring network is now in operation, using the Volz photometer, which measures the intensity of direct sunlight at wavelengths defined by narrow-band interference filters (Flowers, McCormick, and Kurfis, 1969). The readings are readily taken by comparatively unskilled observers and do not rely on the existence of completely cloudless skies. The cost of the instrument itself is on the order of $100, and the time consumed is only a few minutes per day. Comparison with centrally maintained standards approximately twice a year is essential.

A global monitoring system should ideally involve all the above-mentioned types of measurement. Widespread use of the Volz photometer would produce an indication of any trends in major sources of particles. Solar radiation measurements would introduce a higher degree of precision in estimating any gobal trends and would allow some discrimination between scattering and absorption—important in estimating the effects on global albedo. Lidar measurements isolate the effects in the important lower stratosphere region. The feasibility of long-term precision monitoring by satellite of global albedo should be examined.

The estimated costs for integration with existing meteorological and other networks, with staff costs estimated on timesharing basis are as follows:

500 Volz Photometer Sites

Equipment	$75,000
Annual running cost	$75,000

10 Solar Radiation Sites

Equipment	$400,000
Annual running cost	$250,000

4 Lidar Sites

Equipment	$400,000
Annual running cost	$250,000
Feasibility study of Albedo Satellite (no hardware)	$100,000

MONITORING PARTICLES AND GASEOUS PRECURSORS

OF PARTICLES BY CHEMICAL METHODS

Sulfate and Nitrate Particles

The organic particles, sulfates, and nitrates are major constituents of atmospheric aerosols of which a proportion is man-made (Junge, 1963; Cadle, 1966, 1970; Robinson and Robbins, 1970). They may affect the climate through their contribution to turbidity, their absorption of radiation, and their activity as condensation nuclei. The human sources of sulfates and nitrates are the combustion of sulfur-containing fuels and nitrogen fixation during combustion. The former produces sulfur dioxide, which is oxidized and hydrated in air to form sulfuric acid droplets. The latter produces nitric oxide (NO), which is oxidized and hydrated to form HNH_3 and nitrates (Cadle and Allen, 1970). There are numerous natural sources of sulfur dioxide and sulfates, such as volcanoes, the oceans, and forest fires. Forest fires and lightning are natural sources of oxides of nitrogen. Organic particles are formed by chemical reactions in smog and are emitted directly by various combustion processes (natural and man-made).

Probably the best way of monitoring sulfates and nitrates, at present, is at surface monitoring stations, collecting particles on polystyrene fiber filters (Cadle and Thuman, 1960). The filters are extracted and the extracts analyzed using colorimetric techniques. The monitoring for organic particles could use glass fiber filtration followed by extraction with benzene, evaporating the benzene, and weighing the residue. Total particulate concentrations are obtained by weighing the filters before and after sampling. Since these substances are atmospheric particles, their residence time in the atmosphere is relatively short and a large sam-

Table 3.4 Costs of a Sampling Network

	Each	Total
	(thousands of dollars)	
1. Cost of Central Laboratory	400*	400*
2. Laboratory for manufacturing polystyrene filters	100	100
3. Sampling equipment	2	200
4. Cost (annual) of operating stations	50	5,000

* This assumes that many other substances would be monitored at the stations and the cost is for all monitoring.

pling network is advisable, perhaps 100 stations. Ten percent accuracy is satisfactory. One sample should be taken at each station every 2 weeks. The costs are itemized in Table 3.4.

Hydrocarbons

Certain types of hydrocarbons react with ozone, or, in the presence of sunlight, with oxygen and nitrogen oxides, to form particles (Cadle, 1966, 1970; Cadle and Allen, 1970).

Hydrocarbons are emitted by both urban and biosphere sources. Methane is not of concern because it is not a source of aerosols; terpenes from forested areas can be oxidized to form aerosols (Cadle, 1966). Only a very small fraction of the hydrocarbons from urban pollution have structures such that they form organic aerosols in significant amounts. Total emission of these olefins probably is only a few percent of the global terpene emissions.

We know very little about the half-life of hydrocarbons in the atmosphere. The reactive ones may have much shorter lifetimes than the others, so perhaps 100 stations are required. One sample should be collected every 2 weeks. Accuracies of ±20 percent may be obtainable if a concentration technique is workable. The most practical monitoring technique would be to sample those higher molecular weight hydrocarbons of interest on a solid substrate such as silica gel and transport these samples back to a central laboratory for analysis on laboratory gas chromatographs equipped with flame ionization detectors. The capital investment per station specifically for hydrocarbons will probably be about $2,500. Some direct aircraft and balloon sampling can also be undertaken.

Nitric Oxide and Nitrogen Dioxide

These oxides are converted rather slowly to particulate nitrates, but probably more important to particulate formation is the role oxides play in the photochemical formation of particles from hydrocarbons. Urban air pollution contributes appreciable amounts of nitrogen oxides. The rates of conversion to particulate nitrate appear to be slow, so conversions occur at least on a regional scale. It has been estimated that the biosphere over land areas contributes about an order of magnitude more nitrogen oxides to the global atmosphere than does urban pollution (Robinson and Robbins, 1970). However, the biosphere values are estimates based on only a few experimental measurements.

Spatial resolution should take into consideration the need to estimate biosphere contributions as well as upper atmosphere mechanisms of conversion of nitrogen oxides to nitrates. One sample every 1 or 2 weeks should be sufficient at each of 100 ground level monitoring sites. Aircraft sampling for nitrogen oxides also should be included in an overall aircraft sampling program. Accuracies of about ±20 percent would appear sufficient for initial monitoring efforts.

Nitrogen dioxide can be determined by colorimetric methods: these methods are suitable for sampling periods up to 2 hours (Intersociety Committee, 1969). Nitric oxide is analyzed by the same procedure as nitrogen dioxide after oxidation by a chemical substrate. The same sampling schedule should be used for nitric oxide as for nitrogen dioxide. The only long-path or remote correlation instrument for nitrogen dioxide now available requires evaluation concurrently with the colorimetric procedures at nonurban sites.

The capital investment specifically for NO and NO_2 would be about $2,500 per station for about 100 stations.

Hydrogen Sulfide and Sulfur Dioxide

Hydrogen sulfide is readily oxidized to sulfur dioxide, which in turn is converted to sulfuric acid aerosol or particulate sulfate (Cadle and Allen, 1970). These sulfur compounds in particulate form could be contributing an appreciable fraction of the increase in atmospheric turbidity reported in recent years.

Concentration levels of hydrogen sulfide on a global basis are very low, probably at or below 0.1 ppb (Junge, 1963). Sulfur dioxide occurs at a few tenths of 1 ppb on a global basis. These low levels also indicate effective conversion to sulfate on a global scale. Sulfur dioxide causes plant damage effects in many local and regional areas around the world.

Hydrogen sulfide is the major biosphere source of sulfur compounds. The conversion of hydrogen sulfide to sulfur dioxide over the land appears to be so rapid that the detection of any hydrogen sulfide over the oceans or polar regions is unlikely. Sulfur dioxide is produced both by oxidation of hydrogen sulfide and from urban pollution by combustion of sulfur-containing fuels and smelting processes. The amounts of sulfur produced from urban sources in the United States will be extensively inventoried as part of the implementation of air pollution control programs.

A network of 100 stations is recommended and an accuracy of 20 percent is desirable if it can be achieved.

No chemical method exists that is acceptable for measuring very low levels of hydrogen sulfide. Sulfur dioxide can be measured by variations on the colorimetric Schiff reagent method referred to as the West-Gaeke procedure in urban air pollution measurements (West and Gaeke, 1956). Collection of air samples in the liquid reagent over periods up to 24 hours can be conducted satisfactorily. Long-term sampling followed by automated analysis is also feasible. If the methods are not automated, the capital investment per field station will be about $1,000. This assumes the existence of the central laboratory mentioned earlier.

Satellite monitoring may someday be feasible. It will probably be several years before it is available, even if it can be made sufficiently sensitive.

Ammonia

Ammonia combines chemically with the sulfuric acid droplets resulting from sulfur dioxide oxidation and hydration to form ammonium sulfate particles. The nature of the particles in the atmosphere has a strong effect on the extent to which they scatter and absorb solar radiation and also their behavior as nuclei.

Both natural and man-operated sources of ammonia exist, but there are many uncertainties. For example, it is not absolutely established whether the oceans are sources or sinks for ammonia (Junge, 1963). The soil can be both a source and a sink for ammonia, depending upon the pH. Alkaline soils tend to release ammonia and acid soils to absorb (react with) it. In general, natural ammonia results from the decomposition of amino acids by bacteria. Man-produced ammonia results from some combustion processes and from farming.

Ammonia is probably short-lived in the atmosphere (a matter of days), so, if possible, it should be monitored at about 100 stations every 2 weeks. Ten percent accuracy should be sufficient.

Simple, effective sampling and analysis techniques involving bubblers, reagents, and colorimeters (photometers) are available that can achieve the desired accuracy. Both sampling and analysis can be performed in the field at little additional cost over that for operating field stations for other trace constituents. As for most other trace atmospheric substances, satellite monitoring technology for ammonia has not been developed but might be preferable to ground stations if available.

Ozone

Ozone in the stratosphere is being monitored continuously and need not be considered here. However, tropospheric ozone is also of interest and is not being monitored. It reaches the lower troposphere from the stratosphere and also is produced by photochemical reactions involving the pollutants, oxides of nitrogen, and hydrocarbons. The latter is one reason for monitoring, since ozone is an indicator of photochemically active pollutants. Also, it reacts chemically with many hydrocarbons, both natural and man-made, to form particles (Cadle, 1966).

Since its half-life in the troposphere is short (hours or days), it should be monitored if possible at 100 stations. Commercially available units for continuous analysis cost a few hundred to a few thousand dollars per unit.

Recommendations

1.   We recommend extending and improving the precision of measurements of the transmission of solar radiation which are already being made (Volz photometer; total and diffuse solar radiation at the surface).

2.   We recommend beginning measurements by lidar (optical radar) methods of the vertical distribution of particles in the atmosphere.

3.   We recommend studying the scientific and economic feasibility of initiating satellite measurements of the albedo of the whole earth capable of detecting trends of the order of fractions of 1 percent per 10 years.

4.   We recommend beginning a continuing survey, with ground and aircraft sampling, of the atmosphere's content of particles and those trace gases that form particles by chemical reactions in the atmosphere. This will require about 10 fixed stations for relatively long-lived atmospheric constituents, and about 100 fixed stations for short-lived constituents.

5.   We recommend about 100 measurement sites for monitoring each of the specific particles and gases by chemical means—including sulfate and nitrate particles, gaseous hydrocarbons, nitric oxide and nitrogen dioxide, hydrogen sulfide and sulfur dioxide, ammonia, and ozone. In general, one monitoring station could measure all particles and gases. However, a total of somewhat more than 100 stations would probably be necessary, since the combination of short residence time and some isolated sources which should be monitored would require stations that monitored only one or a few of the constituents.

MODIFICATION OF THE COMPOSITION OF THE
LOWER STRATOSPHERE

The lower stratosphere is a quiescent region of the atmosphere. Residence time of gases there is on the order of a year. The purging effect of precipitation is also absent. Material formed in or injected into the stratosphere may therefore accumulate in higher concentrations than elsewhere in the atmosphere (Commission for Air Chemistry and Radioactivity, 1970; Martell, 1970).

It has been estimated that by 1985–1990, 500 supersonic transport (SST) aircraft will be operating, and computation shows that they will change the steady-state water content of the stratosphere by up to 10 percent (Machta, 1970). Higher transient local concentrations may lead to the formation of clouds. The aircraft will also produce particles, directly as carbon soot and indirectly following reactions between emitted sulfur, nitrogen oxides, and hydrocarbons, and the ambient ozone. Carbon monoxide is believed to be destroyed in the lower stratosphere by oxidation to CO_2 in a complex photochemical reaction involving H_2O. Sulfur dioxide of tropospheric and man-made origin may reach the stratosphere and be oxidized to sulfate aerosol (Cadle et al., 1969); Cadle et al., 1970).

The emissions from the supersonic transport represent a relatively massive environmental change. Here is one instance where we can anticipate a potential unwanted side effect of technological advance. We feel it would be unwise not to institute monitoring now, before the change begins. Monitoring is required in the vicinity of projected major air routes, where the greatest concentration of contaminants is expected, as well as in regions remote from such routes.

WATER VAPOR IN THE STRATOSPHERE

If, as estimated, in the 1985–1990 time period, 500 SSTs fly 7 hours per day in the stratosphere, and each engine produces about 41,400 lb/hr of water vapor, the SSTs will increase the 3 ppm concentration of H_2O in the stratosphere to 3.2 ppm H_2O (see report of Work Group on Climatic Effects). This may have an effect on world temperature.

Several monitoring techniques based on direct sensing are at present available. The instrumentation is borne by balloons, rockets, or airplanes. Airplanes offer the best opportunity for systematic monitoring (Cadle et al., 1969; Cadle et al., 1970), but few types have the required performance in altitude and endurance. The RB57 aircraft of the U.S. Air Weather Service is suitable. Several types of instrumentation might be used. Mastenbrook has used balloonborne frost-point recorders, while instruments used on aircraft include frost-point hygrometers and others based on changes in the electrical properties of a surface as a result of the adsorption of water vapor.

Eventually, satellites may be able to do this monitoring. The needed technology is under development.

The costs for monitoring by aircraft are difficult to estimate, but if flights are available anyway, as at present, the capital investment could be as little as $10,000 for instrumenting two aircraft. The instrumentation would need to be carefully calibrated, and the calibrations repeatedly checked. Thus, instrument maintenance and analysis might cost $20,000 per year. Obviously, if special flights had to be financed, the costs would increase considerably.

Until trends are detected, 4 flights per year in the Northern and Southern Hemispheres should suffice. Since the instruments are continuous-reading devices, a given flight could cross both well-traveled and little-traveled regions. The accuracy should probably be about 0.2 ppm.

STRATOSPHERIC PARTICULATE MATTER

Particulate matter can be monitored on the same aircraft flights used for water vapor. However, the sampling will then be by filters, rather than continuous. Measurement accuracy will probably have to be ±1 ppb by weight.

It may be appropriate to use two types of filters on each flight. Glass-fiber filters can be extracted with organic solvents following a flight to furnish concentrations of organic particles (largely from the SSTs), while the extremely high purity polystyrene fiber filters of the type now being prepared for such purposes at the National Center for Atmospheric Research may be used for sampling for $SO_4^=$, Si, and NO_3^- (Cadle, 1960). The Si measurements are included to indicate the amounts of soil mineral matter present. By weighing filters before and after flights, total particle loadings can be obtained.

Remote sensing by lidar (optical radar) can also be done, and probably ground-based lidar stations should be used to supplement the aircraft flights (see earlier in this section).

We are advised that SST operation may lead to the production of stratospheric clouds in heavily traveled areas in subarctic regions (see report of Work Group on Climatic Effects). Lidar appears to be the most promising device for monitoring this effect. While we estimate that 4 lidar sites would be adequate to supplement aircraft observations in the period prior to operational introduction of the SST, we anticipate the need for increased use of lidar monitoring as traffic density increases. In its present state of development lidar does not provide an absolute measure of particle concentration. Study of the lidar monitoring technique is recommended. As the detailed properties and quantities of the particles expected to result from SST operation are classified by theoretical and field study, the feasibility of their detection and monitoring by ground- and satellite-based remote sensing should be reviewed continuously.

The main capital investment in the initial phase will be for

(1) the clean rooms and other equipment for manufacturing the polystyrene fiber filters ($100,000); (2) a laboratory for conducting the analyses ($400,000*); and (3) the 4 lidars ($400,000). The annual operation excluding aircraft flights would cost on the order of $50,000 for analysis and $200,000 for the lidar sites.

CARBON MONOXIDE

Until recently there was no explanation of the apparently steady global atmospheric content of CO in spite of the contribution from vehicular emissions (Robinson and Robbins, 1969) and isotopic carbon measurements on carbon monoxide which could be interpreted to indicate that biosphere sources of carbon monoxide may produce much higher amounts of carbon monoxide than arise from urban pollution (Stevens, 1970). It is now thought that a removal process occurs in the stratosphere.

Carbon monoxide levels over unpolluted land and oceanic sites have been averaging 0.1 to 0.2 ppm in 1969–1970 (Robinson and Robbins, 1969). Whether or not the steady-state concentration is increasing, determinations are needed of the effect of increased emission on the regions where CO is removed. If, as seems likely, a significant fraction of the carbon monoxide is being transported into the ozone layer of the stratosphere, the increase in the importance of the $CO - OH$ reaction could be influencing the overall ozone reaction mechanism in the stratosphere (Hesstvedt, 1970).

Since carbon monoxide is long-lived (a few years) in the troposphere, monitoring is recommended at about 10 remote stations. Aircraft sampling could assist in supplementing this network by low-level sampling. Time resolution of one sample over 2 weeks per site should be adequate. An accuracy of at least ± 5 to ± 10 percent is essential to detect trends. No suitable wet chemical analytical methods are available. A highly sensitive and specific gas chromatographic technique is available for immediate use. A modification of the mercury-vapor-type analyzer has been used for carbon monoxide analysis in unpolluted areas. No instrument exists at present for remote measurements of carbon monoxide. The capital investment, other than that already included in the

---

* This assumes that the laboratory will also be used to analyze samples collected at 10 to 100 ground-based stations, for servicing instruments, and for data processing.

central laboratory, would be $10,000 at each field site if gas-chromatographic techniques or the mercury-vapor-type analyzer were used.

Recommendations

We recommend early implementation of a special monitoring program to be comprised of

1. Measurement by aircraft and balloon of the water vapor content of the lower stratosphere. The height range of particular interest is 55,000 to 70,000 feet. The area coverage required is global, but with special emphasis on areas where it is proposed that the SST should fly.
2. Sampling by aircraft of stratospheric particles with subsequent physical and chemical analysis.
3. Immediate development of techniques for monitoring trace amounts of gaseous hydrocarbons, sulfur dioxide, and oxides or nitrogen in the stratosphere.
4. Monitoring by lidar of optical scattering in the lower stratosphere, again with emphasis on the region in which heavy traffic is planned.
5. Monitoring of tropospheric carbon monoxide concentration.

Monitoring Surface Changes

POTENTIAL SURFACE CHANGES

As the human population of the earth increases, greater and greater changes in the surface features of the planet take place. Forests are cleared for use as farmland, farmland is buried under the concrete of cities and roads. All the fossil and nuclear energy converted by man is ultimately dissipated as heat into the atmosphere and surface waters. These and other changes to the physical geography might ultimately affect the climate of the planet. Examples of land use or surface changes that, on a large enough scale, could have a global effect are the following:

1. Change of vegetation type. Changes from tropical forest to agricultural land may have a significant effect on the ability of the biosphere to absorb part of the carbon dioxide released to the atmosphere as a result of the burning of fossil fuel. It is also possible that the forests play a significant role in the transfer of water from the surface to the atmosphere in the tropics. Large-scale

changes in the amount of forestland in these regions might have a significant effect on the general circulation of the atmosphere (see report of Work Group on Climatic Effects).

2.   The spreading of deserts. New deserts may develop as a result of the diversion of groundwater, or existing deserts may change size as a result of man's activities on or near the deserts. Changes in desert size might affect the climate in two possible ways: (1) through the change in reflectivity of the region, and (2) through the greatly increased amounts of dust introduced into the atmosphere by the winds.

3.   The filling of bays and estuaries. The tidal marshes that border bays and estuaries are extremely important to the growth of many marine species. For example, species as diverse as the shrimp or bluefish spend a critical part of their life cycles in these regions. As the marshes are filled to provide space for housing and other activities, the species that are dependent on these areas are lost, leading to the loss of important ocean fisheries. The trend should be monitored before a multitude of local encroachments has produced an irreversible global effect.

4.   Release of stored energy. This adds to the solar energy that drives the atmospheric circulation. In certain circumstances, it might lead to global as well as local climatic change. Much of the heat might be injected into rivers or estuaries, with serious effects on the life-forms existing there.

MONITORING TECHNIQUES

Land-use statistics can be derived from a multitude of sources, but over much of the world the data are probably inadequate. No system exists at present for compiling these data in a central file, for standardizing the data, and for filling in gaps in the data where they are known to exist (see report of Work Group 2). The major problems here seem to be data accuracy and obtaining data from the remote regions of the world.

A second approach to the monitoring of land-use changes is to utilize an airplane or an earth-orbiting satellite which can observe all or part of the earth. Satellite-type sensors, designed for land-use studies, are under development and are scheduled for flight during the next few years. By using multispectral scanning devices, instrument outputs that are compatible with large, high-speed digital computers can be obtained. The outputs can be interpreted not only as to whether the scene is soil, cropland, or forest but also as to type of soil, crop, or forest. Ground resolutions better than those required for climatological monitoring can be achieved now (Goldberg, 1969). Since these climatological surveys do not need the short time resolution that is required of these scanners for agricultural uses, it is quite unlikely that clouds would pose a significant problem. Surveys of 1-year duration should probably be flown at 5- or 10-year intervals. For this category the technology exists and will be in use in the near future. Costs for the instruments are of the order of $1 million (since the development costs have already been absorbed for the most part). It is quite likely that a survey could be conducted during the next few years using existing sensors and satellites. For use on a more limited geographical scale—for example, estuary studies—aircraft are of great utility.

THERMAL POLLUTION

Thermal pollution might play a significant role in certain fisheries through its effect on the usefulness of the estuary as a spawning or nursery ground for fish that are found offshore in their mature state (Snider, 1968). Since thermal pollution is caused primarily by the discharge of heat from large generating plants, it can be monitored (at least in most countries) by monitoring the level of output and sources of electrical power. Estimates (possibly crude but quite meaningful) of the effect of a plant can be made from its power output and from knowledge of its method of heat dissipation. This information can often be ascertained simply from a knowledge of plant location. Monitoring thermal pollution, then, can be most easily done by a study of national statistics.

Recommendation

We recommend monitoring surface changes by attention to land use and economic statistics and by the study of the output of satellite observations of land use and heat balance.

Considerations for Implementation

No unified action whose purpose is to change the course of man's impact upon the environment can be begun without some basis for an agreement that the change is necessary.

As we have reported on our ability to identify and measure

global environmental problems, we have often referred to limitations that hinder an attempt to discover or present facts and conclusions that would provide that basis for agreement. We have said that sufficient base-line data do not exist; that existing mechanisms for keeping track of man's industrial activities do not include an "environmental effects" component; that we lack an adequate model of the workings of an environmental system; that we need better instrumentation; that more monitoring stations are needed.

Though we have said that such deficiencies should be remedied and have, in many cases, suggested specific steps toward a solution, we have generally not explored the organizational possibilities for overcoming the deficiencies. Nor have we attempted to survey the numerous organizations—governmental and nongovernmental, national and international—which are already engaged in the monitoring, measurement, and research activities we have discussed.

Yet, it is our opinion that global problems are most efficiently studied and monitored on a global basis by a combination of national and international effort. We have viewed our work as the first stage of a larger exercise in exploring the dimensions, both scientific and political, of that effort. In what follows we recommend a general direction of inquiry for that exercise.

As we have said before in this report, we think that the most effective method for gathering information would be an integrated global network of fixed and mobile stations, each equipped to sample and analyze various pollutants in various subsystems of the environment. We also think that satellites can be important elements of such a network.

But we do not think that sufficient knowledge exists to establish such a system. A large amount of preliminary work must be done before any such attempt is made. We need comprehensive base-line surveys of the environment. We need a comprehensive survey of current monitoring activities and facilities around the world. We face long discussions among all those involved in measurement and monitoring of the global environment on objectives, on what should be monitored, on standards for measurement and analysis of data, and on financial support. Assuming that any system which is set up will involve some combination of national and international activities, we also need agreements concerning divisions of responsibility and control within the system.

Such a list of requirements—one which is far from comprehensive—manifests the difficulties that stand in the way of any attempt at establishing a global network, difficulties that do not even begin to take into account the genuine differences in priority given to "the problem of the environment" by the various nations of the world.

Given the problems of negotiating, planning, and setting up such a network, it may be many years before it is providing decision makers with comprehensive, integrated information about the global environment. Because of this time lag, and because of the immediate need for new kinds of information about the problems which SCEP has identified, we stress that the setting up of this network should not impede continuous and necessary expansion of current measuring and monitoring activities.

Therefore, we recommend that an immediate study of global monitoring should be instituted to examine the scientific and political feasibility of integration and to set out steps for establishing an optimal system. One major component of that study should be a consideration of how existing and expanded monitoring activities, particularly those whose main value lies in continuity and homogeneity, can be merged into the integrated system with a minimum loss of data during the transition.

We believe that the proposed study might consider the following organization principles as it develops a structure for a monitoring system:

1.   As many national and international organizations as possible should be participating members of the system. Participants should be willing to modify their own monitoring activities in relation to the integrated monitoring plan.

2.   The participants should define the problems to be monitored and evaluated.

3.   The global monitoring system should be supervised by an agent of the participants—a truly international body, composed primarily of scientists and engineers, which would be responsible

for determining how the measurements and monitoring should be carried out, but not with *what* problem to monitor.

a. The supervisory body should not be separated either from scientific research or from the political scene; it should be able to give and take advice in both areas; its tasks and outputs must be very clearly defined. It should be obligated to maintain communications with the "user" of the monitoring information and to respond to "user" needs.

b. The supervisory body should be initially charged with as great a responsibility as possible for determining the process by which the monitoring system would be established—the fewer constraints imposed on technical operations by the initial agreements setting up the system, the better.

c. The supervisory body should have a technically competent monitoring staff to fill gaps and provide instruction.

4. The global system should contain a center or centers concerned with the indefinite maintenance of physical, chemical, and biological standards and with the control of procedures.

5. The global system should contain a "real time" data analysis mechanism. This would assure, for example, proper maintenance of measuring standards by allowing for prompt feedback to monitoring units in terms of modification of measurement parameters, levels of accuracy, and frequency of observation.

6. The implementation of the global system by the supervisory body should proceed by

a. The preparation of feasibility studies clarifying the objectives for measurement, examining the applicability of instrumentation, and completing a cost-benefit analysis of alternative approaches for solving the given problems.

b. The preparation of comprehensive plans which would, among other things:

(1) Establish fiscal, hardware, and man-power requirements.

(2) Clarify the nature of the problem, whether it was to monitor a well-defined environmental problem, assess the magnitude of a suspected problem, or provide information needed to understand the global environment.

(3) Define standards and procedures for sampling, data processing, and analysis.

(4) Define the output of the system.

Recommendations

1.   We recommend an immediate study of global monitoring to examine the scientific and political feasibility of integration of existing and planned monitoring programs and to set out steps necessary to establish an optimal system.

2.   We recommend the expansion of current measuring and monitoring activities in accordance with our recommendations in the rest of this report to satisfy the immediate need for new kinds of information about the problems SCEP has identified.

3.   We recommend that, because some components of what might ultimately be an integrated global monitoring system are so obviously needed, study and implementation of them be attempted independently of the investigation of an optimal global monitoring system. We specifically recommend a study of the possibility of setting up international physical, chemical, and biological measurement standards, to be administered through a monitoring standards center with a "real time" data analysis capability, allowing for prompt feedback to monitoring units in terms of such things as measurement parameters, levels of accuracy, frequency of observations, and other factors.

References

Ayres, R. U., and Kneese, A. V., 1968. Environmental pollution, *Federal Programs for the Development of Human Resources,* Vol. 2, a report submitted to the Subcommittee on Economic Progress of the Joint Economic Committee, U.S. Congress (Washington, D.C.: U.S. Government Printing Office).

Bolin, B., and Bischof, W., 1970. Variations in the carbon dioxide content of the atmosphere, *Tellus,* forthcoming.

Cadle, R. D., 1966. *Particles in the Atmosphere and Space* (New York: Reinhold).

Cadle, R. D., 1970. Atmospheric chemistry and aerosols, background paper prepared for SCEP (unpublished).

Cadle, R. D., and Allen, E. R., 1970. Atmospheric photochemistry, *Science, 167.*

Cadle, R. D., Bleck, R., Shedlovsky, J. P., Blifford, I. H., Rosinski, J., and Lazrus, A. L., 1969. Trace constituents in the vicinity of jet streams, *Journal of Applied Meteorology, 8.*

Cadle, R. D., Lazrus, A. L., Pollock, W. H., and Shedlovsky, J. P., 1970. The chemical composition of aerosol particles in the tropical stratosphere, *Proceedings of the American Meteorological Society Symposium on Tropical Meteorology* (unpublished).

Cadle, R. D., and Thurman, W. C., 1960. Filters from submicron-diameter organic fibers, *Industrial and Engineering Chemistry, 52.*

Callender, G. S., 1961. Temperature fluctuations and trends over the earth, *Quarterly Journal of the Royal Meteorological Society, 87.*

Carver, T. C., 1970. Estuarine monitoring program, *Report of the Subcommittee on Pesticides of the Cabinet Committee on the Environment* (Washington, D.C.: U.S. Government Printing Office).

Citron, R., 1970. *National and International Environmental Monitoring Programs* (Cambridge, Massachusetts: Smithsonian Institution), forthcoming.

Collis, R. T. H., 1966. Lidar: a new atmospheric probe, *Quarterly Journal of the Royal Meteorological Society, 92.*

Commission for Air Chemistry and Radioactivity of the International Association of Meteorology and Atmospheric Physics of the International Union of Geodesy and Geophysics, 1970. *Report to the World Meteorological Organization on Station Networks for Worldwide Pollutants.*

Feltz, H. R., 1969. *Monitoring Program for the Assessment of Pesticides in the Hydrologic Environment* (Washington, D.C.: Water Resources Division, U.S. Geological Survey).

Flowers, E. C., McCormick, R. A., and Kurfis, K. R., 1969. Atmospheric turbidity over the United States, 1961–66, *Journal of Applied Meteorology, 8.*

Goldberg, I., 1969. Design considerations for a multi-spectral scanner for ERTS, *Proceedings of the Purdue Centennial Year Symposium on Information Processing* (unpublished).

Hesstvedt, E., 1970. Vertical distribution of CO near the tropopause, *Nature, 225.*

Intersociety Committee on Methods for Ambient Air Sampling and Analysis, 1969. Tentative method of analysis for nitrogen dioxide content of the atmosphere, *Health Laboratory Science, 6.*

Junge, C. E., 1963. *Air Chemistry and Radioactivity* (New York: Academic Press).

Keeling, C. D., 1970. Is carbon dioxide from fossil fuel changing man's environment?, *Proceedings of the American Philosophical Society, 114.*

McCormick, R. A., and Ludwig, J. H., 1967. Climate modification by atmospheric aerosols, *Science, 156.*

Machta, L., 1970. Stratospheric water vapor, background paper prepared for SCEP (unpublished).

Martell, E. A., 1970. Pollution of. the upper atmosphere, background paper prepared for SCEP (unpublished).

Minzner, R. A., and Oberholtzer, J. D., 1970. Space applications instrumentation systems, National Aeronautics and Space Administration Technical Report C-136.

Mitchell, J. M., 1969. Climatic change—an inescapable consequence of our dynamic environment (paper delivered at the American Association for the Advancement of Science, Boston).

National Air Pollution Control Administration (NAPCA), 1969. *Air Quality Criteria for Particulate Matter* (Washington, D.C.: NAPCA).

Pales, J. C., and Keeling, C. D., 1965. The concentration of atmospheric carbon dioxide in Hawaii, *Journal of Geophysical Research, 70.*

Robinson, E., and Robbins, R. C., 1969. Sources, abundance, and rate of gaseous atmospheric pollutants (Menlo Park, California: Stanford Research Institute).

Robinson, E., and Robbins, R. C., 1970. Gaseous nitrogen compound pollutants from urban and natural sources, *Journal of the Air Pollution Control Association, 20.*

Robinson, G. D., 1962. Absorption of radiation by atmospheric aerosols, as revealed by measurements at the ground, *Archiv für Meteorologie, Geophysik, und Bioklimatologie, B.12.*

Rohlich, G., ed., 1969. *Eutrophication: Causes, Consequences, Correctives* (Washington, D.C.: National Academy of Sciences). See especially Chapter IV, Detection and measurement of eutrophication.

Sand, P. F., 1970. National pesticide monitoring program, *Report of the Subcommittee on Pesticides of the Cabinet Committee on the Environment* (Washington, D.C.: U.S. Government Printing Office).

Shapley, H., ed., 1953. *Climatic Change: Evidence, Causes, and Effects* (Cambridge, Massachusetts: Harvard University Press).

Snider, G. R., 1968. Nuclear power versus fisheries (paper presented at the annual meeting of the Isaac Walton League, Portland, Oregon).

Stevens, C. M., 1970. Natural and man-produced emissions of carbon monoxide (unpublished).

West, P. W., and Gaeke, G. C., 1956. Fixation of sulfur dioxide as disulfitomercurate (II) and subsequent colorimetric estimation, *Analytical Chemistry, 28.*

Woodwell, G. M., and Whittaker, R. H., 1968. Primary production in terrestrial ecosystems, *American Zoologist, 8.*

The true impact of man's activities on the climate can be assessed only after a fuller understanding of climate and climatic change is achieved. The five papers in this section provide a semitechnical view of the factors involved in the determinaton of climate and provide a glimpse of how man might affect the climate through the introduction of pollutants. This latter area will be treated in much more detail in Parts V through IX of this volume.

The first paper in the series was prepared as a background paper for SCEP by Dr. William W. Kellogg. The intention of this discussion was to convey a very general message concerning the implications of man's activities on the climate and the present uncertainties concerning the climate determining mechanisms. It is an appropriate introduction and summary of this complex subject, and Dr. Kellogg has noted which papers in this volume treat in depth some of the more important themes.

Dr. J. Murray Mitchell, Jr., a noted theorist, prepared the second paper in this series for SCEP and elaborated on many of the ideas in the paper during his participation in the Study. In it, he outlines several aspects of climatic change and indicates how the problem of climatic change bears on an approach to the isolation of human influences on climate. The development of mathematical models which he so strongly endorses is discussed in Part III of this volume.

Dr. Kellogg makes the point that the "driving force" of the complex atmosphere-ocean system which determines climate is the radiation that is absorbed and emitted by the planet. The paper by Professor London and Dr. Sasamori reviews in some detail the present state of knowledge of this radiative energy budget. In it, the authors discuss and compare new theoretical calculations and inferences which have been made possible by improvements in radiation theory and by increased information both from satellites and *in situ* measurements. Presented at an international conference just before SCEP, this paper was an important background document. Professor London also participated in the Study for the full month of July.

The fourth paper in this series is a review of the role of major pollutants in atmospheric processes which is part of a report prepared by Dr. G. D. Robinson and his colleagues for the National Air Pollution Control Administration (NAPCA) in Spring

1970. In addition to a brief discussion of the emission rates and sources of pollutants including carbon dioxide, particles, carbon monoxide, sulfur dioxide, nitrogen oxides, ozone, water, and heat, the paper relates these pollutants to many of the atmospheric processes discussed in the previous papers.

A summary of the present state of knowledge of air pollution effects on the climate is provided in the final paper by Dr. Mitchell. This paper, combined with Dr. Robinson's paper, presents a semitechnical overview of where potential problems lie with respect to man-made pollutants and the climate.

4

Predicting the Climate   William W. Kellogg

Introduction to the Central Problem
It has become evident that man can change the entire atmosphere
of his planet in certain subtle ways and that he can modify large
regions in rather obvious ways—for example, with smoke and
smog. There are now new dimensions to his leverage on the atmosphere as he flies his large jet aircraft ever higher in the stratosphere, and the booster rockets of the Saturn class introduce hundreds of tons of exhaust into the thin reaches of the upper atmosphere.

The *effects* of these changes on the environment are diverse.
In addition to the obvious and sometimes acute effects of pollution in the cities of the world, there is the haunting realization
that man may be able to change the climate of the planet Earth.
This, I believe, is one of the most important questions of our time,
and it must certainly rank near the top of the priority list in atmospheric science.

The General Circulation of the Atmosphere
In recent years it has been possible to create fairly realistic numerical models of the global atmosphere that behave very much
the way the real atmosphere does. The atmosphere is a great heat
engine that runs on solar energy, taking advantage of the greater
amount of heat that reaches the equatorial zone. The function of
the heat engine is to transport this heat from the equator to a pole,
in the process of which the atmosphere moves with the patterns
of the winds that we see on any weather map.

Modeling of the atmospheric heat engine is complicated by
the fact that there is another more sluggish but very massive heat
engine, namely, the oceans. The ocean circulation is coupled to
that of the atmosphere, and about one-fifth as much heat is transported from equator to pole in the oceans as in the atmosphere
(Stewart, 1969).

The key to this ocean-atmosphere system, the ultimate driving force, is the solar radiation that is absorbed, mostly in the
equatorial regions, and the infrared radiation that is emitted back
to space, at all latitudes. One cannot consider the heat involved

Prepared for SCEP.

in radiation without also considering the internal heat that is released into the atmosphere by the condensation of water vapor. In fact, about one-sixth of the heat that is transported from the equator to the mid-latitudes is in the form of the latent heat of water vapor, heat that is released whenever it rains or snows.

Experimental general circulation models of the atmosphere that have been run on large computers (at the National Center for Atmospheric Research, the Geophysical Fluid Dynamics Laboratory of ESSA, and UCLA) also take into account the effect of the mountain ranges of the world, the rotation of the earth, and the complex processes that exchange heat and moisture and momentum vertically by means of convection, particularly in the tropics. All these processes can be related to each other by a set of differential equations that involve time, and a model is made to "run" by integrating all these equations in small time-steps. The result is a model or picture of a moving fluid system that behaves very much like the real atmosphere. (See Chapters 10, 11, 17, and 18.)

With these general circulation models we can, in principle, do *experiments* to see how the atmosphere would change with time if there were a change, for example, in the ability of the atmosphere to transmit solar radiation due to smoke and haze and smog, or how it would change if there were a growth or shrinking of the size of the great polar ice caps.

Actually, however, we are still a long way from realizing a model that is adequate for such experiments in "climatic change." The current general circulation models are designed to show the hour-to-hour, day-to-day, and week-to-week changes, and we would run out of computer time if we used them to study really long-term changes. Long-term changes in this system would certainly involve changes in the ocean, so it would not be enough to consider the circulations of the atmosphere alone. Nevertheless, there is hope that we will be able to develop theoretical numerical models with which we can do experiments to study the atmosphere-ocean climate, and how it will change with changes in the heat available to the system. These models will come as the result of considerable effort in developing quasi-statistical shortcuts and the availability of larger computers than we have now. (See Chapters 5, 9, 10, 11, 12, 17, and 18.)

Radiation Balance the Key

As mentioned, radiation is the ultimate source of energy to drive the complex atmosphere-ocean system. So, what do we know about this important factor? First, in order to give an idea of the role that radiation plays in keeping the system in motion, we can do a simple calculation on the rate of energy input from the sun as compared to the amount of energy that the atmosphere contains at any given time. The solar radiation absorbed by the system is about 600 calories per cm² per day, and the average total thermal heat energy of the atmosphere is about 6×10^4 calories per cm². This means that if we imagine the solar radiation to be cut off abruptly, some 10 percent of the energy of the atmosphere would disappear in a period of ten days. This is enough to cause a very appreciable change in the circulation. Such a rough calculation shows that the atmosphere will respond to a change of heat input in less than a week.

The solar radiation that reaches the planet earth is partly reflected back into space, partly absorbed by the atmosphere, and mostly absorbed by the surface. In the 1940s it was estimated that about 40 percent of the solar radiation was reflected back to space, but more recent estimates based largely on satellite observations have lowered it to about 30 percent. This average reflectivity is referred to as "the earth's albedo." The fact that there has been such a large uncertainty in the albedo of the earth is in itself testimony to our uncertainty about the amount of energy available to the system, though our knowledge is certainly improving. (See Chapters 6 and 34.)

Another extremely important variable is the cloud cover, since clouds are generally very much more highly reflecting than the surface of the earth, and an increase in cloud cover therefore increases the albedo. The same is true of snow and ice, which also have a high reflectivity. To show the effect of this, a change in the average cloudiness of the equatorial zone of 5 percent (which would now go unnoticed) would change the total albedo of the earth by about 1.5 percent, and this is an appreciable decrease in the energy available to drive the atmosphere-ocean system.

It is clear that the same effect as a decrease in amount of cloud cover would be achieved by a decrease in the reflectivity of the clouds, since either would decrease the net albedo of the earth.

Curiously enough, clouds moving over regions of the earth with industrial pollution such as Europe show a decrease in reflectivity from about 0.95 (for pure water clouds more than half a kilometer thick) to 0.8 or 0.85. This much reduction in cloud reflectivity would have about the same effect as a 5 percent reduction in cloud amount if it were to occur throughout the tropics—though this is entirely hypothetical and very unlikely.

To take this energy calculation one step further, Budyko (1970) and Sellers (1970) have argued that it would take only a change of 1.6 percent to 2.0 percent in the solar radiation available to the earth to lead to an unstable condition in which the snow cover could advance to cover the continents all the way to the equator, and in the process of doing this the albedo would be raised by the greater snow cover to the point where the oceans would eventually freeze. Lest this rather frightening calculation be taken too seriously, it should be mentioned that we have no evidence that there is in fact a mechanism for a change of as much as 1.5 percent—probably never in the history of the earth—and furthermore the model used by Budyko is highly idealized. Nevertheless, this illustrates the delicacy of the thermal balance of our planet. (See Chapters 12 and 17.)

The aerosols that fill the atmosphere, that is, natural haze, dust, smoke, smog, and so on, probably play an important role in the radiation balance of the earth, but this is one of the great uncertainties in the theory of how the atmosphere behaves. Probably aerosols in cloudless air increase the albedo to some extent, and they absorb sunlight themselves. Also, as we have noted, they can change the reflectivity of clouds. We are quite certain that the variations in the solar radiation absorbed by the atmosphere and surface due to changes in the turbidity, or total aerosol content, of the atmosphere are significant. Furthermore, as will be pointed out later, aerosols in the atmosphere can be greatly affected by man and volcanic activity.

On the other side of the ledger on which we keep track of the heat in and out of the atmosphere-ocean heat engine is the loss to space of terrestrial heat by infrared radiation. Over a period of a year or so the amount of radiation lost by infrared radiation must almost exactly balance the amount of solar radiation that is absorbed by the earth and its atmosphere. Any unbalance would result in a net heating or cooling of the planet. (See Chapter 6.)

As a general principle, any substance in the atmosphere, be it a gas or an aerosol, that absorbs infrared radiation will slow the cooling of the surface. The reason for this is that the energy radiated from the surface is absorbed by the absorbing substance in the atmosphere, and this heats the atmosphere which in turn radiates back down to the ground. In effect then an absorbing layer acts as a sort of a radiation blanket, and its presence will result in a higher surface temperature.

An auxiliary effect of this absorbing blanket will be an increase in the stability of the lower part of the atmosphere, between the surface and the absorbing layer, and this increase in the stability will reduce convection in the lower layers. The ability of the atmosphere to stir itself by convection is the source of much of the cumulus clouds that we see in the lower layers, and a decrease in this activity would also decrease precipitation.

There are two main classes of infrared absorbers in the atmosphere: trace gases (water vapor and carbon dioxide being the most important in the lower atmosphere) and aerosols of all kinds, including clouds. Various estimates have been made of the effect of increasing carbon dioxide in the atmosphere, since it is known that man has in fact been able to raise the total amount of carbon dioxide in the atmosphere by burning fossil fuels. Since 1900 the amount of carbon dioxide in the atmosphere on the average has increased 10 to 15 percent, and this trend has usually been cited to account for the observed rise in the average surface temperature of 0.4°C up to 1940. The theoretical calculations of Manabe and Weatherald (1967) indicate that a *doubling* of the CO_2 content in the atmosphere has the effect of raising the temperature of the atmosphere (whose relative humidity is assumed to be fixed) by about 2°C. This is, of course, a very appreciable change. (See Chapters 17 and 19.)

The role of aerosols in the radiative balance cannot be calculated with anything like the certainty of that for carbon dioxide. Various estimates have been made of the effect of aerosols, and with conflicting results. The most obvious effect of aerosols is to increase the scattering of sunlight in the atmosphere and also to absorb sunlight, the two effects being about equal. Thus, in England Robinson (1970) reports an average decrease in the amount of sunlight reaching the surface due to aerosols in the air of 25 percent, and presumably at least half of this amount went into heating the atmosphere. (See Chapters 22 and 27.) In very clear air, such as that found in the polar regions, the effect of aerosols is much less, but in the tropical zone the turbidity of the atmosphere, probably due mostly to natural haze from vegetation, is very high all the time.

It can be seen from what has been said that aerosols should be taken into account in any calculation of the radiative balance of the earth-atmosphere system, but the fact is that we do not yet know how to do this with any certainty. Furthermore, there is the very practical question of how man-made aerosols compete with natural aerosols. The haze that is observed in many parts of the world far from any industrial sources comes mostly from the organic material produced by vegetation, with large contributions from ocean sea salt and dust raised from dry ground. (See Chapter 24.) At certain times volcanic activity in the tropics produces a worldwide increase of the aerosol content of the high atmosphere. (See Chapter 37.) It is estimated by Budyko (1967), for example, that the solar radiation reaching the ground after the eruption of Agung in Bali (March 1963) reduced the radiation in the Soviet Union by about 5 percent, a significant attenuation whose total effect on the global balance is not clear. (See Chapter 38.) In contrast to these natural aerosols, it is clear that man has overwhelmed nature in certain parts of the world where industrial smog and smoke has a very evident effect on the clarity of the atmosphere. Observations in a few cities (Washington, D.C., and Uccle, Belgium, for example) have documented the increase in the turbidity of the atmosphere and the decrease in the amount of solar radiation reaching the surface over the past few decades, even though great progress has been made in the United States and Europe in cutting down on the production of smoke from coal-burning heat sources.

One further complication, and a possible effect of man-made contaminants in the atmosphere, is the observed reduction of the albedo of clouds already cited due to contaminants absorbed in the cloud droplets. This effect should also be taken into account in a complete calculation of the radiation budget and of man's effects on it.

Steps to Be Taken

In view of the large uncertainties in so many factors connected with the radiation balance of the earth, and the possibility that man is playing a significant role in affecting the radiation balance by his introduction of aerosols and his increase in the CO_2 content, it is necessary to intensify our studies of the effect of these factors on the climate. The key to such studies is the development of adequate climatological models on which experiments can be run. Thus, one would study the change in the average temperature of various parts of the globe for certain changes in the optical characteristics of the atmosphere due to aerosols and carbon dioxide. There are many "feedbacks" in this system, and the model should take them all into account as far as possible. A major feedback that we have already referred to is the changing ice and snow cover of the polar regions, and another is the change of cloud cover, the two probably reacting oppositely to a change in average temperature. (See Chapters 5, 9, 10, 11, 17, and 22.)

Since the oceans play such an important role in the long-term heat balance of the system, a climatic model must certainly include calculations of the ocean circulations, even though these are mostly secondary to the atmosphere circulations in the sense that the atmosphere drives the ocean currents. While the oceans do not move as fast as the atmosphere, the tremendous heat capacity of the ocean more than offsets its slow movement. Progress has been made in modeling the ocean circulation in a number of places, notably the Geophysical Fluid Dynamics Laboratory of ESSA, the National Center for Atmospheric Research, Florida State University, and The RAND Corporation. The trick will be to eventually combine the atmosphere and ocean circulations into one model. (See Chapter 17.)

It is not enough to develop a theory without being thoroughly aware of the changes that are actually taking place in the real atmosphere. For this reason it will be necessary to continue to monitor the *climate,* which is being done in a number of stations throughout the world. In addition to the usual parameters that are observed, such as temperature, wind, precipitation, and so on, the *composition* of the atmosphere and its *turbidity* will have to be monitored better than is now done. This is not a trivial task, since any quantitative measurement of trace gases requires fairly elaborate techniques, and the measurements that describe the aerosol content of the atmosphere should give a picture of the optical properties of these aerosols as well as the amount. One must know how it affects the incoming solar radiation and the outgoing infrared radiation. This has not been done adequately, except on a very few occasions with special equipment. (See Chapters 5 and 27.)

Satellites have been so useful in a number of ways in obtaining new information about the global atmosphere that we should look to them for help in the monitoring task, but so far the observations have not been quantitative enough to be applicable, with one exception—amount of cloud cover. Cloud cover can and should be monitored by satellites. Satellites can monitor snow and ice cover as well, though there is a problem here during the polar night when pictures cannot be taken in the usual way. (See Chapters 6, 34, 42, and 43.)

The situation is improving rapidly with regard to satellite observations of cloud and ice cover, since the High Resolution Infrared Radiometers (of the sort on Nimbus IV and ITOS-1) can obtain pictures day and night and even give an indication of the height of cloud tops. Nimbus-F (scheduled to be launched by NASA in 1973) may have a sophisticated and absolutely calibrated radiation experiment that could mark the beginning of a direct quantitative measure of the total heat budget of the earth. Measurements of lower atmospheric composition (or pollution) from satellites have been proposed, but at this time seem further in the future. (Ozone, a trace gas found mostly in the stratosphere and upper troposphere, has been measured.)

There is one other matter referred to in the Introduction that deserves attention, the possible change in the radiative characteristics of the upper atmosphere due to high flying jet aircraft and higher flying rockets. The latter can probably be dismissed, because even the most extreme assumptions about numbers of big Saturn-class rockets being launched lead to negligible changes (Kellogg, 1964). The contribution of jets to water vapor and aerosols in the stratosphere may be trivial, and the studies of the National Academy of Sciences Panel on Weather and Climate Modification (1966) and of Manabe and Wetherald (1967) strongly suggest that it is. However, the matter deserves another careful look. (The SCEP Report [1970] constitutes such a look.)

Putting the Subject into Perspective

The gist of what I have been saying can be summarized as follows: The atmosphere-ocean system depends on the heat available to run it, and this is the result of a delicate balance between heat received from the sun and reradiated to space. There are ways to disturb this balance, and the ice ages of the past attest to the fact that nature sometimes does in fact alter it. Man could possibly do the same, and this possibility is so awesome that it deserves to be studied with all the tools at our disposal. If we are doing something drastic to our planet we should understand what it is, for then perhaps we could mend our ways if the alternative seems to be disaster.

However, there has been much hand-waving of late, and the "prophets of doom" have taken the spotlight of public attention. Virtually none of these people who speak of the "doom" of our earthly environment are scientists who have a profound knowledge of the way the atmosphere behaves; and atmospheric scientists have not been able to make very convincing rebuttals so far.

The "spaceship earth" (to use another current phrase) actually has a remarkably stable life-support system, and man is unlikely, I believe, to be able to move it very far from its equilibrium. To mention a few facts: Aerosols, of the sort that man or nature creates, stay in the lower atmosphere only about a week, on the average. Thus, the U.S. industrial pollution hardly has time to reach Europe before it is washed from the air. Furthermore, the natural sources of contamination from vegetation, volcanoes, the oceans, and the deserts still far outweigh all of man's contributions, taken on a global scale. Another point to keep in mind is the balance built in by our highly variable clouds. An increase in mean temperature would probably cause an increase in moisture and cloudiness, and this would reflect more solar radiation back to space. Such a negative feedback, forcing the situation back to its stable equilibrium, is only one of several that we are beginning to identify in the complex atmosphere-ocean system.

These, then, are some of the reasons for questioning the glibly pessimistic pronouncements about the imminent collapse of our terrestrial environment. Geophysical scientists should reply to them with thoughtful forecasts of trends in man's environment, based on the best information and the best theories that we can put together. Unfortunately, we still have a great deal of homework to do, and we have been late getting started.

References

Budyko, M. I., and Pivovarova, Z. I., 1967. Vliianie Vulkanicheskikh izverzhenil-na prikhodiashchuiu k poverkhnosh Zemli solnechnuiu radiatsiiu (Effect of volcanic eruptions on the incoming solar radiation), *Meteorologiia i Gidrologiia, 10:* 3–7 (Moscow).

Budyko, M. I., 1970. "Comments on a global climatic model based on the energy balance of the earth-atmosphere system," *Journal of Applied Meteorology, 9.*

Kellogg, W. W., 1964. Pollution of the upper atmosphere by rockets, *Space Science Reviews, 3.*

Manabe, S., and Wetherald, R. T., 1967. Thermal equilibrium of the atmosphere with a given distribution of relative humidity, *Journal of Atmospheric Sciences, 24.*

National Academy of Sciences National Research Council, 1966. *Weather and Climate Modification,* Publication 1350 (Washington, D.C.: National Academy of Sciences).

Robinson, G. D., 1970. Some meteorological aspects of radiation and radiation measurement, *Advances in Geophysics,* edited by H. E. Landsberg and J. Van Miegham (London: Academic Press).

Sellers, E. D., 1970. Reply to comments of M. I. Budyko, *Journal of Applied Meteorology, 9.*

Study of Critical Environmental Problems (SCEP), 1970. *Man's Impact on the Global Environment* (Cambridge, Massachusetts: The M.I.T. Press).

Stewart, R. W., 1969. The atmosphere and the ocean, *Scientific American, 221.*

5

The Problem of Climatic   J. Murray Mitchell, Jr.
Change and Its Causes

The purpose of this background paper is to identify the fundamental nature of the problem of climatic change and its causation,
as I see it, and to indicate how the problem of climatic change
bears on our approach to the isolation of human influences on
climate. It is *not* the purpose of this chapter to detail our knowledge of past climatic changes, which is well documented elsewhere
(see Lamb, 1969). Nor is it to enumerate the various theories advanced to account for the climatic changes of the past. Inasmuch
as many of these theories are speculative and as yet unverifiable,
a discussion of them would be of dubious benefit to our deliberations here (see Mitchell, 1965, 1968; Sellers, 1965).

Climatic Change as a Fundamental Attribute of Climate
It is useful to introduce the problem of climatic change by considering the definition of climate. Practical definitions of the term
"climate" vary in their specifics from one authority to another.
All are alike, however, in distinguishing between climate and
weather (and between climatology and meteorology) on the basis
that climate refers to "average" atmospheric behavior whereas
weather refers to individual atmospheric events and developments.
By "average" is meant average statistical properties in all respects,
including means, extremes, joint frequency distributions, time
series structure, and so on. On the face of it, then, it might seem
that we are left with a very simple decision: what time interval
to choose to average the observed weather into "the climate."

Were atmospheric behavior to proceed randomly in time, the
problem of defining climate would reduce to a straightforward
exercise in statistical sampling. We could make our estimate of
climate as precise as we wish merely by choosing an averaging interval that is sufficiently long. One difficulty immediately arises
here because our knowledge of past atmospheric behavior becomes
less and less detailed (and less and less reliable) the further back
in time we go. But there is another more important difficulty.
If our knowledge of past climates is imprecise, it is at least good
enough to establish that long-term atmospheric behavior does *not*

Prepared for SCEP.

proceed randomly in time. Changes of climate from one geological epoch to another, and apparently also those from one millennium to another, are clearly too large in amplitude to be explained as random excursions from modern norms.

When one examines modern reconstructions of the paleoclimatic record, one might be led to suppose that geological changes of climate—such as those associated with the alternating glacials and interglacials of the Pleistocene Ice Age—are smoothly varying functions of time, readily distinguishable from the much more rapid variability of year-to-year changes of atmospheric state. In other words, one might suppose that each part of a glacial cycle has its own well-defined climate, just as each season of the year is revealed by modern meteorological data to have its own well-defined climate. In such a case, the averaging interval needed to obtain a stable estimate of present-day climate should be long enough to suppress year-to-year sampling variability, but short in comparison to the duration of a glacial cycle.

If we succumbed to the foregoing rationale for defining climate, we would very probably be living in a fool's paradise. The reason is simple enough: the apparent regularity of atmospheric changes in the geological past is only an illusion, attributable to the inadequate resolving power of paleoclimatic indicators. Most such indicators act to one degree or another as low-pass filters of the actual climatic chronology. If our more recent experience is any guide, based on relatively higher-pass filters such as tree rings, varves, ice cap stratigraphy, and pollen analysis applicable to postglacial time, the state of the atmosphere has varied on most if not all shorter scales of time as well.

In other words, the variance spectrum of changes of atmospheric state is strongly "reddened," with low-frequency changes accounting for relatively large proportions of the total variance (in the broadband sense). At the same time, important gaps in the spectrum of climatic change have yet to be identified, and it is possible that such gaps do not exist. Taken together these circumstances imply that there may be no such thing as an "optimum" averaging interval, and therefore no assurance that we can define (let alone measure) a unique, "best" estimate of what constitutes average behavior of the atmosphere.

To summarize, atmospheric state is known to vary on many

scales of time, and it cannot be ruled out from present knowledge that it varies on all scales of time (from billions of years all the way down to periods so short that they are better defined as meteorological variability). Thus it can be argued that the very concept of climate is sterile as a physical descriptor of the real world as long as it adheres to the classical concept of something static. In any event, present-day climate is best described in terms of a *transient* adjustment of atmospheric mean state to the present terrestrial environment.

The Problem of Causes

If climate is inherently variable as I have suggested, different interpretations can be lent to the variability.

One interpretation is the conventional one which I choose to call the "slave" concept of climatic change. This embodies the idea that the average atmospheric state is virtually indistinguishable from an equilibrium state which in turn is uniquely consistent with the earth-environmental conditions at the time, and that when the earth-environmental conditions change the atmosphere requires a relatively short time to adjust to its new equilibrium state.

Another interpretation is one which I choose to call the "conspirator" concept of climatic change. This concept considers that the average atmospheric state is influenced as much by its own past history as by contemporary earth-environmental conditions: that there may be more than one equilibrium state that is consistent with those environmental conditions, and the choice of equilibrium state approximated by the actual atmospheric state at any given time depends upon the antecedent history of the actual state.

The distinction between these two concepts is a sharp one for long-period climatic change, such as the change from Tertiary to Quaternary times. On such a time scale, the dynamic and thermodynamic time constants of atmospheric processes are infinitesimal, even if one chose to include the oceans and the polar ice caps as coupled "atmospheric" processes. As now seems plausible, earth-environmental changes included gradual sea floor spreading and continental drift, together with a gradual increase of average continental elevation. It is usually assumed that the climate acted in

keeping with the "slave" concept throughout, and that after a certain point in the course of continental drift was reached (isolation
of the Arctic Ocean?) the equilibrium climate was transformed in
a deterministic manner from a glacial inhibiting pattern to a glacial stimulating pattern. On the other hand, it is possible to argue,
following Lorenz (1970), that the actual climate of the Quaternary
was not necessarily preordained by its contemporary environmental state; that the evolution of climate to its Quaternary mode was
not a deterministic evolution but a probabilistic one that might
have turned out very differently under identical conditions of continental drift and other environmental change. The different
Quaternary outcomes (two or more) would have followed from
differences in the precise course of the climate itself, due either to
transient environmental disturbances or perhaps to "random"
excursions of atmospheric state along the way.

With regard to relatively rapid climatic change, however, the
distinction between the "slave" and the "conspirator" concepts of
change is much more subtle in character, and perhaps unrecognizable within present bounds of either theory or observation. The
reason for this is to be found in the intimate dynamic and thermodynamic coupling that exists between the atmosphere and the
oceans, and to a lesser extent in the coupling between the atmosphere, the oceans, and the polar ice caps. These couplings introduce long time constants into the changes of atmospheric state,
and result in various forms of autovariation in the total system,
on the time scale of decades and centuries. In the course of such
autovariation, the atmosphere itself may be said to obey the
"slave" principle. But in a relatively limited period of years the
coupled atmosphere-ocean system would exhibit changes of state
that are not independent of its initial state, and therefore is more
properly described as obeying the "conspirator" principle. To
complicate matters further, it is conceivable that the autovariation of the atmosphere-ocean system is riding on top of a transient
of the Lorenz type already mentioned.

In the presence of Lorenz-type transients, the effect of systematic environmental changes on present-day climate (changes,
for example, involving secular increases of CO_2 or other consequences of human activities) might be so badly confounded as to
be totally unrecognizable. Even without Lorenz-type transients,

however, atmosphere-ocean autovariation could effectively obscure the effect of systematic environmental changes that we are seeking to discover.

Rationale for the Isolation of Human from Natural Factors in Climatic Change

We are led to the practical question, what rationale are we to follow in establishing the climatic effects of systematic environmental change on the scale of decades and centuries? More specifically, how do we go about the task of isolating the contribution of man's activities to twentieth-century climatic change?

First of all, I would suggest that there is no real possibility of detecting Lorenz-type transients in present-day climate, so we will have to proceed on the assumption that they are not now occurring, nor are they likely to be induced in the foreseeable future by further environmental change from human activities.

Second, I would suggest that we do not hesitate to use presently available estimates of the climatic effects of atmospheric pollution and other forms of environmental change, as an interim guide in assessing the potential climatic hazards of various human activities. At the same time, however, we should keep in mind that such estimates are of a highly tentative nature, and we should take pains not to put undue confidence in such estimates. In this connection there are two important points to consider:

1. Most present estimates of the climatic impact of human activities are based on relatively simple hydrostatic heat balance models (as refined, for example, by Manabe and used by him to estimate the thermal effect of variable CO_2, stratospheric water vapor, surface albedo, and so on). Manabe himself has often stressed the limitations of such models. Most important among these limitations are the following: (a) they do not take account of atmospheric dynamics other than purely local convective mixing; and (b) they do not take into account changes of atmospheric variables other than the variable that is explicitly controlled as a parameter of the calculation plus water vapor in those experiments stipulating a constant relative humidity.

2. Climatic changes caused by natural agencies, and those possibly caused by human agencies, are not necessarily additive. For example, by analysis of past data on CO_2 accumulation in the at-

mosphere, roughly 50 percent of all fossil CO_2 added to the atmosphere appears to have been retained there. Using published U.N. projections of future fossil CO_2 production, together with a constant 50 percent retention ratio, it can be predicted that by the year 2000 A.D. the total atmospheric CO_2 load will have exceeded its nineteenth-century base line by more than 25 percent. (See Chapter 17.) As pointed out by Lester Machta, however, recent atmospheric CO_2 measurements at Mauna Loa and other locations indicate that the atmospheric CO_2 retention ratio has been steadily dropping since 1958, to a present value of only about 35 percent. It may be significant that the 50 percent retention figure applied to a time when world average temperatures were rising and that the observed decline since 1958 applies to a time when world average temperatures have been falling. It is conceivable, though certainly not proven, that the reversing trend of world climate in recent years has somehow altered the rate at which the oceans can absorb fossil CO_2. If this is the case, we are witnessing an interactive effect whereby climatic changes produced by one agency (presumably a natural one) are at least temporarily reducing the climatic impact of another agency (in this case an inadvertent human agency). Such interactive effects are very poorly understood, and yet they may be a very important element in the evolution of present-day climate.

To return to our question, what rationale should we follow in our study of contemporary climatic change and of human influences on climate, we are left with little choice. We have to rely on the development of advanced mathematical models of the global atmosphere, suitable for long-term integration to generate stable climatological statistics and capable of simulating many dynamic and thermodynamic processes in the atmosphere and at the earth's surface. Relatively sophisticated models of these kinds have already been developed. At least one of these has been expanded to deal with coupled atmosphere-ocean systems. Experiments with such models have begun to lay a solid foundation for a quantitative theory of global climate, and have elucidated the climate-controlling influence of the general atmospheric and oceanic circulations. There appears to be no limit to the refinement possible in such models, other than the limit imposed by computer capacity and speed.

The manner in which such numerical experiments bear on the study of climatic change is essentially twofold:

1.   The experiments verify that a wide range of environmental factors have a bearing on the global pattern of atmospheric circulation and climate. They confirm that the most important factors in this respect are (a) solar emittance; (b) the geometry of the earth-sun system including the orbital and axial motions of the earth; (c) the distribution of oceans and land masses; (d) the state of the ocean surface which along with the juxtaposed atmospheric state governs the fluxes of energy, moisture and momentum across the surface; (e) the state of the land surfaces with respect to albedo, thermal capacity, water and ice cover, relief, and aerodynamic roughness; and (f) the gaseous and aerosol composition of the atmosphere itself. To the extent that all of these factors may vary with time, either slowly or rapidly, in response to forces other than the contemporary atmospheric state itself, all such factors are automatically to be regarded as potential causes of climatic change.

2.   In the numerical experiments, it is possible to simulate the behavior of circulation and climate as a function of *arbitrarily chosen* boundary conditions and atmospheric constituency which enter the experiments as controllable parameters. This makes it possible to vary any of the environmental factors listed above and determine how the circulation and climate respond. In this way various theories of climatic change can be tested in terms of their meteorological consistency. With the further refinement of joint atmosphere-ocean models, the more realistic modeling of continents and ocean basins, and the introduction of ice cap interactions in the models, the range of factors in climatic causation that are amenable to this kind of study will eventually become almost exhaustive of all reasonable possibilities. Factors related to human activities would, of course, be included.

As necessary as such model experiments may be to the study of climatic change, it is important to realize that they are not sufficient to solve the problem of climatic change. It is not enough that we develop the ability to measure the response of climate to varying environmental conditions. If we are to decipher past climatic changes or to predict future changes, it is necessary to determine which environmental controls of climate have been (or will be) doing the varying, at what rate, and in what direction. At present there is a deplorable lack of understanding about the variability of our environment. There can be no guarantee that the necessary understanding will ever be acquired in full, for that in turn may depend on unknowable past events and unpredictable future events. But we should learn what we can, for in no other way can we be certain whether the climatic changes of the twentieth century are or are not causally related to man's activities.

References

Lamb, H. H., 1969. Climatic fluctuations, *World Survey of Climatology*, Vol. 2, H. E. Landsberg, editor in chief (Amsterdam: Elsevier Publishing Co.), pp. 173–249.

Lorenz, E. N., 1970. Climatic change as a mathematical problem, *Journal of Applied Meteorology, 9* (reprinted in this volume, Chapter 9).

Mitchell, J. M., Jr., 1965. Theoretical paleoclimatology, *The Quaternary of the United States*, edited by H. E. Wright, Jr. and David G. Frey (Princeton: Princeton University Press), pp. 881–901.

Mitchell, J. M., Jr., ed., 1968. Causes of climatic change, *Meteorological Monographs*, Vol. 8, No. 30 (Boston: American Meteorological Society).

Sellers, W. D., 1965. *Physical Climatology* (Chicago: University of Chicago Press), pp. 197–228.

6

Radiative Energy Budget
of the Atmosphere

Julius London and
T. Sasamori

Abstract

Improvements in radiation theory and increased information, both from satellites and *in situ* observations, of the cloudiness and general structure of the atmosphere provided some motivation for revision of earlier calculations of components of the earth's radiation budget. The new calculations show more energy absorbed and reradiated by the earth-atmosphere system than was derived from previous studies. Also, the presently computed value for the average annual planetary albedo of the earth (about 33 percent with interhemispheric variations of approximately ±1 percent) is less than earlier estimates. Within the accuracy of the results, the incoming radiation and the outgoing radiation for the entire earth-atmosphere system are in approximate balance.

In recent years satellite observations have generally verified the latitudinal and seasonal patterns of the net components of the radiation field as derived from theoretical considerations. The earth's planetary albedo as inferred from these satellite observations is of the order of 29 percent. The largest discrepancy between satellite and theoretical results is for equatorial and tropical regions.

The theoretical calculations and inferences from satellite observations will be compared and discussed.

Introduction

The two basic factors governing the overall circulation of the atmosphere are the nature of the energy input (that is, the amount and distribution of the energy sources and sinks) and the magnitude of the earth's rotation. All other parameters are secondary at most. The earth's rotation is fixed, but the distribution of energy sources and sinks depends on the large-scale variations of the different components involved in the atmospheric radiation budget.

It would be extremely satisfying if we could start from the known planetary orbital parameters with accurate knowledge of the solar constant, atmospheric composition, and a physical de-

Proceedings of the COSPAR Symposium, Leningrad, May 1970. To be published in *Space Research, XI.*

scription of the lower boundary to compute theoretically all of the radiative properties of the atmosphere. Although notable attempts were made in this direction by Milankovitch (1969) and others, a complete solution to this problem is not likely at the present time since the atmosphere is a highly nonlinear physical system and radiation represents an important process in this interacting system. Moreover, the earth-atmosphere system contains nonpredictable (at present) elements that affect its composition and structure. For these reasons, it is necessary that we constantly combine experimental observations of radiation fluxes in the atmosphere with theoretical computation of these fluxes so that we can understand and explain the significant radiation processes and predict their overall input to the thermohydrodynamic system.

Since it is not possible to solve completely the radiation problem for the earth-atmosphere system, we can probably make the most beneficial progress by dividing the problem into a set of suitable subproblems involving different time and space scales. In that way, the various relevant radiative properties of the atmosphere can be parameterized by use of convenient empirical approximations which are still physically meaningful for the specific class or scale of the problem to be studied. In addition, laboratory measurements as well as ground-based, airborne, and satellite observations of the radiative flux, both spectral and total, would suggest the types of proper approximations, and more important, would provide the necessary check on the computations so that the theory is not allowed to stray very far from nature itself.

For instance, ground-based measurements of the extinction of incoming solar radiation give significant information on atmospheric turbidity which, in turn, can be used to suggest modifications to theoretical calculations of atmospheric scattering and absorption. Balloonborne measurements of both infrared and short-wave components of the upward and downward radiation flux and flux divergence are absolutely essential at the present time as an additional check on the correctness of detailed radiation calculations and their approximations.

In addition to the general verification of these theoretical calculations, satellite observations make a unique contribution to our understanding of atmospheric energetics. Apart from their use as important atmospheric probes, satellites provide us with direct observations on a global and relative real-time scale of the distribution of various radiation parameters at the upper boundary of the atmosphere—and in combination with surface measurements, at the lower boundary (see, for instance, Vonder Haar and Hanson, 1969). Accurate satellite measurements can provide global distributions of the earth-atmosphere albedo as well as the absorption, emission, and net radiation balance of the total earth plus atmosphere system (Raschke, Möller, and Bandeen, 1968).

Although the distribution of the net radiation balance provides the initial energy source for atmospheric motions, subsequent interactions between motions and radiation proceed through a number of nonlinear processes. For instance, in the energy relationship which is part of the interrelated system of equations describing the general circulation of the atmosphere (see, for instance, Lorenz, 1967) the local temperature change for a nonsaturated atmosphere is given as a function of the radiative flux divergence and the transport terms which, in turn, depend on the motion and temperature distribution within the system. Atmospheric motions, on the other hand, are functions of the static stability, among other things. But the static stability itself depends, in large part, on the radiation characteristics of the atmosphere. Some of these problems dealing with the interaction of various radiative components with the dynamics of the atmosphere have been studied by Manabe, Smagorinsky, and Strickler (1965) and more recently by Washington and Kasahara (1970) among others, but certainly much remains yet to be done.

The advent of meteorological satellites has made the need for theoretical computations based on the various spectroscopic and structural characteristics of the atmosphere even more apparent. Prior to the launching of the various satellites, theoretical results were evaluated, in general, according to their reasonableness. But now these results must pass the test of verification as well. Where the theory and observations agree, it is possible that they are both correct. Where they do not agree, however, it is necessary that the reason for the difference be investigated and resolved if either system is to be useful. It should again be emphasized that theoretical calculations are not made obsolete by satellite observations, but, on the contrary, continue to be important as a primary method of our understanding the basic radiative characteristics of the atmosphere and, therefore, can be used for prediction purposes—which is certainly the primary goal of all scientific investigations.

At this point it might be useful to interject a few words about scales and approximations. For short time and space scales, the radiation fluxes immediately incident at the lower boundary of the atmosphere are of primary importance, and the atmosphere can be considered homogeneous in time and horizontal distance (except for variations in cloudiness). In this case the net radiative flux divergence in the free atmosphere is relatively slow and minor as compared with other physical processes. As the time and distance scales increase, however, radiative processes in the atmosphere increase in importance and should be included somehow as part of the energy equation. It is found that under certain circumstances it is possible to depict the radiative flux divergence in the atmosphere free of clouds by relatively simple linear approximations, involving a temperature (and, in some cases, water vapor) dependence (London and Sasamori, 1968). Although cloudy conditions represent a more complex picture, it is probable that even these could yield to some suitable approximations. As the time scale increases beyond the seasonal, and the region of interest becomes global, accurate evaluation of the various radiation components increases in importance. We need now to know the value of the solar constant and its possible variations. We also need to know how changes in the composition of the atmosphere (such as carbon dioxide) could influence the vertical temperature structure and therefore the thermodynamic stability of the atmosphere. The greatest needs, however, as have become apparent to all who work in the field of atmospheric radiation, are correct determination—through experimental and theoretical programs—of the scattering and absorption characteristics of clouds and aerosols. Although much progress on this problem is presently being made (see, for instance, Plass and Kattawar, 1969), it is hoped that satisfactory solutions will be forthcoming in the next few years in a form that could be included directly into the dynamical equations.

The discussion below summarizes recent results of calculations of the various components of the mean annual latitudinal variations in radiation fluxes for both hemispheres based on recent data on the atmospheric structure and distribution of cloudiness. The computed annual average radiation balance gave a slight excess of total emission over absorption in each hemisphere. Since the assumed structural characteristics of the atmosphere represented a long-term average, some of the components were adjusted slightly to give a global radiative balance on an annual basis. The maximum adjustment was of the order of 1 percent increase in the absorption. Details of the methods of calculation of the individual radiative flux components are discussed by Sasamori and London (forthcoming).

The Global Radiation Budget
The average annual computed global radiation budget is shown schematically in Figure 6.1. The numbers represent the percentage of average incoming solar radiation at the top of the atmosphere. Because of the method of calculation, small changes in the accepted value of the solar constant (as discussed, for instance, Drum-

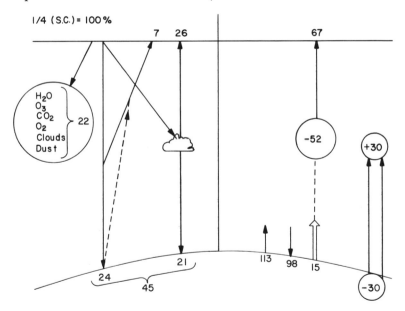

ANNUAL AVERAGE GROSS RADIATION BUDGET

Figure 6.1 The average annual global radiation budget for the earth-atmosphere system
Numbers are given as percent of the average solar energy at the top of the atmosphere.

mond, 1970) would result in only slight adjustment to these figures. The amounts shown on the left represent the various components of the short wave flux; the amounts in the center are the infrared flux; and the value on the right gives the nonradiative energy exchange between the earth's surface and the atmosphere.

Incoming solar radiation is either absorbed within the atmosphere, transmitted to and absorbed at the ground surface, or reflected back to space from the surface, the atmosphere, and from clouds. Absorption of solar radiation within the atmosphere is by the various trace gases, principally water vapor, carbon dioxide, ozone, and molecular oxygen, and by dust and haze. Additional absorption takes place within clouds by the liquid water droplets and ice crystals in the clouds. Computations of these various components of absorption indicate that the average total absorption accounts for approximately 22 percent of the incoming solar energy where over half of this absorption is due to water vapor. The total atmospheric absorption derived from present calculations is somewhat higher than earlier estimates (for just the Northern Hemisphere) given by London (1957). This is due primarily to inclusion of absorption of solar radiation by O_2 and CO_2 in the present calculations and increased absorption within clouds. Approximately 45 percent of the incoming solar radiation is absorbed at the earth's surface with slightly more than half of this energy coming as direct radiation.

The planetary albedo of the earth was computed separately for the clear and cloudy portions of the earth-atmosphere system. The clear part represents the albedo due to surface and atmospheric reflection, while the cloudy part comes from the reflection due to various types of clouds. The total planetary albedo is computed to be 33 percent with the major contribution due to the relatively high cloud reflectivity. Since the cloud cover over the earth is approximately 50 percent it can be seen that an overcast planet having the cloud characteristics of the earth would have an albedo of about 50 percent whereas a cloud free planet, but with surface and atmospheric structure similar to that of earth, would have an albedo of about 15 percent. It should be noted that except for a few isolated regions (such as the ice-covered plateau of central Greenland) satellite observations (Nimbus II) as discussed by Raschke and Bandeen (1970) give just this range of the geographic distributions of the planetary albedo. Their estimates of the overall global albedo based on an assumed value of the solar constant of 1.95 cal cm^{-2} min^{-1}, is 31 percent—not very different from the calculated albedo as shown in Figure 6.1.

As is well known, the presence of infrared active gases in the atmosphere produces a relatively high surface temperature which in turn results in a positive infrared radiation balance at the earth's surface. This net upward infrared surface flux is, on a global basis, approximately 15 percent of the average incoming solar radiation. Even so, the atmosphere acts as a source for infrared energy primarily as a result of the radiative flux divergence due to water vapor and, in the higher layers, due to carbon dioxide. In this way, an amount equal to approximately one-half of the average incoming radiation is used in cooling the atmosphere by infrared radiation. This results in an average infrared cooling of about 1.5°C day^{-1}. The upward flux from the atmosphere together with that from the earth's surface results in terrestrial emission at the top of 0.335 cal cm^{-2} min^{-1} (67 percent of the incoming solar radiation for an assumed solar constant of 2.0 ly min^{-1}), which then gives an equivalent brightness temperature for the earth of about 253°K, a value very close to estimates from recent satellite observations (see, for instance, Raschke and Bandeen, 1970).

Calculations made for the radiative components of each hemisphere separately show that, on an annual basis, each hemisphere is in near equilibrium but that the Southern Hemisphere has a higher planetary albedo (and therefore less absorption by the earth-atmosphere system) and lower emission to space than the Northern Hemisphere. This result is largely based on the derived distribution of cloudiness which gives more cloudiness in the Southern Hemisphere, and therefore a higher contribution of cloud reflectivity to the planetary albedo and a slightly colder surface for infrared emission to space. Obviously, better cloud statistics derived either from more complete and representative ground-based observations or from carefully calibrated high resolution satellite observations are needed to narrow down some of the uncertainties in these numbers.

The energy of solar radiation directly absorbed in the atmosphere is not sufficient to make up for the large infrared radiative cooling. This is accomplished by two somewhat related nonradiative processes. Energy absorbed at the earth's surface is used, in part, to heat the skin layer of air near the boundary and then by convective processes is distributed upward through the lower atmosphere. Most of the solar energy absorbed at the earth's surface, however, is used to evaporate water over the oceans. The water vapor is transported upward in the atmosphere where condensation may occur, clouds form, and the latent heat of condensation is released to the atmosphere. Although this process of transferring heat from the surface to the atmosphere is not as efficient as that involving conduction from the earth's surface, it is by far the more important. Approximately three-quarters of the nonradiative energy transfer from the surface to the atmosphere takes place through evaporation and condensation. Moreover, this process is relatively rapid (as compared, for instance, to radiative processes in the free atmosphere) and therefore contributes a major energy source for the development and maintenance of atmospheric circulation systems involving short to moderate time and space scales.

Radiation Fluxes at the Top and Bottom of the Atmosphere

The solar energy absorbed by the earth-atmosphere system is a function of the sun's zenith angle and the distribution of different atmospheric parameters such as cloudiness, water vapor, and so on, and thus varies with space and time. Except for regions of high albedo, the total absorption is a maximum at the subsolar point and a minimum at places of high solar zenith angles. On an annual basis this means that, in general, the maximum absorption is found near the equator and minimum absorption at the poles of both hemispheres.

The results of calculations of the absorption by the earth-atmosphere system as a function of the atmospheric variables at each latitude are shown in Figure 6.2. As mentioned earlier, the absorption values have been adjusted upward slightly to maintain an energy balance for the entire globe.

Infrared emission also depends on the structure of the atmosphere. But in this case the dependence is primarily on the temperature distribution. Since most of the outgoing infrared flux originates within the atmosphere, and latitudinal tempera-

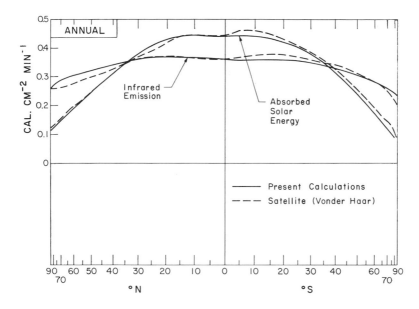

Figure 6.2 The latitudinal distribution of annual average absorbed and emitted radiation for the earth-atmosphere system (according to present calculations and satellite observations)

ture differences in the free air are relatively small, the average variation in infrared emission is also relatively small. As a result, calculation of the annual radiative components shows that the absorbed solar energy is larger than the outgoing infrared flux at all latitudes from about 33° N to about 37° S. Poleward of these latitudes the reverse is true (that is, the outgoing infrared flux is larger than the absorbed solar energy). The implications of this distribution of radiation excess and deficit, insofar as it affects the necessary integrated horizontal energy transport in maintaining an overall balance, will be discussed later.

Since their initial launching, meteorological satellites have had as a prime task the measurement of various components of the earth's radiation budget. Many of these measurements have recently been summarized by Vonder Haar and Suomi (1970). The absorbed energy as derived from the satellite observations is computed as a function of the satellite determined albedo and the solar energy available at the top of the atmosphere. The satellite results for the annual latitudinal distribution of absorption and emission are shown by the dashed lines in Figure 6.2.

Comparison of the satellite results with those derived from computations based on climatological data shows, in general, very good agreement. The small differences in each of the two sets of curves are well within the limits of accuracy of either determination. The slightly lower absorption and emission values calculated for the Southern Hemisphere, however, might be the result of an overestimate of the assumed cloudiness amount in the Southern Hemisphere. Unfortunately, it is still not possible to derive a completely accurate climatic cloud distribution either from ground-based or from present satellite observations. This is particularly true for cloud delineation by type and height. Better cloud data are presently being accumulated and more adequate cloud statistics will certainly be available soon.

On the average, about half of the incoming solar radiation at the top of the atmosphere is transmitted to the earth's surface. The actual insolation received at the surface depends, of course, primarily on the cloudiness—although water vapor and aerosol absorption cannot be neglected. The calculated values of insolation at the earth's surface as well as the net infrared emission both show a slight minimum near the equator but otherwise flat maxima in the northern and southern subtropics (Figure 6.3).

Poleward of latitude 30° the insolation and the net radiation balance at the surface decreases rapidly with latitude in both hemispheres. Indeed, as a result of the high surface albedo in the Southern Hemisphere south of 70°, the net radiation balance becomes negative (that is, more total radiation is emitted than absorbed by the surface). The observed latitude distributions of insolation and net radiation balance as given by Budyko (1963) are also shown in Figure 6.3 for comparison. Although the computed values for the insolation in the Southern Hemisphere are slightly higher than those given by Budyko, the distributions of net radiation balance are, in general, quite close. The discrepancy between computed and observed values of the insolation at the surface is largest near the equator, and it may be that there is increased haze absorption in the atmosphere that was not completely accounted for in our calculations. Further observations of the effect of haze on the absorption of solar radiation—particularly in the lower atmosphere—are needed to resolve this difference.

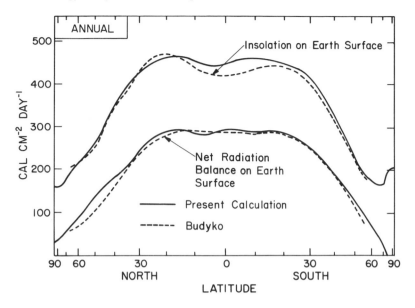

Figure 6.3 The calculated and observed annual average latitudinal distribution of insolation (top curves) and net radiation balance (bottom curves) at the surface

Radiative Cooling and Required Energy Transport

Within the atmosphere absorption of solar radiation results in an average heating of about 0.6°C day^{-1}, whereas infrared flux divergence between the surface and top of the atmosphere gives an average cooling of about 1.5°C day^{-1}. The net annual average radiative temperature change for the global atmosphere is therefore close to 0.9°C day^{-1}. It has long been recognized that this is the approximate value needed to maintain the energy budget for the atmospheric heat engine (see, for instance, Brewer, 1950). As has already been pointed out, about 75 percent of the heating necessary to balance this net radiative loss is supplied by condensation of water vapor within the atmosphere (primarily in the region 2 to 5 km above the earth's surface).

The latitudinal distributions of heating by absorption of solar radiation, cooling by infrared flux divergence, and net radiative temperature change for the entire atmosphere is given in Figure 6.4.

There is slightly less heating and cooling indicated in the

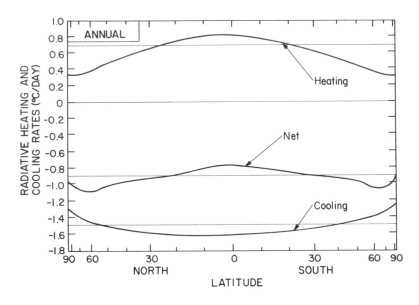

Figure 6.4 The distribution of calculated radiative heating and cooling within the atmosphere

tropical Southern Hemisphere probably as a result of the higher assumed cloudiness in that region (giving a higher albedo and a cooler radiation surface). The net radiative divergence, however, is reasonably symmetric about the equator. At equatorial latitudes there is a net cooling of the order of about 0.8°C day⁻¹ increasing to about 1.1°C day⁻¹ at polar latitudes. This variation represents a measure of the annual average available potential energy between the equator and poles that is contributed by all radiation processes in the atmosphere.

　　The difference between the solar radiation absorbed in the earth-atmosphere system and the infrared emission to space at the top of the atmosphere represents the net accumulation of energy in the system. On an annual basis and over a period of years there must be a balance between the incoming and outgoing radiation, at least to within the accuracy of calculations and/or observations, for the entire globe. At each time and geographic location, however, there is generally an imbalance between these two radiative components, and this imbalance can have large variations as has been shown by recent satellite observations (see, for instance, Raschke and Bandeen, 1970). When averaged over longitude, many of the small-scale variations are smoothed out, and the average latitudinal distribution of the radiative absorption and emission remains. These distributions were given for the annual average calculated and observed values in Figure 6.2.

Since the annual mean temperature change at any latitude when averaged over a number of years must be very small, the net energy accumulation (loss) resulting from the radiation excess (or deficit) in any latitude belt must be transported across the latitude belt by the combined ocean-atmosphere circulation. The radiation imbalance must then be equal to zero when integrated from pole to pole. However, as already indicated, there is a radiation excess (more absorption than emission in the earth-atmosphere system) between 33° N and 37° S, and a deficit beyond. The accumulated energy must therefore be transported poleward by the atmospheric and ocean circulation systems. This transport reaches a maximum in each hemisphere where the radiation imbalance changes sign. The calculated total energy transport that is required to maintain an overall heat balance is shown in Figure 6.5. For comparison purposes the total combined ocean-atmosphere energy transport required by the satellite observations for

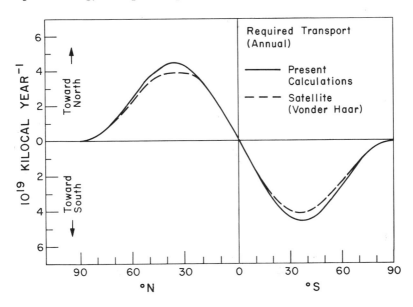

Figure 6.5 The average annual energy transport by the atmosphere-ocean circulation system to satisfy balance requirements

the annual case (as discussed by Vonder Haar and Suomi, 1970) is also shown.

The maximum poleward heat transport, when averaged over the year, is about 4×10^{19}kcal year^{-1} (slightly less according to the satellite data) and occurs between 30 and 35° latitude. A surprising result that is shown by both calculations and satellites is the zero transport across the equator and the apparent symmetry of both hemispheres despite the difference in cloudiness and the marked difference in land and ocean surface between the two hemispheres. It may be that there are small differences in the radiative components that are masked by the uncertainties in both the climatological data upon which the calculations are based and in the assumptions used in reducing the satellite observations. The similarities of the overall radiation regimes can have an important effect in modeling a global circulation pattern.

This is certainly not a final model of the atmospheric radiation budget. Improvements can be made as we acquire more complete information on the climatic distribution of atmospheric variables—in particular the distribution of clouds by type and height—and develop more accurate methods for treating the absorption and transmission characteristics of clouds and aerosols in the real atmosphere. Meanwhile, satellite observations can be used to complement the theoretical calculations based on climatic data. Both, of course, are indispensable for the type of dynamic calculations recently discussed by Manabe (1969).

Acknowledgments
We would like to thank Mr. Douglas V. Hoyt for his considerable aid in programming a good part of this problem. We are grateful also to the National Center for Atmospheric Research, which is sponsored by the National Science Foundation, for use of their computing facilities. This work was supported in part by the Atmospheric Sciences Section, National Science Foundation through Grant NSF GA-10838.

References
Brewer, A. W., 1950. *Centenary Proceedings of the Royal Meteorological Society* (London: Royal Meteorological Society), pp. 31–32.

Budyko, M. I. (ed.), 1963. *Atlas of the Heat Balance of the Earth* (Moscow).

Drummond, A. J., 1970. *Advances in Geophysics, 14*, edited by H. E. Landsberg and J. van Meigham (New York: Academic Press), p. 1.

London, J., 1957. A study of the atmospheric heat balance, New York University College of Engineering, Research Division, Department of Meteorology and Oceanography, Final Report on Contract No. AF 19 (122)–165.

London, J., and Sasamori, T., 1968. *Proceedings of the Radiation Symposium, Bergen, Norway, 1968* (Geneva: World Meteorological Organization), forthcoming.

Lorenz, E. N., 1967. *The Nature and Theory of the General Circulation of the Atmosphere* (Geneva: World Meteorological Organization).

Manabe, S., 1969. Climate and ocean circulation. Part I—The atmospheric circulation and the hydrology of the earth's surface, *Monthly Weather Review, 97:* 739.

Manabe, S., Smagorinsky, J., and Strickler, R. F., 1965. Simulated climatology of a general circulation model with a hydrologic cycle, *Monthly Weather Review, 93:* 769.

Milankovitch, M., 1969. *Canon of Insolation and the Ice-Age Problem* (Jerusalem: Program for Scientific Translations).

Plass, G. N., and Kattawar, G. W., 1969. Radiative transfer in an atmosphere-ocean system, *Applied Optics, 8:* 455.

Raschke, E., and Bandeen, W. R., 1970. The radiation balance of the planet earth from radiation measurements of the satellite Nimbus II, *Journal of Applied Meteorology, 9:* 215.

Raschke, E., Möller, F., and Bandeen, W. R., 1968. *Swedish Meteorological and Hydrological Institute Publication No. 28, Series B:* 42.

Sasamori, T., and London, J., 1971. *Meteorology of the Southern Hemisphere* (Boston: American Meteorological Society).

Vonder Haar, T., and Hanson, K. J., 1969. Absorption of solar radiation in tropical regions, *Journal of Atmospheric Sciences, 26:* 662.

Vonder Haar, T., and Suomi, V., 1970. Measurements of the earth's radiation budget from satellites during a five-year period, forthcoming.

Washington, W. M., and Kasahara, A., 1970. A January simulation experiment with the two-layer version of the NCAR global circulation model, *Monthly Weather Review, 98:* 559.

7

The Major Pollutants: G. D. Robinson
Their Emission and Role
in the Atmosphere

A recent report by E. Robinson and E. C. Robbins (1) gives an excellent summary of the incidence, chemistry, and physics of the major pollutant gases. This is the main source of the following brief summary of some of the facts which seem particularly relevant in planning an investigation of long-term effects.

Carbon Dioxide

This is not the place to discuss in depth the literature on the carbon dioxide content of the atmosphere. It has rather cursory treatment in Robinson and Robbins' report. The comprehensive work of Bolin and Keeling (2) has been updated in a publication by Bolin and Bischof which became available after completion of the text of this report. The Bolin and Keeling paper, with others which are referenced in it, covers the major factors of interest without perhaps doing full justice to the role of the biosphere in the CO_2 turnover. The broad facts which appear to be established are that there is an increasing annual production of CO_2 from fossil fuel combustion of about 1×10^{10} tons per year at present, and that there has been for many years a steady increase in the atmospheric CO_2 concentration which in the mid-1960s was around 7×10^{-7} per year, accounting for about half the annual production of CO_2 by combustion. There is recent evidence (14) that this factor may have fallen to about one-third the annual production. There is an exchange of CO_2 between atmosphere and biosphere—continental and oceanic—involving a quantity of CO_2 greater than that produced by fossil fuel combustion. There is a very large oceanic reservoir of dissolved CO_2 and CO_2 "fixed" as carbonate, with a chemistry, involving silicates and phosphates as well as carbonates, which has been studied in some detail but which is not yet able to account *quantitatively* for the ocean-atmosphere exchange of CO_2 and its geographical variation or for the fate of that fraction of the CO_2 produced by

Reprinted from G. D. Robinson, 1970. *Long-Term Effects of Air-Pollution—A Survey*, prepared for the National Air Pollution Control Administration (Hartford: Center for the Environment and Man, Inc.), pp. 3–9.

combustion which does not appear to be retained in the atmosphere.

Estimates of the quantity of CO_2 fixed in the biosphere are about 25 percent of the total atmospheric content. There is a detectable seasonal effect of the biosphere on the atmospheric content, but nothing is known about secular change.

A recently reported observation (Seiler and Junge [3], quoting Georgii) introduces a complication, and if confirmed, may require a re-examination of the mechanism of the atmospheric transport of the CO_2 produced or exchanged at the surface. This is an apparent change in the CO_2 content of about 0.2 percent, "a small but distinct difference," as the tropopause is crossed—the stratospheric content being lower. Apart from this, the relatively small systematic temporal and spatial variations in CO_2 content are so satisfactorily explained by Bolin and Keeling's simple diffusive transport model that we might expect a comprehensive general circulation model to account for them quantitatively.

Particulate Content of the Atmosphere

There are recent summaries and bibliographies by Horak (4) Shah (5) and Twomey and Wojciechowski (6) dealing with the particulate content of the atmosphere. A few notes concerning the role of pollutant aerosols are added here. Radiative effects are discussed in Section 5* and effects on precipitation processes appear elsewhere in this volume (13).

We cannot entirely separate "aerosol pollution" from "natural aerosol" even by definition—the New Zealand meteorologist who remarked that he had never seen clean air north of 40°S did not necessarily mean unpolluted air. Avoiding pedantic classification, we can recognize three broad types of natural aersol—volcanic dust, wind-raised dust, and water-soluble nuclei. The first two categories are clearly enough definable but soluble nuclei are produced by all combustion processes, as well as by seaspray and by chemical reactions in an "unpolluted" atmosphere, and the extent to which any population is "natural" must always be in some doubt.

There is evidence of a fairly homogeneous population of sol-

* All section or appendix references in this paper refer to the original report.

uble nuclei in air near the surface in the world's remaining empty spaces, and growing evidence of a similar state of affairs in the stratosphere. Some of the evidence is indirect—it comprises the measurements of atmospheric electrical conductivity and the earth's potential gradient which were commonly made on scientific expeditions in the early years of the century. Conductivity decreases, and the potential gradient increases, with the nucleus and particle content of the air because the particles capture small ions and reduce ionic mobility. Atmospheric potential gradient, indicating the concentration of condensation nuclei, is an excellent man-detector or rather combustion-detector. (Radioactive fallout has confused this issue since World War II.) There are some more direct observations of particulate content of air over the oceans but realization of the importance of size distribution and the practicability of determining it readily are developments of the last 20 years, and we may in time find the electrical observations made by early Antarctic explorers and the cruises of the "Carnegie" a useful indication of the particulate content of air in the earliest days of the population explosion. There is evidence in recent NCAR observations in Panama and over the adjacent sea that the smaller "natural" soluble nuclei, Aitken nuclei of radius <0.1 μm, are largely $(NH_4)_2SO_4$ particles formed by oxidation of H_2S or SO_2 in the presence of an excess (around 10 ppb) of free NH_3.

The larger soluble nuclei act as cloud condensation nuclei (13). They appear to have a "natural" concentration over the oceans around 100 cm^{-3}, but their number is much more variable in populated continental regions—from several hundreds to several thousands per cm^3. Nevertheless, according to Squires (7), man's contribution to the number of cloud condensation nuclei is only a few percent.

It is very difficult to estimate the lifetime of soluble particles in the atmosphere. Twomey (8) has estimated that air masses moving from continent to ocean achieve the typical maritime concentration in about three days. A similar time seems to be available for the processes, whatever they may be, which produce the extreme clarity of Arctic air. There is no entirely satisfying explanation of this cleansing process.

Wind-raised and windblown dust, the other particulate component of the troposphere not the result of industrial processes, may be natural or artificial, in the sense of resulting from cultivation. Windblown material from desert land and semiarid or temporarily arid cultivation seems to be more pervasive than was once suspected: indeed it is suggested in Appendix 1 (on the basis of an unpublished survey) that only a few percent of the total mass of atmospheric particulate is man-made. It is often detected by its effects on solar radiation and it appears to be a major source of ice nuclei. Together with stratospheric fallout and industrial particles, it is to some extent preserved in the annual layers of permanent snowfields so that there is a possibility of investigating its secular changes in the past. Russian workers claim to have traced the industrial development of the USSR in the snows of the Caucasus.

A more recent development, potentially of considerable biological importance, is the long-distance transport of persistent insecticides, either on raised dust or directly on the base used in crop dusting operations. Much pollutant insecticide is waterborne but it seems that a proportion, not negligible in some localities, has been an air pollutant at one stage of its unfortunately long life.

The modification of the earth's albedo by volcanic dust is one of the mechanisms postulated to cause climatic change. A recent incident—the stratospheric dust cloud following an eruption on Bali in 1963—has been well documented and is the first such occurrence which has been extensively investigated quantitatively. Dust content of the middle stratosphere appears to have temporarily increased by an order of magnitude, with a half-life of a year or two. The extent to which the sulfate aerosol discovered in the stratosphere by Junge is of volcanic origin is not clear. This aerosol, which appears to have a maximum concentration in the height range 15 km to 25 km has been directly sampled and chemically analyzed and has been detected optically by several techniques—searchlights, lasers, and scattering from the solar beam at twilight. There is as yet no convincing explanation for this concentration, and although it seems unlikely, pollutant SO_2 may have some part in its formation. For this reason, it seems advisable that it should be extensively observed for a number of years to determine any secular trend.

Carbon Monoxide

There is no reason to expect CO to produce any direct geophysical changes at the volume concentration of about 10^{-7} which appears to be its present atmospheric level in regions remote from pollution sources. Furthermore, there is no evidence that this concentration is increasing. Atwater has confirmed that at this concentration possible radiative effects are far less than our uncertainty of the radiative effects of H_2O, CO_2, cloud and aerosol. CO cannot be ignored in the context of long-term changes, if only because we do not know how it is removed from the atmosphere, or whether or not there is a natural source comparable in magnitude with the very considerable pollution source. Robinson and Robbins (1) summarize the data on pollutant emission and distribution in the atmosphere—we see from these data that CO production by combustion processes, about 2×10^8 tons per year, is 1 to 2 percent of CO_2 production so that if CO is removed from the atmosphere by a process leading ultimately to its oxidation to CO_2, it will add negligibly to the CO_2 pollution load.

There is a possibility of biological effects connected with CO. It has low solubility, and there are some reports of saturated or even supersaturated ocean water and large concentrations associated with certain aquatic plants, and there has been speculation on the possibility of biological sources, as well as of biological sinks of the gas.

Recent observations (3) and theories (9, 10) may, if confirmed, remove at least some of the uncertainties. The observations, made on commercial aircraft, indicate a sudden decrease of CO content on passing from the troposphere to the stratosphere; the theory explains this as the result of a series of photochemical reactions in which CO is involved in the O_3-H_2O photochemistry in the lower stratosphere and ultimately oxidized to CO_2 with destruction of O_3. The theory has only been briefly reported and there are some puzzling features which may be resolved in a fuller publication. If the theory is quantitatively sound detailed observations of CO concentrations in the region of the troposphere will be of considerable meteorological interest, but not in the context of long-term geophysical change. (If CO were removed by a different photochemical oxidation in the lower ionosphere there might be cause for concern, since in this region photochemistry and ionization are interrelated.)

Whilst the fate of CO remains unexplained, it is a reasonable subject for research in the context of long-term effects of pollution, if only because of the magnitude of CO emissions and the social and economic implications of substantial control. At present it seems reasonable to wait for clarification of the suggested photochemical removal process, though this may well require future measurements in the tropopause region.

Sulfur Dioxide

There is no reason to expect geophysical influence of pollutant SO_2 in its gaseous form. It now seems well established that the volume concentration of SO_2 in the upper atmosphere and over the oceans and remote land areas is as low as 2×10^{-10} (1, 15). At this concentration its radiative influence is negligible. As with CO, interest from the point of view of long-term effects centers on the methods by which it is removed from the atmosphere, but unlike the situation in respect of CO, the difficulty is not to find plausible removal mechanisms but to decide which among many possibilities are significant. From the geophysical point of view, the interest is in the formation and persistence of sulfate aerosol; from the biological point of view, in the accumulation in soil and surface waters of sulfuric and sulfurous acids and their salts following rain and fallout.

SO_2 is a short-lived atmospheric constituent. Pollutant production of 1.5×10^6 tons per year and a concentration of 2×10^{-10} require a mean residence time (as SO_2) of about 2.5 days. Indirect estimates of the lifetime of SO_2 in highly polluted regions are much shorter, ranging from Meetham's 12 hours (southern England 1950), to a half-life of 1 to 3 hours estimated for the State of Connecticut (11). There have been reports of half-lives as short as 20 minutes observed by European investigators, but details are not yet generally available. There is little doubt that variations in pollutant aerosol concentration have some influence on the variation of these figures, which suggest that acidic rain-out and fallout could be a problem at distances varying from a few tens to a few hundreds of miles downwind from major pollution sources (i.e., about 24 hours' travel).

Persistent geophysical effects are more likely to be associated with the more persistent sulfate aerosol than with the gas. There is some evidence that over some seas and tropical land areas, the condensation nuclei are largely $(NH_4)_2SO_4$, associated with a free NH_3 content of about 10 ppb (16), and Junge identified $(NH_4)_2$-SO_4 as an important constituent of the stratospheric aerosol. It is not clear to what extent pollutant SO_2 contributes to the "background" 0.2 ppb of SO_2 in remote areas, or to the formation of the SO_4 ion in these areas and in the upper atmosphere. There is a sufficient biological source of H_2S, and plausible oxidation mechanisms, to account for current levels of the "background" sulfate aerosol without the intervention of pollutant SO_2. The residence time of sulfur in the form of sulfate aerosol is currently estimated as being on the order of tens of days in the troposphere, and hundreds of days in the stratosphere. A question of major geophysical importance in the very long term is whether there is a mechanism for net transport of pollutant SO_2, even in a very small proportion of the total output, into the stratosphere. Alternatively, or in addition, is there a mechanism for transport of sulfate aerosol from troposphere to stratosphere in amounts even very slightly greater than the loss by fallout or mixing?

Nitrogen Oxides

N_2O is a long-lived, fully-mixed constituent of the atmosphere. It is not significantly an industrial pollutant. NO is a combustion product, particularly of combustion at high temperatures, and is readily oxidized in the atmosphere to NO_2 by reaction with O_3—a reaction so fast that even at the low (order 10^{-8}) concentrations of O_3 in the lower atmosphere the NO_2 concentration is significant. NO_2 is radiatively important—it absorbs solar radiation in the visible (with photochemical decomposition). It may, in fact, be responsible for a considerable part of the absorption of solar radiation in urban atmospheres elsewhere attributed to aerosol (12). It is a key constituent of photochemical smog. Because of its high reactivity, it does not seem to present problems on the global scale, but it could have significance in chronic and long-term biological effects in the vicinity of large cities.

Photochemical reactions involving NO are of prime importance in the very high atmosphere, and touch significantly on human activity (communications) in their control of ionization in the lower region of the ionosphere. Again, because of high reactivity, it seems unlikely that pollutant NO or NO_2 from the surface could be transported to the ionosphere.

Ozone

Ozone is a constituent of the unpolluted atmosphere, being formed and destroyed photochemically. It is a highly reactive gas, particularly liable to destruction on surfaces. It is considered that the normal tropospheric concentration, of order 10^{-8}, represents an equilibrium between transport from a predominantly stratospheric source and destruction at the surface. One very fast homogeneous reaction $O_3 + NO \rightarrow O_2 + NO_2$ may be important even in air which might otherwise be considered unpolluted.

Ozone is also a pollutant of major importance and is unusual in being a secondary pollutant formed in a photochemically initiated chain reaction—the "photochemical smog" process. In smog conditions ozone concentrations may be as high as 5×10^{-7}. Smog does not seem to be a significant factor on the global scale, but it certainly affects city climates and appears to be responsible for widespread chronic and acute damage to plants in the areas which it affects, so that the possibility of secondary "long-term" biological effects must be considered. The associated aerosol—polymerized organic nitrates—appears also to be removed locally, but the possibility that a very small proportion of it is long-lived and mixed on the global scale cannot be dismissed.

Speculations concerning artificial climate modification have included some on possible manipulation of the stratospheric ozone layer—"punching a hole in the ozone" is a phrase which has been used. Since the layer is a most effective u-v filter, its removal would be biologically disastrous. The proposed mechanism is by introduction of a substance very reactive with O_3 (e.g., NO). Published discussion follows the consequences of such an introduction on the stratospheric photochemistry only to the first step of the chain and it is difficult to see how anything more than a change in the distribution of the O_3 with height could result. Normal pollutant materials to be considered in this context are sulfate and other aerosol, and oxides of nitrogen, and the possibility now arises that the development of supersonic transport might introduce both NO and hydrocarbons into the lower section of the ozone layer. The first sign of a "threat" of modification of the O_3 distribution might be an increase in the stratospheric content of these materials.

Water and Heat as Air Pollutants

"Waste" heat is an inevitable consequence of the utilization of the energy stored in fossil fuel or the atom, and it dissipates in the atmosphere, surface layers of the earth, and surface waters. An estimate of the level at which it *might* induce global climate change is not difficult to make. The average rate of absorption of solar energy is about 250 watts m^{-2}, and simple climatic models suggest that variations of 1 percent of this might have serious consequences. This is only a few watts per square meter but is the equivalent of about half a million 1000 MW generating plants distributed throughout the world and more than 100 times man's current level of energy conversion. We, therefore, need not yet consider "heat pollution" as a global problem. It is, however, already significant locally—the air temperature of large cities is notably modified, by direct heating, from that of the surrounding countryside, individual smoke-stacks frequently generate cumulus clouds, and precipitation patterns may be modified. It is conceivable that, if some current predictions of population trends are sound, "thermal air pollution" might have more than local climatic effect through modification of such significant features as the sharp temperature gradient near the eastern seaboard of North America in winter. Current atmospheric models could be used to investigate these possibilities by introduction of an artificial surface heat source. From the geophysical point of view, the exact mode of disposal of the waste heat is probably not very important, but persistent local climatic and biological effects could be quite different from dry cooling towers, wet towers, or cooling to bodies of water. "Water pollution" is already producing unpleasant modification of fog and drizzle frequencies near badly sited wet cooling towers, and "thermal pollution" of streams is leading to biological changes.

Some concern has been expressed over the introduction of H_2O into the high troposphere and lower stratosphere by aircraft.

There have been suggestions of a significant increase in cirrus cloud amounts in some localities, resulting from aircraft condensation trails and some confirmation of this has recently been found by search of standard meteorological records (17). It has been argued that circumstances in which condensation trials persist or grow, and in which cloud would not otherwise form, are meteorological rarities since they require high relative humidity and absence of appreciable vertical motion, either up or down.

In the lower stratosphere, where supersonic transport aircraft will operate, the existing H_2O concentration is low—2×10^{-6}. This is almost certainly a level set by atmospheric dynamics—it corresponds to saturation at the coldest tropospheric temperature. Residence time of water in the lower stratosphere is probably about one year. With this residence time computation shows that commercial operation of *one* large transport aircraft in the lower stratosphere would increase the water content by a factor 10^{-4}. Some traffic projections have suggested that in 20 years' time there may be several hundred such aircraft in operation so that the H_2O concentration could be changed by several percent. It would, therefore, be prudent to monitor the water content of the lower stratosphere, and to investigate theoretically the radiative and other effects of a few percent increase in concentration.

References

1. E. Robinson and R. C. Robbins, "Sources, abundance and fate of gaseous atmospheric pollutants," 1968, American Petroleum Institute, New York, N.Y.

2. B. Bolin and C. D. Keeling, "Large-scale atmospheric mixing as deduced from the seasonal and meridional variations of CO_2," *J. Geophys. Res., 68*, pp. 3899–3920, 1963.

3. W. Seiler and C. Junge, "Decrease of CO mixing ratio above the polar tropopause," *Tellus, 21*, pp. 447–449, 1969.

4. H. G. Horak, "Aerosol concentration and extinction in the earth's atmosphere," Los Almos Report LA-4023, Clearinghouse, 1969.

5. G. M. Shah, "Aerosols in the stratosphere," *Canadian Aero. and Space J., 15*, pp. 321–326, 1969.

6. S. Twomey and T. A. Wojciechowski, "Observations of the geographical variation of cloud nuclei," *J. Atmos. Sci., 26*, pp. 684–688, 1969.

7. P. A. Squires, "An estimate of the anthropogenic production of cloud nuclei," *J. Rech. Atmos., 2*, pp. 297–308, 1966.

8. S. Twomey, "On the nature and origin of natural cloud nuclei," *Bull. Obs. Puy de Dôme, 1*, pp. 1–19, 1960.

9. E. Hesstvedt, "Vertical distribution of CO near the tropopause," *Nature, 225*, p. 50, 1970.

10. J. Pressman and P. Warneck, "The stratosphere as a chemical sink for CO," *J. Atmos. Sci., 27*, pp. 155–163, 1970.

11. G. R. Hilst, "The sensitivities of air quality prediction to input errors and uncertainties," NAPCA Symposium on multiple source urban diffusion models, Chapel Hill, North Carolina, 1969. (To be published).

12. G. D. Robinson, "Absorption of solar radiation by atmospheric aerosol as revealed by measurement at the ground," *Arch. Meteorol. Geophys. Biokl.* B12, pp. 19–40, 1962.

13. Chapter 32.

14. L. Machta, personal communication.

15. J. B. Pate, personal communication.

16. J. B. Pate, personal communication.

17. L. Machta, personal communication.

8

Summary of the Problem   J. Murray Mitchell, Jr.
of Air Pollution Effects
on the Climate

Effects on Urban Climate

Nature of Urban-Rural Differences of Climate
It has long been recognized that cities—which represent the most concentrated form of environmental modification by man—differ appreciably from rural areas as to local climatic conditions. Urban climatic anomalies extend to temperature, humidity, cloudiness, visibility, radiation, wind speed, and apparently precipitation as well. The typical direction and magnitude of such anomalies in an average large city is as indicated in Table 8.1 (after Landsberg; see Peterson, 1969).

Air Pollution as a Contributory Cause
In accounting for these observed anomalies of urban climate, the role played by air pollution is not altogether clear. Other causative factors include (1) the local addition of heat by combustion and other energy expenditures in the city, or so-called *thermal pollution;* (2) the effect of buildings and pavement in altering the disposition of solar heat; and (3) disturbances in the natural wind produced by the mechanical turbulence and obstacle effects of large buildings. Each of these factors gives rise to various climatic effects similar to the presumed effects of locally concentrated air pollution and are thus difficult to distinguish.

Urban air typically contains many gaseous and particulate materials, in concentrations substantially higher than those found in "clean" rural air (Ludwig, Morgan, and McMullen, 1970). Those materials thought to have significant effects on climate include water vapor, carbon dioxide, combustion particles, and possibly nitrogen dioxide. Other pollutants, for example sulfur oxides, nitrogen oxides other than NO_2, ozone, and metallic compounds, may also produce indirect climatic effects of one kind or another; these, however, are thought to be relatively minor and limited to special meteorological conditions.

Urban heat sources (thermal pollution) account for part of

Adapted from "The Effect of Man's Activities on Climate," paper presented at the American Geophysical Union Symposium on the Environmental Challenge, Washington, D.C., April 22, 1970.

Table 8.1 Climatic Changes Produced by Cities

Element	Comparison with rural environs
Temperature	
Annual mean	1.0 to 1.5°F higher
Winter minima	2.0 to 3.0°F higher
Relative humidity	
Annual mean	6% lower
Winter	2% lower
Summer	8% lower
Dust particles	10 times more
Cloudiness	
Clouds	5 to 10% more
Fog, winter	100% more
Fog, summer	30% more
Radiation	
Total on horizontal surface	15 to 20% less
Ultraviolet, winter	30% less
Ultraviolet, summer	5% less
Wind speed	
Annual mean	20 to 30% lower
Extreme gusts	10 to 20% lower
Calms	5 to 20% more
Precipitation	
Amounts	5 to 10% more*
Days with <0.2 inch	10% more*

* Precipitation effects are relatively uncertain (see text).

the observed excess of temperature in cities (the so-called urban "heat island"), especially in winter when the energy used for space heating and for transportation is comparable per unit city area to the energy received from the sun. The locally added water vapor and carbon dioxide may also contribute to the urban heat island, especially at night, by their ability to absorb and reradiate long-wave thermal radiation escaping from the surface toward the overhead sky. In other words, these gases may contribute to a locally intensified "greenhouse effect."

Effects of Atmospheric Particles

The basic effects of airborne particulate materials (aerosol) are to scatter and absorb solar radiation, to deplete solar radiation reaching the urban surface, and to reduce visibility through the atmosphere (U.S. Department of Health, Education, and Welfare, 1969). Further effects on urban temperature and other climatic conditions are uncertain; these depend on many physical properties of the aerosol which must be known in detail, and which tend to vary greatly from season to season and from city to city according to the type of local industry, heating fuel, and other factors. Such properties include the number density and size distribution of the particles, the vertical distribution of the particles, and the chemical makeup of the particles which governs the relative strengths of absorption and scattering by the aerosol.

Urban Pollution and Rainfall

The efficacy of urban pollutants in modifying precipitation, both in and downwind of cities, is not yet clearly established. Observed anomalous patterns of rainfall in cities are difficult to interpret with reliability, partly because of the extreme variability of natural precipitation which erodes the statistical significance of the observed patterns, and partly because of systematic gauge catch or sampling errors which are largely unavoidable in urban rainfall measurements. Furthermore, those patterns that do appear to be real may be attributable to various thermal or mechanical factors in the urban environment not necessarily related to local excesses of pollution. On the other hand, certain air pollutants (or byproducts of chemical reactions taking place in polluted atmospheres) are credited with adding significant concentrations of either *ice nuclei* or *cloud condensation nuclei* to the natural background levels of such nuclei. In certain meteorological conditions the addition of such nuclei to urban air might alter the precipitability of clouds either in the immediate vicinity or downwind of the city (Robinson, 1970). Observational evidence of such an effect is equivocal at best. Some confirmation, however, is to be found in a study by Frederick (1970) which shows a rather systematic tendency for cold-season precipitation at twenty-two urban stations in the eastern United States to average several percent greater on weekdays than on weekends. Further investigations of this problem are highly desirable.

Effects on Global Climate

Global Trends of Climate

Meteorological data reveal a systematic fluctuation of global climate in the past century (Lamb, 1969; Mitchell, 1961, 1963, 1970). This fluctuation has consisted in part of a net worldwide warming of about 0.6°C between the 1880s and the 1940s, followed thereafter by a net cooling which to date (1970) has accumulated to about 0.3°C (see Figure 8.1). The temperature fluctuation is thought to reflect a change in the planetary heat budget; it has been accompanied also by changes in the large-scale atmospheric circulation and in other climatic elements. It is likely that the bulk of the fluctuation is ascribable to natural causes, for example variable volcanic dust loading of the upper atmosphere (Table 8.2 and Figure 8.2). A part of the fluctuation, however, may be related to increasing atmospheric pollution in the period.

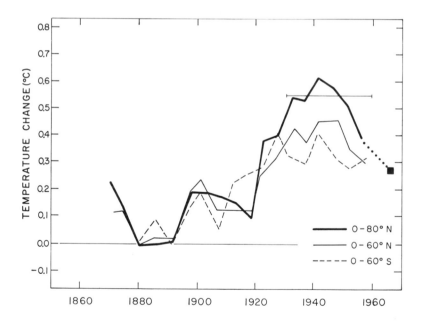

Figure 8.1 Changes of mean annual temperature, 1870–1967, integrated over various latitude bands of the earth
Data after 1960, shown for 0–80°N band only, indicates continuation of cooling trend that began in 1940s.

Source: Mitchell, 1970.

Table 8.2 Volcanic Eruptions since 1855

Date	Name and location			Severity class*
1855	Cotopaxi, Ecuador	1°S	78°W	1–1/2
1856	Awu (Awoe)	3.5°N	125.5°E	2
1861	Makjan, Molucca Is.	.5°N	127.5°E	2
1870	Ceboruco, Mexico	21°N	105°W	3
1872	Vesuvius	41°N	14°E	3
1872	Merapi, Java	7.5°S	110°E	3
1875	Askja (Vatna Jökull), Iceland	65°N	17°W	2
1877	Cotopaxi, Ecuador	1°S	78°W	3
1883	Krakatoa	6°S	105.5°E	1
1883	St. Augustine, Alaska	59.5°N	153.5°W	3
1883	Bogoslov, Aleutians	54°N	168°W	3
1885	Falcon Island	20°S	175°W	3
1886	Tarawera, N.Z.	38.5°S	176.5°E	2
1886	Niafu, Tonga Is.	16°S	175.5°W	3
1888	Bandai San, Japan	38°N	140°E	2
1888	Ritter Is.	5.5°S	148°E	2
1890	Bogoslov, Aleutians	54°N	168°W	3
1892	Awu (Awoe)	3.5°N	125.5°E	2
1902	Mont Pelée, Martinique	15°N	61°W	2
1902	Soufrière, St. Vincent	13.5°N	61°W	2
1902–04	Santa Maria, Guatemala	14.5°N	92°W	1–1/3
1907	Shtyubelya, Kamchatka	52°N	157.7°E	2
1911	Taal, Luzon	14°N	121°E	3
1912	Katmai, Alaska	58°N	155°W	2
1913	Colima, Mexico	19.5°N	104°W	3
1914	Sakurashima, Japan	31.5°N	131°E	3
1921	Andes (Chile-Arg. border)	≃30°S	≃70°W	3
1929	Asama, Japan	36.5°N	138.5°E	4
1931	Kluchev, Kamchatka	56°N	160.5°E	4
1932	Quizapu, Chile	35.5°S	70.5°W	3
1947	Hekla, Iceland	64°N	19.5°W	2
1953	Mt. Spurr, Alaska	61°N	153°W	2
1955	Ranco Puyehue, Chile	40°S	72°W	3
1956	Bezymyannaya, Kamchatka	56°N	160.5°E	2
1960	Puntiagudo et al., So. Chile	39–45°S	72°W	3
1963	Gunung Agung, Bali	8.5°S	115.5°E	1–1/2
1963–65	Surtsey, Iceland	63°N	20.5°W	3
1966	Awu (Awoe)	3.5°N	125.5°E	2
1968	Fernandina I., Galapagos	0.5°S	92°W	2

* Severity class reflects the estimated order of magnitude of the total mass of material ejected.
Class 1: 10–1 km³ ∼ 10^{10} metric tons
Class 2: 1–0.1 km³ ∼ 10^9 metric tons
Class 3: 0.1–0.01 km³ ∼ 10^8 metric tons
Class 4: 0.01–0.001 km³ ∼ 10^7 metric tons
Fractional classes represent compromises believed to be appropriate. Estimates derived from data in Lamb (1970).

Air Pollutants Potentially Involved

Of the many substances entering the atmosphere as a byproduct of human activities, specifically two—carbon dioxide and particulate materials—are of special concern as possible causes of disturbances in global climate. With regard to CO_2, the average longevity (residence time) of a molecule of this gas in the atmosphere is several years. This is a long enough time for new atmospheric CO_2 supplied through combustion of fossil fuels (now exceeding 10^{10} tons per year worldwide) to become well mixed by air circulations into all parts of the atmosphere. Precision CO_2 measurements in remote sites (Hawaii and Antarctica) verify that such mixing has taken place very efficiently and that the global atmospheric reservoir of CO_2 has been growing by about 0.2 percent per year since 1958 when the observations began.

With regard to particulate materials (or aerosol), average residence times of those that enter the atmosphere near ground level are measured in days or weeks, rather than years as in the case of CO_2. However, the sources of particles are so ubiquitous that these materials also tend to be distributed through very large expanses of the atmosphere, albeit in much lower concentrations than those found in the immediate vicinity of urban or industrial centers. Evidence for the geographically extensive spread of particles, including most continents and the North Atlantic Ocean, has accrued in recent years from a variety of sources. The worldwide total atmospheric particulate loading derived from all human activities is estimated to have increased by a full order of magnitude in the past century to a present value (smoke-sized particles only) of the order of 10^6 tons (see Figure 8.2). (See Chapter 24.)

Effect of Pollutants on Atmospheric Heat Budget and Temperature

Both CO_2 and atmospheric particulates (aerosol) are of concern to climate because both are capable of altering the terrestrial heat budget. In the case of CO_2, the gas absorbs long-wave thermal radiation emanating from the earth's surface toward space and reemits part of that radiation back toward the surface again. Thus it tends to warm the climate near the ground by a mechanism usually referred to as the "greenhouse effect." In the case of atmospheric particles, the situation is more complicated. The particles absorb a part of the solar radiation passing through

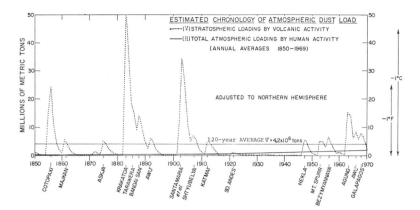

Figure 8.2 Estimated chronology of worldwide atmospheric particulate loading by volcanic activity and by human activity, 1850–1970
Average of volcanic loading added for comparison (thin solid line). Rough calibration of volcanic loading in terms of planetary average temperature effect is shown outside margin at right. Volcanic chronology derived from data in Table 8.2.

Source: Mitchell, 1970.

the atmosphere toward the earth's surface (warming effect), and scatter another part back toward space (cooling effect). Depending (1) on the relative efficiencies of the aerosol in absorbing and backscattering solar radiation, (2) on the altitude of the aerosol, and (3) on the albedo and evaporable water content of the underlying surface; the *net* thermal effect near the ground may be one of either warming or cooling (Mitchell, 1971). In most studies of the problem, it has been tacitly assumed that the backscatter effect of aerosol dominates over the absorption effect, inferring a net cooling of surface climate. Surface cooling is clearly indicated in the case of high-altitude aerosol such as volcanic dust veils in the stratosphere. In the case of low-altitude aerosol, such as that associated with most human activities, however, recent investigations indicate that the surface effect is more likely to involve a warming instead (Mitchell, 1971). Confirmation of the latter conclusion, with the aid of better observations of the optical properties of aerosol than are presently available, is highly desirable. (See Chapter 22.)

The best available determination of the temperature effect of changes of atmospheric CO_2 is that of Manabe and colleagues (see Chapter 17). Assuming constant relative humidity in the atmosphere, and conditions of surface albedo, cloudiness, solar radiation intensity, and other parameters chosen as typical of middle latitudes in an equinoxial season, these investigators calculated that a 10 percent increase of CO_2 concentration (which approximates to the total CO_2 increase since the mid-nineteenth century) would result in a warming of about 0.3°C at all levels in the lower atmosphere, and a substantial cooling at higher stratospheric levels (above about 10 km).

Quantitative estimates of the temperature changes attributable to the long-term increase of atmospheric particulates from human sources have been attempted (see Mitchell, 1970). These, however, are based on very insecure assumptions and therefore will not be cited here. As noted earlier, the temperature effect of such low-altitude aerosol is now thought likely to be one of warming and not cooling and may therefore supplement the warming effect of fossil CO_2 increases. Thus it is reasonable to ascribe to air pollution at least a part of the global climatic warming of 0.6°C between the 1880s and 1940s, but not the global cooling that set in after the 1940s.

Indirect Large-Scale Climatic Effects of Pollutants

It should be emphasized that the ultimate climatic response of global atmospheric increases of CO_2 and particles may extend beyond the basic thermal response discussed above. Concomitant changes of cloudiness, for example those conceivably stimulated by the temperature changes or by possible cloud-nucleating effects of particles, might introduce important revisions to the ultimate thermal adjustment of global climate to the pollution changes. Moreover, the basic thermal effects of pollution might at some point either past or future begin to modify the pattern of the global atmospheric circulation itself. The circulation change, in turn, could then alter the worldwide pattern of climate in such a manner as to cause contrary temperature changes as well as other climatic aberrations in some geographical areas. It will be necessary to await the development of more advanced numerical modeling techniques for the study of atmospheric phenomena, now on the way, before such matters can be investigated intelligently.

In perspective, it should be added that the global climatic fluctuation of the past century is not unlike similar fluctuations known to have occurred in earlier centuries and millennia. From this viewpoint we are amply justified in attributing all past climatic change to natural geophysical forces lying beyond man's control. At the same time, the impact of air pollution on global climate, along with other effects of human activities, clearly threatens to become more competitive with natural climate-driving forces in the future. The problem clearly merits very close monitoring and improved understanding.

References

Frederick, R. H., 1970. Preliminary results of a study of precipitation by day-of-the-week over eastern United States, preprint, Second National Conference on Weather Modification (Boston: American Meteorological Society), pp. 209–214.

Lamb, H. H., 1969. Climatic fluctuations, *World Survey of Climatology*, Vol. 2, H. E. Landsberg, editor in chief (Amsterdam: Elsevier Publishing Co.), pp. 173–249.

Lamb, H. H., 1970. Volcanic dust in the atmosphere, with a chronology and assessment of its meteorological significance, *Philosophical Transactions of the Royal Society of London, A, 266:* 425–533.

Ludwig, J. H., Morgan, G. B., and McMullen, T. B., 1970. Trends in urban air quality, *EOS (American Geophysical Union), 51,* 468–475 (reprinted in this volume, Chapter 25).

Mitchell, J. M., Jr., 1961. Recent secular changes of global temperature, *Annals New York Academy of Sciences,* Art. 1, *95:* 235–250.

Mitchell, J. M., Jr., 1963. On the world-wide pattern of secular temperature change, *Changes of Climate,* Arid Zone Research 20 (Paris: U.N. Educational, Scientific, and Cultural Organization), pp. 161–181.

Mitchell, J. M., Jr., 1970. A preliminary evaluation of atmospheric pollution as a cause of the global temperature fluctuation of the past century, *Global Effects of Environmental Pollution* (New York: Springer-Verlag [D. Reidel]), pp. 139–155.

Mitchell, J. M., Jr., 1971. The effect of atmospheric aerosol on climate, with special reference to temperature near the earth's surface, *Journal of Applied Meteorology, 10* (in press).

Peterson, J. T., 1969. *The Climate of Cities, A Survey of Recent Literature,* Publication No. AP-59 (Raleigh, North Carolina: National Air Pollution Control Administration).

Robinson, G. D., 1970. *Long-term Effects of Air Pollution—A Survey* (Hartford, Connecticut: Center for the Environment and Man, Inc.).

U.S. Department of Health, Education, and Welfare, 1969. *Air Quality Criteria for Particulate Matter,* Publication No. AP-49 (Raleigh, North Carolina: National Air Pollution Control Administration).

Many aspects of the very complex interactions that determine the weather and the climate can probably be understood only through the development of and experimentation with mathematical models of these phenomena. The mathematical and physical principles involved in such models were set forth in the early 1920s by L. F. Richardson, but these models have become practical only with the advent of very high-speed computers with large data storage capabilities. The four papers in this section discuss the concepts involved in climate model development and review some of the capabilities of the more advanced models which presently exist.

In the first paper, Professor E. N. Lorenz reviews the mathematical approaches that can be taken to develop an understanding of the climate and of the nature of climatic changes. He then relates these procedures to the development of climate models. Although Professor Lorenz did not participate personally in SCEP, his influence was felt when theories of climate change and atmospheric models were being discussed. The paper republished in this volume appeared just prior to the beginning of SCEP and served as an important document in the deliberations of the Work Group on Climatic Effects.

During those deliberations in July, Professor Hans Panofsky prepared a working paper for other participants in which he explained in nonmathematical terms the formidable array of concepts that are involved in formulating a mathematical model of climate. This paper, which follows that of Professor Lorenz, also outlines the limitations of such models. The ideas in Professor Panofsky's paper represent his own synthesis and interpretation of many discussions at Williamstown, particularly those with Dr. Joseph Smagorinsky who was a part-time participant of SCEP.

The third paper in this series was prepared as a background paper for SCEP by Dr. Smagorinsky, one of the pioneers in research, on the bases for numerical models of general atmospheric circulation. In it he defines the components of the problem of modeling large-scale atmospheric circulation, discusses the current status of these models, and lists three broad areas that will require extensive upgrading in the future.

The final paper is another section of the report which Dr. G. D. Robinson and his colleagues prepared for the National Air

Pollution Control Administration (NAPCA) in Spring 1970, and which was an important background document for SCEP. This paper furnishes a critical review and evaluation of several modeling concepts and of the models of L. R. Rakipova (and Budyko) and Messrs. Sawyer and Sellers.

All four of these papers conclude that computer modeling is of critical importance if climatic changes are to be understood and if the impact of man's activities on the climate is ever to be assessed. Yet, all also conclude that much more effort and knowledge will be required before these goals will be realized. Further insight into how these have been used and might be used in the future is provided in Part V of this volume in the papers of Drs. Manabe and Washington.

9

Climatic Change as a   Edward N. Lorenz
Mathematical Problem

Abstract

Formulating reasonable hypotheses regarding climatic change requires physical insight and ingenuity, but subsequently testing these hypotheses demands quantitative computation. Many features of today's climate have been reproduced by mathematical models (equations arranged for numerical solution by digital computers), similar to those used in weather prediction. Models currently in use generally predict only the atmosphere, and prespecify the state of its environment (oceans, land surfaces, sun, etc.). Newer models, where certain environmental conditions enter as additional dependent variables, should be suitable for testing climatic-change hypotheses. Aspects of the atmosphere which play no role in these hypotheses may be highly simplified. A supermodel where virtually all not-strictly-constant features of the atmosphere and its environment enter as variables may ultimately lead to an acceptable theory of climatic change.

Introduction

The problem of climatic change occupies but one corner of the field of climatology. Yet, perhaps because it requires its followers to visualize an age when things did not all look as they do today, it has succeeded in attracting the imagination and effort of many scholars who might have looked upon general climatology as something rather prosaic. Probably for the same reason, it is highly conducive to speculation, and hypotheses easily outnumber established results. When, some years from now, someone will see fit to assemble the body of knowledge which may properly be called the theory of climatic change, the greater part of this knowledge will likely consist of facts and results which are not known today.

The complete problem of climatic change entails several distinct sub-problems. First, there is the observational task of establishing that changes of climate actually have occurred—by no means a trivial undertaking—and of determining the nature and extent of these changes. At the other extreme, there is the theoreti-

Reprinted with permission of the American Meteorological Society from *Journal of Applied Meteorology, 9* (1970), pp. 325–329.

cal task of determining just what changes in climate would take place as a result of specified hypothetical causes. An intermediate problem is that of identifying the principal cause or causes of those changes in climate which have actually happened.

The first of these tasks is fairly well in hand, although it is by no means completed. During the past century or two, routine meteorological measurements have revealed certain progressive changes, such as a general warming trend during the first half of the twentieth century. Earlier historical times have seen changes in vegetation of the sort which evidently demand changes in rainfall or temperature regimes. However, the most spectacular changes are presumably those which accompanied the advance and retreat of the prehistoric continental glaciers. We feel confident that only a climate different from today's could have produced and maintained the great ice sheets, while, conversely, the presence of the ice must have produced and maintained a climate different from today's. When, however, we ask how greatly the ancient temperature and precipitation patterns differed from the current ones, we find no general agreement.

In the matter of determining the response of the climate to specified influences, we are still in the speculative era. Moreover, it seems unlikely that we shall obtain new results in which we can place very much confidence until we have perfected a more quantitative approach. We shall presently consider this matter in greater detail.

The intermediate problem, that of properly identifying the causes of well-established changes of climate, is the one whose solution would seem to advance our general knowledge the most. It is mainly this problem which will concern us in this discussion.

At first glance it might not appear that mathematics would play an important role in attacking the problem. What would seem to be called for is physical insight and ingenuity. We prefer the point of view that physical insight is indeed required, especially in the formulation of hypotheses, but that the ultimate choice among the numerous hypotheses which have been proposed, and the many more which presumably will be appearing, must be based upon mathematical considerations.

In the following we shall first describe a mathematical procedure which has been and is being successfully applied to the more general problem of climate. We shall then indicate the types of modification needed to make the procedure applicable to the problem of climatic change, and the resulting degree of success to be anticipated.

Mathematical Models of Climate

In a mathematical treatment, we may define the climate as the collection of all long-term statistical properties of the state of the atmosphere. We may represent the instantaneous state of the atmosphere by the three-dimensional fields of temperature, pressure, density, and wind velocity, and water in its gaseous, liquid, and solid phases. The variations of these fields as time progresses are governed by familiar physical laws; we may express these as mathematical equations which specify the time derivative of each atmospheric variable as a function of the state of the atmosphere and its environment.

If we are to solve these equations, we must know how the environment will behave. Frequently, we simplify the problem by assuming that the state of the environment is known. If we do not wish to do this, we may introduce additional variables describing the state of the ocean surface and land surface and other environmental features, and formulate appropriate additional equations governing their behavior.

If long-term statistics are understood to mean statistics taken over an infinite span of time, there is sometimes just one set of statistics compatible with a particular system of equations. In this event the system is said to be *transitive,* and the set of statistics constitutes the climate. It is also possible that two or more sets of statistics are compatible with a given system of equations. The system is then said to be *intransitive,* and the various sets of statistics constitute alternative, physically possible climates. The selection of a particular climate by a real physical system is then perhaps fortuitous. We do not know whether the atmosphere, or, more appropriately, the atmosphere-ocean-earth system, is transitive or intransitive, although there are some reasons for believing that it is transitive.

Analytical procedures for determining the climate or climates from a system of governing equations include the derivation and solution of new equations whose dependent variables are statistics, and the evaluation of statistics from analytic solutions of the original equations. However, the extreme nonlinearity of the atmospheric equations renders these procedures unfeasible. There remains the possibility of solving the equations numerically, and compiling statistics; it is this procedure which is currently in use.

In the numerical method, the continuous fields of atmospheric variables are replaced by their values at a pre-chosen three-dimensional grid of points. Partial derivatives in the governing equations are replaced by finite differences, and integrals are replaced by sums. Initial conditions are chosen, often arbitrarily, and the equations are solved in a stepwise manner with the aid of a digital computer.

Our concept of climate now requires a slight modification. Numerical solutions necessarily extend over finite spans of time, and infinite statistics cannot be compiled. However, this apparent shortcoming proves to be an advantage when we come to the problem of climatic change. By their very nature, statistics taken over an infinite time span do not vary as time progresses, and changes of climate defined by such statistics are non-existent. Statistics taken over long but finite spans of time are more in keeping with the concept of climate which we wish to pursue.

It has become common practice to refer to a particular system of governing equations, together with a specific procedure for solving it, as a mathematical *model* of the atmosphere. Likewise, the process of obtaining a particular solution for a special purpose is often called a numerical *experiment*.

It must not be supposed that such an approach to the problem of climate was or could have been developed overnight, once computers had become available. For one thing, the development of mathematical models has accompanied the development of computers rather than following it, and has always been limited by the size and speed of the computers available. Of greater importance, certain technical questions had to be answered before the models would work properly. What finite difference operator must replace a partial derivative, for example, if spurious sources or sinks of energy are to be excluded? If a grid contains only a few thousand points, how are systems like thunderstorms to be taken into account, when an individual thunderstorm may occupy no more than one ten-millionth of the atmosphere? What aspects of the atmosphere and its environment may be considered irrelevant, and completely disregarded, in order to reduce the problem to manageable size, and what properties must be retained?

These and similar questions have by now been at least partially answered. Yet the very size of the task has placed it almost beyond the reach of the individual worker who happens to have a computer at his disposal. We find instead that much of the progress has come from the efforts of several groups addressing themselves to the specific problem. We mention one such group, the Geophysical Fluid Dynamics Laboratory of ESSA, which has been headed by Dr. Joseph Smagorinsky since its founding during the 1950s. Its staff presently includes some two dozen scientists, some of whom are specialists in certain physical or mathematical aspects of the subject. If it has not become big science, it has certainly left the realm of little science.

During its lifetime this group has constructed and tested a succession of models. The earlier ones (Smagorinsky, 1963) used a few thousand grid points, and disregarded the presence of water in its various phases. One of the more recent models (Miyakoda et al., 1969) uses about 50,000 grid points, and contains a complete hydrological cycle. Many of the principal climatological features are fairly realistically reproduced.

Nevertheless, some important aspects have yet to be introduced. For example, the model does not produce its own clouds; liquid water is assumed to fall out immediately as rain. The amounts of absorption, emission and reflection of radiation by clouds, which exert a profound influence upon climate, are taken to be the amounts which would accompany a climatological normal distribution of clouds. Likewise, the influence of the atmosphere upon the oceans is omitted, and climatological normal sea surface temperatures are assumed. In short, the behavior of the environment is assumed to be known in advance. The omission of these and other aspects does not stem from a lack of regard for their importance, nor from an inability to incorporate them; it has simply not been possible to do everything at once while retaining confidence that one is doing it correctly.

Models and Climatic Change

We now come to the central question. How can one use mathematical models to study climatic change, and, in particular, to identify and establish the principal causes of climatic change? It is not certain that we can presently use them at all, at least insofar as formulating hypotheses is concerned. For example, in a number of current hypotheses (e.g., Donn and Ewing, 1968), increases and decreases in the extent of sea ice, and the subsequent influence of the ice upon atmospheric conditions, play an essential role. In current mathematical models the presence of sea ice, when it is recognized at all, is represented by constants rather than dependent variables. If an investigator has chosen to regard sea ice as a constant feature, no amount of mathematical finesse will reveal to him the positive results which he might have obtained by treating it as a variable feature instead. Likewise, in one hypothesis (Weyl, 1968), variations of oceanic salinity are assumed to exert their control upon the amount of sea ice, which in turn influences the atmosphere. An investigator using a model where salinity is assumed constant, or, more likely, where it is disregarded altogether, even though variations of sea ice are included, could never have arrived at such a hypothesis.

In view of the manner in which mathematical models have evolved, and in view of our failure to have yet incorporated every feature which we *know* to be relevant, it is inconceivable that in the near future we shall construct a model possessing every feature which could possibly be relevant, i.e., which treats every not-strictly-constant feature of the atmosphere and its environment as a dependent variable. We therefore ought not to look upon a mathematical model as a means of by-passing the physical imagination needed to formulate hypotheses. We should, however, regard a model as a valuable tool for *testing* hypotheses. For this purpose, we can and must incorporate into our model each individual feature, such as variable sea ice or salinity, suspected of being important.

Such testing seems essential if the hypotheses are not simply to remain hypotheses forever. For example, one might argue convincingly that a decrease in evaporation from the ocean would bring about a decrease in surface salinity, which would inhibit vertical overturning and thereby favor the formation of sea ice, which would in turn bring about increased reflection of solar radiation, and thereby lower the atmospheric temperature. Such reasoning could be completely sound, and yet not be particularly relevant to the problem of climatic change, if the decrease in temperature arising from a given decrease in evaporation should prove to be negligibly small, or if the decreased evaporation should simultaneously initiate a second chain of events which would favor a rise in temperature. Yet all the essential features of this reasoning *can* be incorporated into a mathematical model, and the step-by-step numerical integration of the equations will then constitute a system of bookkeeping for the ensuing temperature changes. We hasten to add that although the chain of events appearing in our example is modeled after certain currently proposed hypotheses, it is not intended to be an accurate presentation of the content of any particular hypothesis. We are not aiming to criticize specific pieces of work for not being numerical; after all, the formulation must come first. Also, we are not accusing all current hypotheses of being non-quantitative; we simply maintain that further exploitation of quantitative procedures is essential.

The manner in which we may put a model to work depends upon the nature of the hypothesis being tested. Some hypotheses regard changes in climate as the direct result of changes in the external environment, i.e., the portion of the environment which is not in turn appreciably influenced by the atmosphere. Changes in the intensity or spectral distribution of solar energy reaching the earth would fit this category. So also, most likely, would changes in the geographical locations of continents, although in the absence of numerical computations one might argue that the continual wind stress over the centuries plays some part in continental migration. Here the simplest procedure would be to compare numerical experiments already performed, using environmental conditions typical of today's with additional experiments to be performed with an altered environment. In essence, one would be assuming that today's climate is the one which would continue to prevail if the external environmental status quo could be preserved.

Other hypotheses, however, involve only the immediate environment, i.e., the part of the environment whose variations result at least partly from atmospheric effects. A typical feature of the immediate environment would be sea ice. In investigating such hypotheses, the physical system, i.e., the thing which is described by the dependent variables, should include all portions of the immediate environment as well as the atmosphere itself. The envisioned climatic changes would then become completely internal.

One might simply perform two or more numerical experiments of limited duration, with different arbitrarily chosen initial conditions, to see whether more than one climate could ensue. However, the possibility of more than one climate is not quite the same as a change of climate, especially when intransitivity looms as a possibility; and in any event it gives no indication of the time required for a change to be realized. A more satisfactory test would be a single experiment of sufficiently long duration to capture the climatic changes.

This ideal procedure has obvious practical drawbacks. Experiments so far performed with the more realistic models have extended over less than a year of simulated time. Climatic-change experiments may require hundreds or even tens of thousands of years. Even with the anticipated continual improvement in computer speed, the envisioned experiments could be prohibitively lengthy.

One is therefore tempted to settle for the performance of a few shorter experiments with differing initial conditions. However, alternative possibilities should not be overlooked. First of all, in many qualitative hypotheses the meteorological portions of the arguments are rather naive. There is no suggestion that the atmosphere is a system requiring several hundred thousand numbers for its proper description. One cannot escape the feeling that 50,000 grid points, although not irrelevant, are somehow redundant. To investigate the plausibility of a hypothesis as it has been formulated, the numerical description of the atmosphere should not have to be more sophisticated than the verbal description entering the hypothesis. A model with a few hundred grid points rather than many thousand is therefore suggested, even though it may be unrealistic in its treatment of aspects of the atmosphere which do not enter the hypothesis. If the hypothesis is sound, the model should reproduce the envisioned climatic changes. The work required to obtain solutions extending over centuries should be no more than that needed to extend more detailed experiments over years.

A further simplification could, if realizable, lead to even greater savings. Current experiments use the equations of short-range weather forecasting, even though they are not short-range experiments, and in the course of generating their climates they recreate the life history of each transient weather system, such as the familiar migratory cyclones and anticyclones. Perhaps there is some way to filter out explicit reference to these systems, while still retaining their overall effects. The numerical solution of the equations could then proceed in time increments of days or even weeks, instead of hours or minutes, and experiments extending over millennia would become feasible.

Meanwhile, it is of interest to ask what would happen if we took the mathematical models which are currently being used to simulate climate, without any modifications to accommodate existing climatic change hypotheses, and performed experiments lasting centuries or more. Would climatic changes be revealed? If we include as one hypothesis of climatic change the proposition that no processes other than those commonly considered in short-range weather forecasting are needed to bring about changes in climate, we would be testing this hypothesis.

The proposition is by no means preposterous. There are extremely simple and also very complicated systems of equations possessing solutions which behave in one manner for an extended period of time, and then change more or less abruptly to another mode of behavior for an equally long time. Such systems have been described as *almost intransitive* (Lorenz, 1968).

There are certain indications against almost-intransitivity as a major cause of climatic change, if the system hypothesized to be almost intransitive is taken to be the atmosphere alone. Enough numerical experiments with different initial conditions have already been performed by different groups to see whether widely differing climates are likely to appear. Invariably, the climate is found to look more like today's climate than an ancient one. It is, of course, possible that some investigators have obtained climates which do not look like today's and have simply assumed that something must have gone wrong, and that their results are not worth publishing. It does seem likely, however, that something favoring the older climates is missing from the experiments.

The situation is quite different if the system includes some portion of the environment, even if this portion is nothing more than the sea-surface temperature field. Almost-intransitivity becomes still more plausible if ocean currents are included. Models which generate their own oceanic properties as well as their own atmospheric properties are only beginning to be explored. When the system under consideration includes not only sea ice but also such features of the continents as snow cover and storage of water in the ground, almost-intransitivity becomes an attractive hypothesis. Indeed, it may be said that any hypothesis which does not invoke changes of the external environment is effectively attributing climatic changes to almost-intransitivity.

There appears to be in the mathematical model of the atmosphere a new and powerful tool for studying the phenomenon of climatic change. Availability of this tool to those who logically should be using it poses a problem; perhaps there should be a center for climatic-change hypothesis testing. It has been said that new hypotheses are being introduced much more rapidly than older ones are being rejected; the next ten years could see a reversal of this trend if the models are properly exploited.

If, instead, we look into the 21st century, and make an optimistic forecast concerning the type of computer which will be available, we find that yet another approach to climatic change may become feasible. We may construct a super-model, including as variables every feature of the atmosphere and its environment which can conceivably have varied over the ages. Included will be such features as the detailed composition of the atmosphere and the oceans, the extent of continental glaciation, and the distribution of vegetation. We can probably omit human activity on the grounds that human tampering was not responsible for *past* climatic changes. When we integrate the equations, if they are correct, we shall necessarily obtain changes in climate, including the great ice ages.

Such a solution may give us little insight as to why the changes took place. However, we can now eliminate various features, singly or in combination, and see whether climatic changes are still produced. In this manner we can eventually say what features or combinations of features *could have* produced the changes. In essence, we shall have reached the day when mathematical procedures will be instrumental in formulating hypotheses as well as testing them. This is a brute-force approach, and undoubtedly involves much computing which a little careful planning could eliminate, but this appears to be the way of modern computations. As to what features *did* produce climatic changes, we shall still have the privilege of arguing.

References

Donn, W. L., and M. Ewing, 1968: The theory of an ice-free Arctic Ocean. *Meteor. Monogr.*, 8, No. 30, 100–105.

Lorenz, E. N., 1968: Climatic determinism. *Meteor. Monogr.*, 8, No. 30, 1–3.

Miyakoda, K., J. Smagorinsky, R. F. Strickler and G. D. Hembree, 1969: Experimental extended prediction with a nine-level hemispheric model. *Mon. Wea. Rev.*, 97, 1–76.

Smagorinsky, J., 1963: General circulation experiments with the primitive equations. I. The basic experiment. *Mon. Wea. Rev.*, 91, 99–164.

Weyl, P., 1968: The role of the oceans in climatic change: A theory of the ice ages. *Meteor. Monogr.*, 8, No. 30, 37–62.

10

Introduction to the Subject　　　Hans A. Panofsky
of Mathematical Modeling
of the Atmosphere

Variables and Equations

The state of the atmosphere is well described by seven variables:
pressure, temperature, density, moisture (usually the fraction of
water vapor by weight, the specific humidity), two horizontal velocity components, and the vertical velocity. The behavior of these
seven variables is governed by seven equations: the equation of
state (gas law), the first law of thermodynamics, three components
of Newton's second law, continuity of mass, and continuity of
water substance (water can change phase, but is not destroyed).

A knowledge of the sets of seven variables is satisfactory for
many purposes, but does not completely specify the state of the
atmosphere and how it will behave. For example, if we know that
the specific humidity has reached a limiting value which is a function of temperature and pressure only, "saturation" occurs. This
fact by itself does not determine whether water vapor will change
to small drops, which remain suspended in the form of clouds, or
coalesce to form raindrops. The size distribution of the drops is
determined by other considerations, for example, the distribution
of particles.

If particles are introduced into the model, additional equations for the behavior of the particles are needed. Not only must
different equations appear for different particles, but somehow the
size distributions of particles must be specifiable by the additional
equations. Therefore, so far no model used in meteorology has
adequately treated the role played by particles.

Most of the terms in the equations (but not all) depend only
on the dependent variables of the atmosphere and on the coordinates. Among the terms that do not satisfy this condition is the
heat added to the air, a term occurring in the first law of thermodynamics. This term depends on the addition or removal of latent
heat, and on the addition or removal of radiation. Latent heat
provides no difficulties if it is assumed that changes of phase between vapor and liquid occur when the theoretical saturation conditions are reached and that water changes to ice at a specific tem-

Prepared during SCEP.

perature. This latter condition is of particularly dubious validity.

But radiation presents some more difficult problems; it depends not only on the position of the sun, which is known, but also on the amount, height, and other characteristics of clouds which cannot yet be handled. Radiation also depends on the distribution of CO_2, usually taken as constant, and on H_2O, which cannot be treated as constant.

In the stratosphere, water substance can often be neglected as a variable, and therefore the continuity equation for water vapor can be omitted; instead, ozone concentration becomes important, and a continuity equation for ozone is needed. This requires knowledge of the concentration of atomic oxygen, which is needed for ozone formation; hence an equation for this constituent has to be added.

There exists in the ocean a set of variables and equations similar to the atmosphere. Instead of water vapor, salinity is the principal variable factor in the composition; and, therefore, an equation for continuity of salt is required. Again, of course, this set of variables and equations is not "complete." For some problems, other variables of composition are needed such as oxygen and CO_2 content.

Integration of the Equations
A mathematical model of the atmosphere, ocean, or atmosphere-ocean system is described by the results of the integration of the relevant equations for specified initial and boundary conditions. The equations are solved for the time rates of change of the variables, the derivatives are replaced by ratios of finite differences, and changes of the variables over a certain time interval Δt are computed. The time interval must be chosen small enough to satisfy the condition that Δt must be smaller than x/c, where x is the distance between points at which the initial conditions are specified (the grid spacing), and c is the velocity of the fastest moving change that results from the integration (the speed of sound, around 300 m/sec). For example, if the equations are applied to observations 300 km apart, Δt would be fifteen minutes, though in practice it is taken as about five minutes. Given an initial estimate of the fields of the variables, the new fields are computed for a time Δt after the initial time; the process is then re-

peated as long as needed. Clearly, the construction of global models with small spatial (and therefore temporal) resolution, continued for a long period, requires a long time even on the fastest computer.

The initial conditions are either chosen from actual observations (for example, in routine numerical weather prediction, NWP) or they may be specified as a simple, artificial state (an atmosphere at rest with uniform thermodynamic characteristics).

The topography of the earth must enter as a lower boundary condition for the atmosphere, and there are further complications at the air-sea or air-land interface. Here, the vertical fluxes of momentum, heat, and mass must be continuous. One difficulty with these conditions is that the vertical fluxes are generally carried by systems of a much smaller scale (much smaller eddies) than the scale of the motions treated by the model. (We will discuss this point further.) Also, the complexity of the earth's surface means that the real boundary has to be greatly simplified to fit the model. Other boundary conditions require continuity of horizontal motion at the interface and that the motion just above a land surface be parallel to that surface.

Scale Considerations and Parameterization

The atmosphere contains motions with scales varying from about 1 mm to thousands of kilometers. Ideally, mathematical models should be constructed from observations every millimeter and with time steps of a fraction of a second. Clearly, this is impossible practically, and models are constructed separately for systems of different scales. Thus, for example, there are models for individual clouds, for local circulations such as sea breezes, for flow over mountains, for local weather developments over Western Europe, for the Northern Hemisphere, or for the globe.

Depending on the system modeled, the equations can be simplified; for example, the Coriolis force due to the earth's rotation can be neglected in cloud models, and the vertical acceleration can be neglected in large-scale models.

If, again, x represents the distance between points at which the initial data are available, there are in all cases motions with scales smaller than x. These motions then cannot be treated explicitly in the models. However, these small-scale motions cannot be neglected. Mathematically small-scale motions interact with large-scale motions, because the basic equations are nonlinear. Physically, the properties of large-scale systems can be changed by small-scale eddies, because the latter can produce mixing; for example, if a warm air mass is placed next to a cold one, the small-scale motions will cool the warm air and warm the cold air.

In practice, it is not necessary to specify the small-scale motions in detail. Instead, the large-scale motions are affected by certain *statistical* properties of the small-scale motions, which describe the fluxes produced by these motions. Such terms are proportional to covariances between small-scale characteristics. For example, the vertical heat flux depends on the covariance between small-scale vertical motions and the small-scale fluctuations of temperature.

The fluxes due to small-scale motions are especially important close to the ground because here the *large-scale* vertical motion is zero or nearly so, thus the small-scale motions furnish the only mechanism of vertical exchange near the boundary.

Since the small-scale motions do not explicitly occur as variables in the equations, we must find ways of relating the relevant statistics of small-scale motions to the large-scale field. This process is called "parameterization" of the small-scale effects. Parameterization is often based on statistics guided by physical intuition. Thus, for example, the vertical flux of moisture is assumed to be proportional to the large-scale vertical gradient of moisture and is directed from moist to dry air. The coefficient occurring in such an equation is called an "exchange coefficient." For models used in day-to-day forecasting, the properties of such coefficients near the ground are well known.

Special Characteristics of Climate Models
Climate models are global numerical models for which the computations are carried out for periods longer than a few days. They generally start with a simple atmosphere, such as a dry, uniform atmosphere at rest. The spacing between grid points is of the order of hundreds of kilometers, and the time step is of the order of five to ten minutes. The calculations are carried out for periods of a year or more.

The purpose of the numerical models is, first, to learn to understand the characteristics of the climate of the atmosphere, including the general circulation. To do this, the statistics of the model after considerable integration time are compared to the statistics of the actual atmospheric parameters. Another purpose is to assess the possibilities for long-range weather prediction. In principle, then, these models could be used in assessing climatic changes due to changes in certain characteristics of the atmosphere; for example, the initial state of the atmosphere could be changed in arbitrary ways, and the effect of these changes on the statistics of temperature and wind could be judged. There are difficulties here which will be discussed in detail later.

Aside from these general similarities of numerical models there are also many differences. For example, the vertical resolution varies widely between models; oceanic properties may be prescribed throughout the modeling period, or oceanic variables may be predicted; the surface may be given in various degrees of simplification; radiation may be modeled accurately or crudely; and the stratosphere may or may not be modeled explicitly. Due to the formidable difficulties described earlier, particles and clouds have not yet been introduced as variables to be predicted.

In principle, it might be desirable to shorten the integration time for climatic numerical models by not predicting detailed weather developments with five to ten minute time steps, but by working with, for instance, monthly average characteristics that can be predicted with much longer time steps so that machine time can be saved. This is being tried by G. B. Tucker of Australia. However, this method runs into a severe difficulty: monthly and other long-time averages eliminate traveling weather systems such as highs, lows, or upper-level waves. But, these systems affect the monthly averages through their ability to provide horizontal mixing. This means that the fluxes produced by these traveling weather systems have to be related to the characteristics of monthly averages. Attempts to do this have not been encouraging, although there is now increased understanding of some statistical characteristics of the traveling systems from the theory of two-dimensional turbulence.

Relevance of Climate Modeling to Environmental Problems

The role of atmosphere-hydrosphere numerical models in dealing with environmental problems is to provide two classes of predictive capability:

1. To determine the large-scale, long-term dispersive and storage characteristics of the combined system (atmosphere and hydrosphere).

2. To assess secondary interactions which would significantly influence the structure and variability of the combined system.

The present level of modeling capability of the atmosphere and oceans will allow meaningful estimates of the dispersion of inert material. Recent experience has shown that verbal arguments may be wholly misleading, for example, the observed slow dispersion of radioactive tungsten released in the lower equatorial stratosphere was the result of an almost balanced opposition of the transports by mean meridional circulation and the large-scale quasi-horizontal eddies.

The second type of problem, to estimate inadvertent climate modification, is a more demanding one. It requires models that are capable of simulating the natural climate and its variability in sufficient detail. Ten years ago modeling techniques were clearly inadequate to answer most questions. Though a great deal of progress has been made, an unqualified potential has not yet been demonstrated. For example, it is improbable that present models can simulate the differences between the global climate 100 years ago and today.

The object in introducing more sophisticated models is to make them more general and physically self-sufficient—that is freer from empirical parameterization derived from present terrestrial observations. In a sense the ultimate objective is to build models that are equally applicable to simulating the climate of Mars, Venus, Jupiter, and the Earth.

To understand the behavior of the atmosphere or the ocean under abnormal conditions we must avoid specifying the imposed influence of one on the other—that is, we must deal with the combined dynamical system in which the parts mutually determine each other's behavior. Oceanic modeling is somewhat behind in sophistication relative to that which we enjoy for the atmosphere, in part due to the critical lack of understanding of the small scale vertical transfer processes which influence the response characteristics of the ocean as a function of depth.

Most of the inadvertent climatic change questions we pose represent variations comparable to the noise level of the natural variability of the atmosphere as well as to the span of uncertainty in current modeling technique. When looking for small differences in the climate, smaller than the annual differences, we are asking very delicate questions of a model, and the simulation must be carried on for a long enough time to develop good statistics. There are apparently anomalies in the real atmosphere that last over a year—we hardly know how long—and some models tend to show the same thing. This poses a practical problem with current models and computers, since it is expensive to run a sophisticated model for five or ten years. (For example, 2.5 hr/model day is what the GFDL model requires using a 250×250 km grid size and 11 levels, running on an IBM-360-91. The NCAR model, which has a 400×400 km grid size and 6 levels, requires about 1 hr/model day on a CDC-6600.)

Modeling the Factors Influencing Climate

Glaciation
The fact that the earth is partially glaciated, rather than fully or not at all, represents a delicately balanced situation. Variations from an ice age to the periods in between must represent relatively small shifts. These shifts are due to very small but systematic differences in the balance between snow in winter and thaw in summer. But it is only recently that annual variability has been attempted with models. With models, a very small change can cause glaciation to the equator, which has never actually occurred. This means that the models are more unstable than the real climate. One reason for this is the omission of clouds as free variables. Clouds act as thermostats, contributing to the relative stability of the real climate.

Aerosols
Except for clouds, aerosols have been ignored—not only in their variability but even their systematic influence on the radiative

transfer. The determination of climatic effects due to changes of the aerosol distribution, quality, and amount due to human activity is therefore not possible in the near future.

Clouds

Even this preponderant aerosol has been dealt with cavalierly. As was mentioned before, water vapor is predicted quantitatively, but not cloud amount. The reason is that cloud storage of water is a negligible element in the water balance, and the size distribution of liquid water is immaterial when saturation occurs. The latent heat released is the same whether the water vapor forms drops, buckets, or sheets of liquid water. This may not be quite as true for cumulus clouds. In either case adequate parameterizations to predict cloud type and amount do not exist, because cloud formation depends on variables not in the problem as formulated at present.

Attempts have been made to determine empirically the stratiform cloud amount from the water vapor distribution, but this has not proved very satisfactory. The corresponding statistical relations for cumulus clouds have not yet been determined at all —which may be particularly critical in the tropics.

The mean effects on the radiative transfer of clouds distributed according to their observed vertical and latitudinal variation is however taken into account. Despite the clamor to relax this parameterization, it is still not known to what extent the dynamics are sensitive to synoptic time and space scale variability of clouds. This is, however, an urgent question that is beginning to be studied.

CO_2

Ideally one would like to change just one external variable or parameter at a time, such as the concentration of atmospheric CO_2, in order to get a result that is interpretable. (Distinguish between "a parameter" and "parameterization.") Since there are so many such external parameters in current models, and since they *should* be interacting (for example, CO_2 and cloud cover and polar ice), it is not possible to do a "proper experiment" now. We need a more self-determined model; we need to take clouds and polar caps into account; we need to take aerosols into account, and so on in order to see what the effect of changing CO_2 would be.

198 Hans A. Panofsky

Water Vapor

To give some insight to the dynamical influence of relatively gross modifications, consider the differences between a moist atmosphere (the normal case) and a completely dry one (the abnormal case). These are somewhat basic to physically conceivable intermediate situations where the atmosphere might be slightly drier or moister than it is presently, due to a different mean temperature or changes in the sources and sinks of water vapor at the earth's surface. This simple analysis is based on a post hoc reflection on two rather complex simulation experiments.

Latent heat released in the atmosphere reduces the effective static stability and thereby alters the baroclinic instability of extratropical disturbances. Hence in a moist extratropical atmosphere as compared to a dry atmosphere:

1. The horizontal scale of disturbances is smaller.
2. The equilibrium meridional temperature gradient is smaller.
3. The disturbance amplitudes are smaller, since the baroclinic waves can accomplish the needed meridional heat transfer in latent as well as sensible form.
4. The meridional angular momentum transfer is correspondingly smaller, and then
5. The surface winds are weaker in order to maintain an angular momentum balance.

Conclusions

The object of these comments is not to discourage expectations for use of theoretical models to simulate atmospheric-ocean response to human activity. It is an attempt to convey a *realistic* perspective as to present capabilities and prospects for the future. It is important to realize that if physically comprehensive models are deemed inadequate to answer some of our questions then certainly one should be wary not to rely on handwaving arguments or back-of-the-envelope calculations. Atmospheric dynamics can introduce both positive and negative feedbacks so that static calculations may be wrong, even in sign.

In order to make judgments of the effects of contaminants on climate, certain aspects of climatic models have to be made so complex as to exceed the capacity of computers. It therefore becomes necessary to look for trade-offs; one must find out what parts of present models can be simplified without important effects on the results. Then such simplified models can be used to study models that are more complex in other ways.

11
Large-Scale Atmospheric Circulation

Joseph Smagorinsky

The Problem and Its Implications

An understanding of the structure and variability of the global atmospheric circulation requires a knowledge of the following:

1. The quality and quantity of radiation coming from the sun.

2. The atmospheric constituents—not only the massive ones, but the thermodynamically active components, such as water vapor, carbon dioxide, ozone, and clouds as well as other particles. Furthermore, one must understand the processes by which these constituents react with the circulations and their radiative properties, that is absorption, transmission, scattering, and reflection.

3. The processes by which the atmosphere interacts with its lower boundary in the transmission of momentum, heat, and water substance over land surfaces as well as sea surfaces. The behavior of the atmosphere cannot be considered independent of its lower boundary beyond a few days. In turn, the lower boundary can react significantly. Even the surface layers of the oceans have important reaction times of less than a week, whereas the deeper ocean comes to play only over longer periods. Hence, from a dynamical viewpoint, the evolution of the atmospheric circulation over long periods requires consideration of a dynamical system whose lower boundary is below the earth's surface.

4. The interactions of the large-scale motions of the atmosphere with the variety of smaller scale motions normally present. If these smaller scales have energy sources of their own, as is the case in the atmosphere, the nature of the interactions are considerably complicated.

Hence, in principle, mathematical models embodying precise statements of the component physical elements and their interactions provide the means for numerically simulating the natural evolution of the large-scale atmosphere and its constituents. Such a capability implies, in turn, potential applications in a number of areas: long-range forecasting, the determination of the large-scale long-term dispersion of man-made pollutants, the interaction of these pollutants in inadvertently altering climate, and the influence of intentionally tampering with boundary conditions

Prepared for SCEP.

to modify artificially the climatic equilibrium. No doubt there are a variety of other applications of a simulation capability to problems that may not yet be evident.

Current Status

Efforts to model the large-scale atmosphere and to simulate its behavior numerically began more than twenty years ago. As additional research groups and institutions in the United States and elsewhere in the world became involved, a steady succession of advances in model sophistication followed. These came from refinements in the numerical methods as well as from improved formulations of the component processes.

Multilevel models today account for a variety of interacting influences and processes: large-scale topographic variations, thermal differences between continents and oceans, variations in roughness characteristics, radiative transfer as a function of an arbitrary distribution of radiatively active constituents, large-scale phase changes of water substance in the precipitation process, interactions with small-scale convectively unstable motions, the thermal consequences of variable water storage in the soil, and the consequences of snow-covered surfaces on the heat balance. More recently, combined atmosphere-ocean models have taken into account the mutual interaction, including the formation and transport of sea ice.

Despite the fact that many of the preceding elements are rather crudely formulated as cogs in the total model, it has been possible to simulate with increasing detail the characteristics of the observed climate—not only the global wind system and temperature distribution from the earth's surface to the mid-stratosphere, but also the precipitation regimes and their role in forming the deserts and major river basins of the world. Attention is beginning to be given to the simulation of climatic response to the annual radiation cycle.

Detailed analyses of such simulations in terms of the flow and transformation of energy from the primary solar source to the ultimate viscous sink show encouragingly good agreement with corresponding analyses of observed atmospheric data. Such models have also been applied to observationally specified atmospheric states in tests of transient predictability. Even within the severe limitations of the models, the data, and the computational inadequacies, it has been possible to simulate and verify large-scale atmospheric evolutions of the order of a week, giving promise that as the known deficiencies are systematically removed, the practical level of the large-scale predictability of the atmosphere can converge upon a theoretical deterministic limitation of several weeks.

Such models have also been used in some other limited applications. For example, an attempt was made to simulate the long-term, large-scale dispersion of inert tracing material, such as radioactive tungsten, which had been released at an instantaneous source in the lower equatorial troposphere. These results have been surprisingly good. Only very limited attempts have been made to apply extant models to test the sensitivity of climate to small external influences. The reason is that one normally seeks to detect departures from fairly delicately balanced states. It is often beyond the current level of capability to simulate an abnormal response that is comparable in magnitude to the natural variability noise level.

The extent of the present large-scale data base is essentially dictated by existing operational observing networks created by the weather forecast services of the world. The existing network is hardly adequate to define the Northern Hemisphere extratropical atmosphere and is completely inadequate in the Southern Hemisphere and in the equatorial tropics. For example, there are only 50 radiosonde stations in the Southern Hemisphere in contrast to approximately 500 in the Northern Hemisphere. The main difficulties arise from the large expanses of open ocean which, by conventional methods, impede the determination of the large-scale components of atmospheric structure responsible for the major energy transformations. This critical deficiency in the global observational data store makes it difficult to define the variability of the atmosphere in enough detail to discern systematic theoretical deficiencies. Furthermore, the data are inadequate for the specification of initial conditions in the calculation of long-range forecasts.

Recent dramatic advances in infrared spectroscopy from satellites promise significant strides in defining the state of the extratropic atmosphere virtually independent of location. However, the motions of the equatorial tropical atmosphere lack strong rotational coupling, making the observational problem there more acute. Independent wind determinations may be needed as well as the information supplied by a Nimbus III (SIRS sensor) type satellite. It is not yet known to what extent balloonborne instrumentation or measurements from ocean buoys will be needed to augment satellite observations, especially in the lower troposphere. It will depend on just how strongly the variable characteristics of the atmosphere are coupled. A more precise knowledge would permit relaxing observational requirements for an adequate definition of its structure.

There is some controversy as to the inherent deterministic limitations of the predictability of the atmosphere, say for scales corresponding to individual extratropical cyclones. The span of controversy ranges from about one to several weeks. Moreover, it is not known at all whether longer-term characteristics of variability of the atmosphere are determinate. For example, is it inherently possible to distinguish the mean conditions over the eastern United States from one January to another in some deterministic sense? In the equatorial tropics there is very little insight at all as to the spectrum of predictability.

Needs for Future Improvements

Broadly, there are three areas which require intensive upgrading, the first two of which are essentially technological:

1.   The establishment of an adequate global observing system (for reasons already discussed).

2.   The acquisition of computers two orders of magnitude faster than those currently available to permit the positive reduction of mathematical errors incurred by inadequate computational resolution. Faster computers will also permit more exhaustive tests of model performance over a much larger range of parameter space to assess the sensitivity of simulations to parameterizations of physical process elements of the model. Faster computers will also provide an ability to undertake the broad range of applications implied by a more sophisticated modeling capability.

3.   The scientific requirements stem from the necessity of refining the formulation of process elements in the models. To cite a few: boundary layer interactions to determine the dependence of the heat, momentum, and water vapor exchange within the

lower kilometer of the atmosphere as a function of the large-scale
structural characteristics; internal turbulence—to determine the
structure and mechanisms responsible for intermittent turbulence
in the "free" atmosphere, which is apparently responsible for the
removal of significant amounts of energy from the large scale, and
possibly also playing a role in the diffusion of heat, momentum,
and water vapor; convection—to determine how cumulus overturning gives rise to the deep vertical transport of heat, water
vapor, and possibly momentum.

We are as yet unaware of consequences of particles, man-made
or natural, either directly on the radiative balance or ultimately
on the dynamics.

In the tropics we have yet to completely understand the instability mechanisms responsible for the formation of weak disturbances or the nature of an apparent second level of instability
which transforms some of these disturbances to intense vortices,
manifested as hurricanes and typhoons. Without an understanding
of the intricacies of the tropics, it is impossible to deal comprehensively or coherently with the global circulation, particularly
with the interactions of the circulation of one hemisphere with
that of the other.

Most all of these critical scientific areas of uncertainty require intensive phenomenological or regional observational studies. These will provide the basic data as foundations for a better
theoretical understanding.

Any one of the three general categories above may at any one
time provide the weakest link in the complex required to advance
a modeling and simulation capability. Obviously then, they must
all be upgraded at compatible rates.

12

Review of Climate Models   G. D. Robinson

Climate and Climate Change

When we come to investigate climatic change, we are faced with an initial difficulty of definition which has far-reaching consequences. The climate of a locality may be described by statistics of certain weather elements. Let us take air temperature over central and southern England, as an example (because observations have been made there for a long time). Craddock (1) has examined about 100 years of the instrumental record of daily mean temperature at Kew (London) and Manley's series of about 250 years of annual mean temperatures of central England, using numerical filters that isolate the variance connected with various periodicities. Eliminating diurnal and annual periodicities and their subharmonics, Figure 12.1 illustrates his findings. There is significant variance at the longest period plotted. The 250-year-long series of observed temperatures is not stationary. The continuous line in Figure 12.1 is the spectrum of a stationary series with a long time scale proposed by Charnock and Robinson (2) as an empirical approximation to many meteorological time-series. It is fitted to Craddock's spectrum at a period of 30 days. If we define climate, as it has been conventionally defined, by the mean and variance of a 30-year series of observations, then the climate of central England has been subject to continuous change for the past 250 years; if we look before the instrumental record, we find clear historical and geological evidence that climate change is a global phenomenon, and was so before man appeared. Early in the present millennium a pastoral community was able to maintain itself in southern Greenland. Twenty thousand years ago ice covered much of Europe and Canada. Climatic change is obviously not necessarily dependent on man's activity. This is not to say that man cannot change, or indeed is not changing climate, but we have insufficient statistics of "natural" changes to allow us to recognize artificial changes on the global or continental scale by statistical techniques, and it is the essence of the problem that we will now never have them.

We, therefore, turn from exclusively statistical to physical

Reprinted from G. D. Robinson, 1970. *Long-Term Effects of Air Pollution— A Survey,* prepared for the National Air Pollution Control Administration (Hartford: Center for the Environment and Man, Inc.), pp. 10–16.

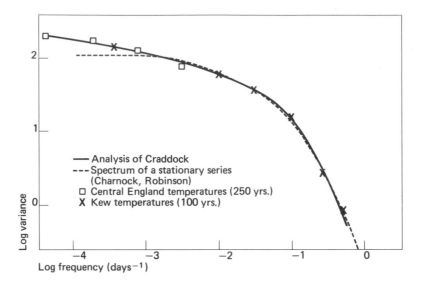

Figure 12.1 Spectrum of temperature variations

methods, though if climate is statistically defined we cannot completely exclude statistical considerations. If we understand in detail the dynamical and thermodynamical equations of the atmosphere and can solve them, we can investigate the consequences of the sort of perturbation which man's activity might introduce into the initial conditions. We have made enormous progress in this direction, best exemplified in the work of Smagorinsky's laboratory (3), but there is a fundamental difficulty which has been most clearly exposed by Lorenz (4). [I first encountered the idea many years ago in an unpublished review by R. C. Sutcliffe and, no doubt, it has been intuitively recognized by most meteorologists; but it is, surprisingly, not a prominent feature of meteorological literature.] The behavior of the atmosphere is expressed as a closed set of equations. This set of equations has numerous solutions, each representing a possible state of the atmosphere. Climate can be mathematically defined as the statistics of the solutions of these equations. If a unique (stationary) set of such statistics exists, the system governed by the equations is said to be transitive. The equations governing the state of the atmosphere certainly include nonlinear differential equations, and it is known that the uniqueness of long-term statistics of solutions of sets of such equations is not assured. Statistically different sets of solutions might, for example, develop from different initial conditions. If a unique set does not exist, the system is said to be intransitive. Lorenz introduces the concept of the "almost-intransitive" system, one for which an infinite set of solutions exists which is independent of the initial conditions but for which large but finite subsets exist which are very dependent on initial conditions. Lorenz postulates that the atmosphere may be an almost-intransitive system. The consequences for the student of climatic change are best stated in his own words—"For one thing, the mere existence of long-term climatic changes cannot by itself be taken as proof of environmental change; alternative explanations are now available. Finally, what about the not unlikely possibility that the atmosphere would be almost-intransitive if the environmental influences were constant, while at the same time external environmental changes actually are taking place? The effect of these changes will then be harder to detect, and causative connection will be more difficult to establish. For example, an environmental change which ought to bring about a 2°C temperature rise might occur just at the time when the temperature was in the process of falling 2°C as a result of almost-intransitivity. The environmental change might then go unnoticed simply because no one would see any reason to look for it." (See also Chapter 9.)

With these considerations in mind, let us take a superficial look at climatic trends in the present century (we will need to look a little more closely later). Over that part of the world where air temperature is comprehensively observed, it is reasonably well established that temperatures increased over a 30- or 40-year period, ending some time betwen 1940 and 1955. It is also reasonably well established that the CO_2 content of the atmosphere increased during the same period, by an amount not inconsistent with the rate of production by consumption of fossil fuels. The two trends were associated, and computation using the radiative transfer equations showed that quantitatively the CO_2 increase could explain the temperature rise. Many scientists did not hesitate to say that it did explain the temperature rise. A few did not hesitate to extrapolate the trend and predict a man-made climatic change which would melt the polar ice and drown many of the world's major cities. More cautious meteorologists recognized the inadvisability of the extrapolation but had to agree that the argument connecting the two trends was physically sound. But by 1960 it was becoming clear that the temperature was falling, and somewhat later it was noted that atmospheric turbidity was increasing, not only in the immediate vicinity of cities but over quite wide areas of Eurasia and North America. The increased turbidity correlated well, geographically, with the emission of pollutant aerosols, and the decreasing temperature was tentatively ascribed to an increased reflection to space of solar radiation by man-made pollution. The argument is plausible but, basically because of lack of knowledge of certain physical properties of the aerosol, it cannot be developed as precisely as was' that concerning CO_2, and perhaps for this reason there has been less talk of a man-made ice age than there was of a man-made deluge.

The relation of these recent events and the conclusions of Lorenz quoted above is obvious. Man has detectably changed the constitution of the atmosphere; in one respect globally, in another at least on a subcontinental scale. At the same time, there have been minor variations of climate. Our present knowledge of atmospheric processes suggests that these changes are what would be expected to follow from man's interference with the atmosphere. But they are also quite compatible with what we know of the statistics appropriate to an atmosphere of undisturbed constitution and they are also compatible with the possible behavior of a dynamical system as complex as the atmosphere. Climate has changed and will change, and man may never know to what extent he has contributed to or inhibited the change. Recognition of this has quite important consequences in the planning of research on the climatic effects of pollution.

Simple Climatic Change Models

There have been some attempts to estimate climatic effects of changes in the input of solar radiation (either by a change of solar constant or of global albedo) by very simple methods. They must be considered, if only because of the nature of some of the conclusions which have been drawn from them. The prototype of such models appears to be that of Sawyer (5, 6) who treated the problem "very crudely as though the atmosphere consisted of two blocks of uniform temperature between which heat was transferred at a rate depending on the temperature difference." On this model a 1 percent decrease in absorbed solar radiation caused a decrease of temperature in the equatorial section of about 0.75°C and in the polar section of about 0.6°C. In a second application of the model, Sawyer reduced the mean annual solar input by 1 percent in regions poleward of 50° latitude and computed a temperature change in this sector of about 0.2°C.

This model is illustrated in Figure 12.2 in which have been inserted values of radiative input consistent with the most modern estimates of the solar constant, the global albedo as measured by satellites, the radiative temperatures measured by satellites, and the estimates of heat-transfer by ocean and atmosphere across latitude 50°. This can be used as an educational toy for simple numerical experiments. The basic assumption of the models under discussion is the relation between advective transport and temperature difference. The toy can be used to illustrate the difficulty of preserving a balance for arbitrary variations of the heat input given a simple, e.g., linear, relation between heat transport and temperature difference. It gives some appreciation of the potentialities and pitfalls of Sawyer's model and its elaborations.

Two of the elaborations, described by Sellers (7) and by Rakipova (8), are directly applicable to the problem of the climatic consequence of a change of albedo. The simpler model (Sellers)

Solar input			Air-ocean transport	Radiative output		
Albedo	cal year^{-1} $\times 10^{-22}$		cal year^{-1} $\times 10^{-22}$	Radiative temp.	cal year^{-1} $\times 10^{-22}$	
0.50	4.2	→		239K	8.2	→
						50°N
0.27	41.6	→	4	259K	37.6	→

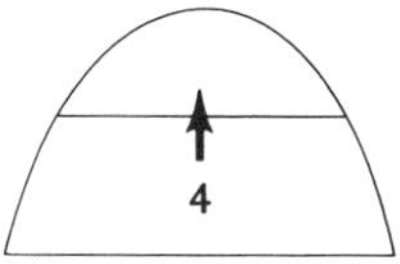

Transfer coefficient 2.0 $\times 10^{21}$ cal year^{-1} deg^{-1}

Whole earth albedo 0.30 Mean radiative temp. 254.5K

Suggested exercise: Hold solar constant and transfer coefficient fixed. Specify a relation (or relations) between albedo and radiative temperature. Arbitrarily change one albedo. Proceed by trial and error to a new balance.

Figure 12.2. Basis of the simplest climate model

carries only sea-level temperature as an (independent) indicator of climate. The type of conclusion which he draws is typified by: "If all other variables are held constant, a decrease in the solar constant by about 2 percent would be sufficient to create another ice age with the ice-caps extending equatorward to 50°." Rakipova's model is more complex in detail (though not in principle) than Sellers's, and carries temperature at various heights in the atmosphere. Rakipova's work exemplifies the approach to this problem by the Leningrad group under M. I. Budyko. One of her conclusions is that a 1 percent decrease in the solar input would result in a temperature decrease varying between $0.3°C$ at the equator to $1.4°C$ at the poles. The apparent difference between this and Sellers's more startling conclusion is caused by a difference in the treatment of the relation between temperature and albedo. This relation is a key factor in the models. According to Sellers, the albedo in all latitudes would be expected to increase as temperature decreases. This is a destabilizing mechanism—a drop in temperature reduces the heat input. The relation proposed is empirical; the physical cause appears to be the high albedo of a snow- and ice-covered surface. The weakness of the relation is that it does not seem adequate to cover the effect of the cloud. Lower temperatures lead to a lower atmospheric water content which might mean less cloud and lower albedo. This type of control of albedo produces a stabilizing mechanism, lower temperatures leading to greater heat input, and the possibility that the planet Earth is behaving as an inefficiently stirred thermostat. In Rakipova's computation, cited earlier, the albedo is not changed, but in the text of her paper she quotes a "warm season" global albedo of 0.410 and a "cold season" global albedo of 0.384. Both figures may be too high, but the overall relation proposed is a stabilizing relation.

The temperature-albedo relation is the sensitive problem in the Sawyer-Sellers-Rakipova type models, and there is some prospect of an empirical solution to it when many more data have been collected from meteorological satellites, but the models have also a major difficulty in principle. Rakipova states this clearly early in her text and does not allow it to inhibit her further. She states, "We are analyzing a purely zonal situation for which (the mean meridional and vertical velocities) $= 0$." One cannot accept this "purely zonal situation" as a reasonable model for investigation of the effects of heat transfer between latitudes. Sellers apparently shares this view but does not appear to circumvent the problem. He states "the inclusion of the mean meridional motion is necessary in order to avoid having to deal with negative diffusivities," but the mean meridional motions he postulates do not satisfy mass continuity and he does not discuss the transports of potential energy consequent on realistic mean meridional circulations.

Detailed Climatic Models

The basis of modern meteorology is mathematical simulation of the atmosphere and the earth's surface as a nearly closed thermodynamic system, and the current trend in such simulation is toward increasingly detailed realism, at the expense of computation loads which stretch the limits of foreseen technology. This type of model was initially conceived as a tool for objective weather forecasting and is often justified economically in this context, but its value in climatology was recognized at an early stage. Reservations concerning the validity of this approach to the detailed forecasting problem (9, 10), even if sound, are not related to the use of the models as generators of climatic statistics. For the present purpose, a minimal description of the methods must suffice. Several groups are engaged in the development and use of comprehensive atmospheric models. Probably the most-advanced and best-documented model is that of Smagorinsky's group at Princeton. The following comments, superficial in the sense that they include only aspects relevant to the present task, are based on a perhaps partial understanding of this model and of that described by Bushby and Timpson (11).

The mathematical basis is the conservation equations for momentum, energy, and matter—partial differential equations, some of which are nonlinear—the integral radiative transfer equation, the perfect gas equation of state, and the thermodynamic equations describing the phase changes of H_2O. Mathematically, solution is only possible by numerical finite difference methods; physically, initial and boundary conditions can be supplied only as averages in time and space. The magnitude of the computation load and the resolution of observations set lower limits to feasible and available spatial definition; when the space scale has been chosen an upper limit to the finite time step which can be used in solution of the momentum equations follows. A spatial resolution greater than about 500 km is probably insufficient in a model which is expected, for example, to indicate changes in the ice- and snow-covered areas of the globe. The corresponding maximum allowable time-step is of order 10 minutes. To seek to use such a model in straightforward continuous simulation to examine a period even as short as the 250 years for which we have some in- strumental record, cannot at present be contemplated. Perhaps this will not still be true 50 years from now but we are concerned with the next few years. In this period the use of models to in- vestigate climate modification will probably be by changing initial and boundary conditions—surface albedo, solar input, etc., or coefficients in the radiative transfer equations (i.e., atmospheric composition) and following the consequences to a quasi-steady state (or to breakdown of the computational scheme).

For the investigation of the effects of pollution, current com- prehensive models have one very serious defect—they do not de- velop and transport cloud systems. Climatological averages of cloud distribution are used in application of the radiative transfer equations, and the water cycle is handled by assuming immediate precipitation of water in excess of saturation—the models simulate rain and snow but not cloud. There is no major difficulty of prin- ciple in modifying models of large-scale processes to incorporate generation, transport, and dissolution of cloud, but this would increase very considerably the complexity of a computation which is already probably the most extensive ever undertaken on a con- tinuing basis. Lack of this modification is not of prime importance in the weather prediction application of large-scale models, but it is a most serious shortcoming when they are used to investigate climatic change, because of the high sensitivity of climatic statis- tics to albedo, the sensitive dependence of albedo on cloud amount and type, and the critical relation of cloud amount and type to atmospheric motion and stability. We see that the major defect of the existing comprehensive models of the atmosphere, in their application to study of climate, is the same as that we noted in the simple models—inadequacy in simulation of the relation of albedo to other parameters of climate.

It is probably safe to assume that on the global and climatic scale, though not on the local scale, the amount and gross radiative properties of cloud do not depend on space and time variations in the nucleating properties of the atmosphere. If this is so, the cloud problem can, at least in principle, be resolved within the logical framework of the model without further empiricism— the dynamic and thermodynamic processes generate and dissolve the cloud; the condensation and evaporation react grossly on atmospheric temperature and motion. The situation in respect of aerosol is rather different. There are both surface and internal sources of both man-made and natural aerosols. These aerosols are transported by atmospheric motions but there is no first-order feedback between aerosol content and motion—the interaction is slow and through the radiative terms. Transport can be handled by the models—on the small scale this is done in "air pollution models," and on a scale of hundreds of kilometers, R. J. Murgatroyd has recently applied a detailed forecasting model (Bushby-Timpson, 10 layer 80 km grid square) to the problem of three-dimensional transport of pollutants. At the time of writing only abstracts of Murgatroyd's work have been seen: the outstanding problem may be the handling, with a tolerable number of layers in the vertical, of the quasi-isentropic nature of the three-dimensional motion and the resultant tendency for locally injected pollutant to be confined, in a stable atmosphere, to gently sloping layers of very limited depth. An approach other than empirical to the specification of a source term for the natural aerosol (e.g., raising of surface dust by the model-generated wind) is a development for the distant future. Man-made sources must always be empirically specified.

Future Development of Climatic Models

The brief overview of simple climatic models did not lead us to any very encouraging conclusions. It is not easy to formulate recommendations for further work with them because one cannot have real confidence, even qualitatively, in the results so far announced. On the other hand, there is now general agreement that modeling of the whole surface-atmosphere complex is essential to an understanding of climatic change consequent on any local or global modification of the atmosphere or surface and inactivity will not advance the cause. A possible approach would be use of the comprehensive global weather models of the type developed by Smagorinsky's group in something like a "Monte Carlo technique" mode to accumulate statistics on the climates of atmospheres of differing constitution, but this expensive undertaking is not likely to commend itself to the very few institutions at present able to contemplate it. The expense and elaboration are sufficient to make this a very inefficient approach, so long as the models do not carry cloud cover as a dependent variable.

It would perhaps be useful to continue development of the simple models as "educational toys"—for example: to attempt to improve the treatment of mean meridional circulations in Sellers' model and then examine the effects of imposing different relations between albedo and temperature. This relation is one of the keys to the modeling of long-term climatic change, and it might be that a new approach with more parameterization of dynamical aspects and more detailed treatment of radiation and cloud physics could be explored. The simple models carry parameterization of all processes to the extreme.

References

1. J. M. Craddock, "A contribution to the study of meteorological time series," M.R.P. 1051, Met. Office London, 1957.

2. H. Charnock and G. D. Robinson, "Spectral estimates from subdivided meteorological series," M.R.P. 1062, Met. Office London, 1957.

3. K. Miyakoda, J. Smagorinsky, R. F. Strickler, and G. D. Hembree, "Experimental extended predictions with a nine-level atmospheric model," *Mo. Wea. Rev., 97,* pp. 1–76, 1969.

4. E. N. Lorenz, "Climatic determinism," *Met. Monog., 5,* pp. 1–3, 1968.

5. J. S. Sawyer, "Notes on the response of the general circulation to changes in the solar constant," Proc. Rome Symposium on Changes of Climate 1961, UNESCO Paris.

6. J. S. Sawyer, "Possible variations of the general circulation of the atmosphere," World Climate from 8000 to 0 BC, pp. 218–229, 1966, *Roy. Met. Soc.,* London.

7. W. D. Sellers, "A global climatic model based on the energy balance of the earth atmosphere system," *J. Appl. Met., 8,* pp. 392–400, 1969.

8. L. R. Rakipova, "Changes in the zonal distribution of atmospheric temperature as a result of active influence on the climate," *Modern problems of climatology,* pp. 388–421, 1966, Leningrad (Translation Clearinghouse AD 670–893).

9. G. D. Robinson, "Some current projects for global meteorological observation and experiment." *Quart. J. R. Meteor. Soc., 93,* pp. 409–418, 1967.

10. E. N. Lorenz, "The predictability of a flow which possesses many scales of motion," *Tellus, 21,* pp. 289–307, 1969.

11. F. H. Bushby and M. S. Timpson, "A 10-level atmospheric model and frontal rain," *Quart. J. R. Met. Soc., 93,* pp. 1–17, 1967.

One of the major difficulties that faced the SCEP participants in their attempt to assess man's impact on the global environment was the poor state of the data base with respect to climatic processes and the rates, routes, reservoirs, and effects of pollutants in the environment. Developing a more complete understanding of the climate will require extensive research and measurements as well as the sophisticated mathematical models discussed in Part III. The data base necessary for determining the interaction of pollutants with atmospheric processes will have to be generated through monitoring.

In the following four papers some of the concepts involved in global monitoring and in establishing monitoring networks are discussed. These papers complement and expand the report of the SCEP Work Group on Monitoring (reprinted in Part I of this volume) with respect to objectives and design considerations. In other sections of this volume monitoring techniques will be discussed in the context of specific pollutants or technologies.

This section begins and ends with papers providing a macro-view of the problems involved in the monitoring of global environmental problems. The middle two papers provide the micro-view of questions that must be answered when actually setting up a monitoring network.

In the first paper Dr. G. D. Robinson notes the ambiguity of the term "monitoring" and the compromises required in allocating resources for monitoring, base-line studies, and research measurements. He also discusses some of the peculiarities of and problems associated with global monitoring systems.

The next two papers were originally prepared to be presented together at a conference in the Spring 1970. They were important resource documents for the Study because they discuss the specific objectives, criteria, guidelines, and design specifications for setting up an air monitoring network. These observations are particularly important because these insights are provided from the actual experience establishing such systems. In fact, the paper by Drs. Morgan and Ozolins originally used an illustrative case based on the National Air Pollution Control Administration (NAPCA) design of a network for the Metropolitan St. Louis Interstate Air Quality Control Region (this discussion does not appear in the paper in this volume). If a global monitoring network is ever established, the design and scientific principles outlined in the papers of Drs. Morgan and Ozolins and of Dr. Saltzman will have to be considered explicitly and carefully.

The final paper in this series is presented to provide an overview of the types of questions that must be addressed in the design of a global monitoring network. Many of these questions were considered by the SCEP Work Group on Monitoring, and those conclusions appear in Part I of this volume. The two major areas addressed by the paper are those of identification and classification of pollutants to be monitored and of determination of the number or density of stations for a global network. This document, which was available during SCEP, was prepared for the World Meteorological Organization (WMO) of the United Nations by the Commission of Air Chemistry and Radioactivity (CACR) which is part of the advisory structure of the International Council of Scientific Unions (ICSU). The members of the CACR are scientists appointed as individuals not as representatives of any government, institution, or national academy. The WMO has made some preliminary recommendations to member countries, based in part on these CACR findings.

Monitoring Global   G. D. Robinson
Environmental Problems

An Attempt to Define "Monitoring"

Monitoring is a word which has meant different things to different people, and it is useful to narrow the range of meaning. It seems unnecessary to use the word, as it has been used by some, as synonymous with measurement or in particular with measurements concerned with natural phenomena. A reasonable description would seem to be "arrangement for the provision of advice or warning concerning some event, usually an unwelcome event." In this sense measurements made in the course of research, even research into problems with a serious impact on mankind, should not be described as monitoring. Less obviously, the description excludes continuing observation designed to allow the preparation of generalized advice, for example, meteorological observations taken to allow provision of routine weather forecasts. The essence of monitoring, conceived in this way, is the existence of a specific problem—for example, meteorological observations taken, processed, and disseminated specifically to detect and warn of the approach of hurricanes would qualify as "monitoring."

It would be wrong to give the impression that all environmentalists agree with this restricted approach to monitoring. It is considered by some to presume too comprehensive a knowledge of environmental problems. On this view there is a place for what has been described as "shotgun monitoring"—widespread measurements and observation, particularly of biological material, designed to disclose hitherto undetected environmental threats. In the present state of knowledge and opinion, environmental monitoring must find some compromise between the two points of view, whose extreme expressions are "measure everything everywhere all the time" and "make the minimum of observation, with no more than the necessary precision, required to assess the status of known threats." The recommendations of the SCEP Monitoring Group are such a compromise.

Peculiarities of Global Monitoring

There are innumerable examples of effective monitoring of local problems; the best-known and most successful ones are of the

Prepared after SCEP for this volume.

timely alarm type, for example, the animal that becomes distressed before the human beings which it protects are too seriously affected, the alarm bell triggered by minute concentrations of smoke. These solutions presuppose a safe base for retreat—the workers have time to leave the mine or the factory while engineers and firemen do what they can to make things safe for return. In a truly global problem there is nowhere to go while the specialists get to work to put things right. The remedy must be worked out inside the system. The minimum requirement of global monitoring is that it should sound a warning in time to allow this, but a monitoring system which not only sounds an alarm but which gives indications useful in working out a solution is to be preferred. The "alarm" will, in general, be a trend in statistics; the ideal is that the statistics should themselves give indications of the causes of the trend which would in turn point the way to remedial action.

The Feasibility of Global Monitoring

It is not obvious a priori that effective monitoring of a global problem is possible. In the first place, it is a general truth that observation in some way disturbs the system observed. The disturbance is frequently negligible. If, in the course of monitoring some global problem, we decide to measure the energy arriving at the earth's surface from the sun, we can certainly neglect the minute change that introduction of our sensor makes in the field of radiation. But what if, in the course of "shotgun monitoring," we attempt to observe and chronicle the ecology of an area. This type of monitoring is advocated by many biologists. If the area were genuine wilderness, the observer's presence would cause reactions—some species have the habit of positively shunning the company of man, however well-intentioned the individual, and we do not know how other species will react to this primary perturbation. There will be changes simply because we have decided to look for changes. If the area is not a genuine wilderness, we will have the greatest difficulty in separating local and global influences, for example, if we attempt to fence off such an area we will invite influx of species hunted in the vicinity. Nothing is more likely to change bird populations than the institution of a bird sanctuary. It may be that these difficulties can be overcome, but clearly the feasibility of effective monitoring of any problem in this way needs to be established.

A second general obstacle to global monitoring lies in the nature of climatic (and probably of ecological) statistics. For the period of the instrumental record, statistics of global climate are not stationary. There is historical and geological evidence of climatic fluctuations on all periodicities, with amplitude increasing with period. In this situation we have no means of testing the significance of a trend. We cannot detect an artificial change as a "signal" by integrating over a period long enough to cancel the natural "noise." It is not that sort of "noise." Paleobiology suggests that ecosystems might be similar in this respect to climate. Meteorologists recognize that the problem of induced climatic change cannot be monitored by observing the climate. We must monitor the perturbing element while continuing our efforts to understand the mechanism of the perturbation. Biologists have not yet universally adopted the analogous viewpoint.

Another situation in which global monitoring, in the sense of production of timely warning, might fail arises if the perturbation we are concerned with has an irreversible impact, or more subtly if there is a "point of no return" in the process. This is perhaps a frequent occurrence in biological systems, and it may be true of some aspects of climate. It is easy to conceive situations in which monitoring, however well devised, could not separate significantly the artificial trend from the natural variance until the process of change was too advanced to be corrected.

The Identification of Critical Global Environmental Problems
There are in the modern world innumerable environmental problems. Many of them are critical in the sense that some unpleasant impact on human life is present or imminent. Most of them are of local concern. A few are obviously global. If a contaminant with a lifetime on the order of years escapes into the atmosphere anywhere in the world, it will mix globally. If it causes a problem, we have a global problem. There is a complication which appears trivial in principle but which causes many difficulties in practice. An atmospheric pollutant may have a short life simply because it reacts quickly with another pollutant or a natural atmospheric

constituent. If the product has a long life we have a global problem with a complex monitoring aspect. The same considerations apply to escape of contaminants into the oceans, but here mixing is much slower and the lifetime of a globally dangerous pollutant must be much longer. Birds and fish mix quickly and if a wide-ranging species has a restricted breeding ground, a local threat to the breeding ground can become a global problem. One of the difficulties facing the SCEP monitoring group was the question of deciding when a multitude of local problems can be considered a global problem. One case which seemed to qualify as a global problem was the modification by pollution, eutrophication, land reclamation, and dredging of numerous estuaries and coastal waters that shelter ocean-going species at some point in their life cycles. Even when such a problem is identified as critical and global, it by no means follows that the most effective approach to monitoring the problem is on the global scale. For example, of varied threats to coastal waters just enumerated, only gross physical manipulation—construction, reclamation, and so on—seems at present more suited to globally than to locally organized monitoring, thanks to the possibilities of satellite observation.

"Base Lines," Research, and Monitoring

In discussions of global monitoring, it is usually possible to come to agreement on the principle that minimum effective effort is desirable; that we cannot, in fact, measure everything everywhere all the time. It is much more difficult to arrive at agreement on what constitutes minimum effective effort, not so much because of conceptual differences as because of a lack of knowledge both of the present state of the environment and of the fundamental behavior of climate and ecosystems. We know that toxic substances have escaped into the oceans. In default of international agreements which will not be easily reached, these substances will continue to escape. We wish to "monitor" their occurrence because they are a threat to life, critical in varying degrees. We do not know how many measurements we must make, and with what precision, because we do not know the present distribution of the materials. If we knew, for example, that the distribution were uniform, and that we could rely on its remaining uniform, we could settle for measurements of the concentration of contaminants in one locality. If we knew the distribution to be uniform but did not know why, we would be wise to set up more than our one monitoring station and simultaneously conduct research into the mechanism of dissemination. If we knew the distribution to be highly non-uniform, we would require a number of monitoring stations but could hardly expect to define an optimum array if we did not know why the distribution had its observed form. We would require first a base-line survey and then research on the distribution process.

In none of the critical global problems identified at SCEP was there sufficient knowledge to allow confident specification of a global monitoring procedure. In each case the recommendations for action included broad base-line surveys, or research into the life cycle of pollutants in the environment, or both.

14
Air Quality Surveillance   George B. Morgan and
                                            Guntis Ozolins

Introduction

The basic design of an air quality monitoring network consists of
the selection of pollutants to be monitored, the determination of
the number and location of sampling sites, the selection of appropriate instrumentation, analytical techniques, and sampling frequencies, and the development of applicable data handling and
analysis procedures. The requirements for and uses of the air
quality data together with practical considerations are the basic
determinants of the resulting monitoring network.

This paper is intended to present the basic concepts of network design for Air Quality Control Regions (AQCR) to provide
guidelines for network development. Twenty-nine AQCRs have
been designated so far by the federal government. The designation
of AQCRs is the first step in the regional approach to air pollution control (Gaulding, 1969). This is followed by the adoption
of ambient air quality standards, development of abatement or
implementation plans necessary to meet the standards, and the
implementation of necessary enforcement. The latter three steps
are the responsibility of the individual states.

A necessary portion of the Implementation Plan is to design
and establish a monitoring program to determine ambient air
quality within the region (National Air Pollution Control Administration [NAPCA], 1966). Overall air pollution surveillance includes surveillance of emissions as well as actual air quality measurements. This outline is concerned with the design and operation of an air quality monitoring network which is necessary to
demonstrate that ambient air quality is progressing toward, con-
forming to, or continuing to meet ambient air quality standards.
In some of the Air Quality Control Regions a variety of monitoring or sampling networks of varying degrees of sophistication
has been in existence for a number of years. These have ranged
from static sampling devices, such as dustfall buckets and sul-
fation candles, to systems composed of continuous air monitoring
stations with data being continually telemetered to a central receiving point. Similarly, the degree of coverage, if any, has varied

Presented at the Eleventh Conference on Methods in Air Pollution and Industrial Hygiene Studies, Berkeley, California, March 1970.

from networks having a few stations to well-designed multiple-station networks.

As a result of the requirements of the Clean Air Act as amended and the subsequent regional control of pollution, it will be necessary to rethink and modify as needed the established objectives, as well as monitoring networks design. With the increasing availability and use of reliable diffusion models, the objectives of air quality monitoring networks, their design and operation can and must be altered to be responsive to this new approach and meet the demands placed upon us by the Clean Air Act as amended. The availability of new and improved instruments, methodology, and data handling procedures now permits a more accurate definition of ambient air quality than ever before. The following are guidelines with respect to the objectives of monitoring, the design of monitoring networks, types of instruments, sampling frequency, and data handling procedures. While these guidelines are currently aimed at monitoring for SO_2 and total suspended particles, the same principles and practices apply to the determination of ambient air quality for other pollutants.

Objectives of Monitoring
An air monitoring network for an Air Quality Control Region must be designed and operated so that it is responsive to the following four objectives:
Objective 1. The network must be capable of measuring and documenting the region's progress toward meeting the adopted ambient air quality standards.
It is necessary that the existing air quality within an entire Region be known and that it can be compared to the adopted air quality standards. Because of the size of the regions and the extreme geographical variability of air pollution levels, it is not economically feasible to operate a sufficiently large network to adequately characterize regional air quality levels. The practical approach is to provide a limited network supplemented by diffusion modeling for extrapolating the data so that it is possible to estimate or predict existing concentrations of a pollutant throughout a region. The initial network must be reevaluated at regular intervals (at least annually) to see that it is doing an adequate job. Certain stations in the existing network may have to be relocated or it may

be just a case of adding more stations. In any event, constant supervision of the network's data output is needed especially at the start of the sampling program. If the network is properly designed and operated, this information will permit year-to-year comparisons on trends and in addition provide feedback on adequacy of adopted control strategy. It is important to be able to depict the changes in air quality as a result of changes in emissions from different source types.

Objective 2. To determine the ambient air quality in nonurban areas of the region.

Most, if not all, Air Quality Control Regions contain areas that are not yet developed and where the pollution is minimal. It is an objective of the monitoring program that the air quality in these areas also be known, especially where future population and industrial expansion is anticipated. The measurement of air quality in nonurban areas that typically are in the periphery can also provide information on the extent to which sources outside the region affect its air quality. In other words, this gives us information as to ambient air quality upwind as well as downwind from an urban area.

Objective 3. To improve the reliability of diffusion models.

Diffusion modeling can be a very important tool in the proper management of regional air resources and is used particularly in the preparation of implementation plans. Modeling can, when properly supported, adequately characterize existing overall regional air quality. More important perhaps is the use of modeling in predicting future levels of pollutants depending upon various control strategies on both a short-term and long-term basis, whether it be in industrial locations, residential areas, center city, or nonurban areas. The increased dependence upon modeling requires continuing availability of ambient air quality data for validation purposes.

Objective 4. To provide air quality data during air pollution episodes.

It is necessary to provide air quality data rapidly during air pollution episodes. The primary requirement is that the data be available as rapidly as possible to permit taking action under the plan. If the episode plan involves forecasting, concurrent meteorological data will also be needed. The U.S. Weather Bureau can assist with necessary data for the description of local meteorology as well as the issuing of high air pollution potential forecasts for the entire country.

Criteria for Locating Monitoring Stations

The placement or location of sampling stations within this limited network must be such that ensuing data can be gainfully employed to meet the four objectives of monitoring. With this in mind, the following criteria are recommended.

Criterion 1. Monitoring stations must be pollution oriented.

It is most important that areas most heavily polluted be identified and monitored. It is in these areas that progress toward meeting ambient air quality standards is most critical.

Criterion 2. Monitoring stations must be population oriented.

A portion of the network must be located according to the population distribution. This is particularly important during times of air pollution alerts and episodes. Such data are also frequently of administrative use in demonstrating concern for the welfare or emotional well-being of the population.

Criterion 3. Sampling stations must be located to provide areawide representation of ambient air quality.

Data must be representative of the entire Air Quality Control Region. Area-wide data are needed for validation of the model as well as to show conformity to the ambient air quality standards. This includes both developed and undeveloped areas within the region. In the nonurban areas increased consideration should be given to those areas where future land development is anticipated.

Criterion 4. Monitoring stations must be source category and/or source oriented.

The primary purpose of these stations is to provide feedback relative to the effectiveness of the adopted control strategies. For example, a control regulation limiting the emissions from domestic use of heavy fuels would require that stations be located where the resulting change best can be appraised.

The air quality monitoring network should then be composed of stations reflecting one or more of these criteria. It should contain stations that are situated primarily to monitor the highest levels in the region, to measure population exposure, to measure the pollution generated by specific classes of sources, and to record the nonurban levels of pollution. *Also, in order to allow comparisons of present and past air quality data and to permit interregional comparisons as well as a check on local and federal monitoring techniques, a center city station should be located adjacent to the National Air Surveillance Network (NASN) station.* In many cases a given station location will be capable in meeting more than one of the listed criteria, that is, a station located in a densely populated area besides measuring population exposure will also monitor the effectiveness of controls on emissions from domestic space heating if such is part of the overall control strategy.

Guidelines for Distribution of Monitoring Stations

In most Air Quality Control Regions it will take from fifteen to twenty-five stations to furnish an adequate amount of air quality data. In unusual circumstances additional stations may be needed to fulfill the criteria. Based upon our past experience, we recommend the following guidelines for the distribution of air quality stations within the region:

1. Heavily polluted or "dirty" areas—in most cases three to five stations will suffice.
2. Nonurban stations—two to four, depending upon the size of the hinterlands.
3. Population oriented stations—three to seven.
4. Source oriented stations—three to five.
5. Comparison oriented (center city) stations—one.
6. Remaining or other necessary stations should be placed where concentration gradient or gradation is greatest as predicted by the diffusion model.

In summary, the development of network designs should be based upon the following factors: (1) past air quality data, (2) isopleth maps from diffusion models for various source, receptor combinations, (3) emission density maps, (4) population distribution maps, including future population patterns, (5) present and future land development maps, and (6) topographical and meteorological considerations.

Monitoring Networks

A very important part of network design is the selection of averaging times, sampling frequencies, specific sampler location, as well as data handling. These specifics are outlined briefly in the next six sections.

Averaging Times

The types of samples, whether continuous or intermittent, depend upon the primary use of the data. To show compliance with or progress toward meeting the standards the sampling equipment must be capable of producing data consistent with the averaging times specified by the ambient air quality standards. For measuring the exposure of population, as well as for emergency episodes, continuous monitoring or data of relatively short averaging times are required. In contrast, for instance, sampling at the nonurban stations can be of a much longer duration.

More specifically, for particulate matter, the basic sampling period is twenty-four hours, whereas for SO_2, it can range from continuous instruments up to twenty-four-hour integrated samples. Similarly, ambient standards for particulate matter will be in terms of twenty-four-hour values (averages and maximums); the ambient standards for SO_2 may be specified in terms of from five-minute values to yearly averages.

Sensors and Methods

The preferred methods of sampling and analysis are those most commonly in use and for which a large body of data is available. When standard methods become available in the near future, they should be used. The recommended sampling method for suspended particulates is the Hi-Vol sampler which collects total suspended particles on an 8×10 inch glass fiber filter at the sampling rate of 50 to 55 cfm. For sulfur dioxide the NAPCA modification of the West-Gaeke method, the flame photometric method, and the gas chromatographic method are all adequate because they are relatively specific and have been shown to be comparable for continuous monitoring. For twenty-four-hour integrated samplers the modified West-Gaeke procedure is preferred.

Sampling Frequency

Twenty-four-hour integrated samples should be collected at a frequency of at least twice weekly in order to be able to predict adequately the maximum concentrations and to define a more reliable average concentration. Sampling should be spread out uniformly over the year, and each day of the week should have equal representation. Sampling on a schedule of every third day will ensure that these conditions are met.

Sampler Locations

In the selection of sampling sites particular consideration should be given to source locations in the immediate vicinity and other parameters that may unduly influence the results. Sampling instruments or ports should be located from 10 to 20 feet above the street and at least 10 feet away from the nearest structure. It is desirable to have the network standardized according to these factors as much as possible. This will result in more accurate measurements by eliminating various interferences and allow for better comparisons between stations.

Regional Distribution of Types of Samplers

To meet the previous listed monitoring objectives, the following guidelines for sampler locations are recommended:

1.   In the heavily polluted spots, a Hi-Vol sampler, a continuous SO_2 instrument, and an American Institute of Steel Industries (AISA) sampler should be located. (The AISI data are useful in air pollution episode situations.)

2.   Nonurban stations should contain as a minimum a Hi-Vol sampler. In many cases, where it is anticipated that land use will change or it is recognized that a problem of SO_2 may exist, 24-hour bubblers should be utilized.

3.   As much as possible, the population-oriented stations should contain a Hi-Vol and a sampler capable of providing short-term averages for SO_2. In populated areas adjacent to major industrial zones, a continuous SO_2 sampler and an AISI tape sampler may be needed.

4.   For source category-oriented stations a Hi-Vol and bubbler are usually sufficient. Where individual large sources predominate, a continuous SO_2 monitor may be necessary.

5.   The comparison-oriented, or center city, station should contain as a minimum a Hi-Vol and a gas bubbler.

6.   For other stations, that is, those to show gradation, a Hi-Vol and bubbler are usually adequate.

Other Types of Monitoring

In addition to the stationary monitoring sites, it frequently may be feasible to operate mobile monitoring stations. Mobile stations with continuous instruments can be used quite advantageously to map urban areas over a short time period. We also envision that in the future large Air Quality Control Regions may find airborne monitoring expedient. While we are not recommending this type of monitoring, we certainly want to point out the fact that it does exist and may be useful.

Data Processing and Presentation

A most important part of the entire monitoring effort is the validation, handling, and analysis of data. It is extremely important that all data be analyzed and be made available quickly and in a standardized format. This means that values are to be expressed in uniform units (metric system) and in useful and systematic averaging periods.

To this end, NAPCA has developed an aerometric data storage and retrieval system (SAROAD) (NAPCA, 1968). This system can easily be instituted and modified to fit the particular regional needs. In addition, for the regions that already have their own data systems, it is relatively simple to convert their format into the SAROAD format for entrance into the National Aerometric Data Bank.

The data system and presentation should be flexible and responsive to meet a number of needs ranging from evaluation of data with respect to the standards to providing inputs for diffusion modeling. For continuous monitoring, the system should be capable of producing statistics on maxima, minima, averages (both geometric and arithmetic), and for various frequency percentiles for the following averaging times: five minute, fifteen minute, one hour, eight hour, twenty-four hour, monthly, and yearly averages.

References

Gaulding, C. L., 1969. Regional approach to air pollution abatement gains momentum, *Environmental Science and Technology, 3.*

National Air Pollution Control Administration (NAPCA), 1966. *Guidelines for the Development of Air Quality Standards and Implementation Plans* (Washington, D.C.: NAPCA).

NAPCA, 1968. *Storage and Retrieval of Air Quality Data (SAROAD), A System Description and Data Coding Manual* (Washington, D.C.: NAPCA).

15
Factors in Air Monitoring Network Design

Bernard E. Saltzman

Air monitoring network design is a compromise between desirable objectives and available resources. The previous chapter, based on much practical experience, details major objectives and outlines recommended designs. The purpose of this chapter is to present some comments on this problem and to introduce scientific considerations that may illuminate some facets.

Statistical Aspects of Sampling

Two primary objectives are to accurately determine air quality within both urban and nonurban areas in order to evaluate degrees of pollution and to measure and document progress toward adopted ambient air quality standards. Because concentrations fluctuate in a statistical manner, the number of samples collected influences the accuracy of the mean value that is determined. Many studies have indicated that the lognormal statistical distribution characterizes concentrations of air pollutants. The standard geometric deviations as listed by Larsen (1969) for various pollutants are given in Table 15.1. The standard geometric deviation and geometric mean are characteristics of the pollutant pattern and are not affected by the number of samples collected.

Table 15.1 Standard Geometric Deviations for One-Hour-Average Concentrations of Various Pollutants in Various Cities, December 1, 1961–December 1, 1964

City	Pollutant						
	CO	Hyc	NO	NO_2	NO_x	Oxi	SO_2
Chicago	1.49	1.37	1.77	1.65	1.66	2.54	1.98
Cincinnati	1.40	1.77	2.82	1.76	2.13	2.25	2.25
Los Angeles	1.50	1.74	2.45	2.09	2.12	2.30	2.07
Philadelphia	1.70	1.80	3.46	1.83	2.69	2.87	2.31
San Francisco	1.53	1.60	2.09	1.85	1.80	2.13	2.54
Washington	1.76	1.77	3.19	1.67	2.46	2.13	2.03
Median	1.52	1.76	2.64	1.80	2.13	2.28	2.16
Minimum	1.40	1.37	1.77	1.65	1.66	2.13	1.98
Maximum	1.76	1.80	3.46	2.09	2.69	2.87	2.54

Source: Larsen, 1969.

Presented at Eleventh Conference on Methods in Air Pollution and Industrial Hygiene Studies, Berkeley, California, March 1970.

They are affected however by sampling time as will be discussed later. The numbers of samples then required to determine the geometric mean within ±5 percent and ±10 percent are given in Table 15.2. For high standard geometric deviations rather large numbers of samples must be collected. These statistical calculations are based upon random sampling. This requires sampling over a sufficiently long period to include many cycles of fluctuations of all periods, some of which may be seasonal. Deviations which are found from these relationships for sampling periods as short as three weeks and other statistical considerations are given elsewhere (Saltzman, 1969; 1970).

Another objective listed is to provide data during air pollution episodes. Once the critical threshold concentrations for undesirable effects are established, the proportion of time that these concentrations are exceeded is a convenient way of expressing the data. Figure 15.1 illustrates how a sampling program can be designed for quality control with minimal analytical effort (Saltzman, 1969; 1970). Its use is illustrated as follows:

Example 1. Design a sampling program so that the probability of acceptance is >0.95 if actually <2 percent of the population exceeds the critical value, but is <0.10 if actually >8 percent do. Answer: See illustration on the upper right of Figure 15.1. Draw a line connecting 2 percent on the left scale with 0.95 on the right scale and another one connecting 8 percent with 0.10. Their intersection in the center area gives values $c = 4$, $n = 98$.

Table 15.2 Numbers of Random Samples Required to Establish the Geometric Mean within Given Degrees of Accuracy

(Confidence Coefficient 0.95)

Standard Geometric Deviation	No. of Samples Required ±5% of GM	±10% of GM
1.1	16	6
1.2	58	17
1.4	180	54
1.6	340	97
1.8	540	160
2.0	670	200
2.5	1300	350
3.0	2000	550
3.5	2400	660

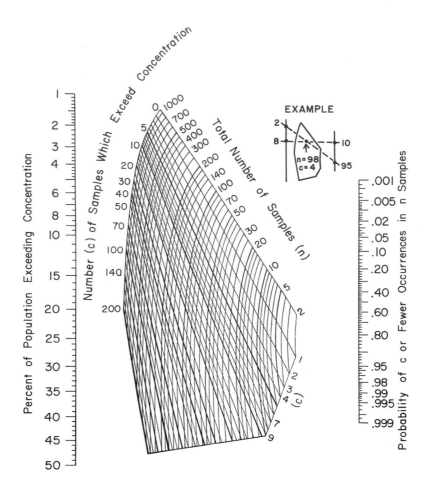

Figure 15.1 Design of a sampling program for quality control

Note: Related variables are connected with a straight line.
For proportions below 1 percent divide numbers on left scale by a convenient integer (such as 10), and multiply numbers on the "n" scale by the same integer. For proportions above 50 percent, the nomogram may be used by regarding the left scale and the "c" scale as representing the numbers of samples *equal to or less than* the critical concentration.

This chart may also be used to estimate the confidence intervals for proportions in the population by drawing lines through the point $(c-\frac{1}{2}, n)$; in this case the right scale represents the probability that the true proportion is greater than the value intercepted on the left scale.

Thus, ninety-eight samples should be collected and no more than four are allowed to exceed the critical concentration.

This figure also may be used to estimate confidence intervals of the proportion of time exceeding the threshold value as illustrated in Example 2.

Example 2. In a group of thirty collected samples, two exceed the specified critical value. What are the 95 percent confidence limits of the proportion in the population? Answer: The point (1 1/2, 30) is located in the central area of the figure. A line drawn between this point and 0.025 on the right scale intersects 19 percent on the left scale as the upper limit; a line from the point through 0.975 on the right scale intersects 1.3 percent on the left scale as the lower limit of the confidence interval. The 99 percent confidence interval may be estimated in a similar manner using 0.005 and 0.995, respectively, on the right scale.

Strategies for Station Location and Operation

It has been mentioned in Chapter 14 that another objective of air monitoring is to improve the reliability of diffusion models. Because pollutants vary both in time and space, accomplishing this requires extremely large numbers of stations and samples. Substantial simplification of present methods in order to reduce the technical effort and expense are needed to conduct such work. In the design of an implementation program to achieve air quality standards, specific data are needed as to the effects on target areas of emissions from known major sources. Because substantial sums of money will be involved, accurate experimental studies at individual stacks using tracer gas techniques (Clemons, Coleman, and Saltzman, 1968) are indicated.

It was pointed out in the previous chapter that some stations should be located to be pollution oriented, some population oriented, some area oriented and some source oriented. Much time and money appears to be wasted by collecting data of essentially zero concentrations. After a preliminary survey is made a more efficient strategy for orienting stations and collecting data is to focus on the significant hot spots, both in time and space, where trouble is expected. Thus, during periods of good atmospheric ventilation fewer samples need be collected, whereas during episodes the sampling effort should be increased. Modifications in design of monitoring instruments would be needed to accomplish this type of cost saving. Stack monitoring devices appear to be the most efficacious means of implementing air quality standards, since they provide information in the minimum time for emergency action, and also a legal record.

Effects of Sampling Time

The length of sampling period does not affect the mean value of the analytical results but does affect the ranges of concentrations obtained. This was studied empirically by Zimmer and Larsen (1965). A theoretical analysis has been made by Saltzman (1965, 1969). The fluctuations of pollutant concentration in air may be regarded as the sum of a series of sine waves of differing amplitudes and periods. These periods may vary from minutes due to wind, to hours or days due to traffic, and to months due to power and heat generation. For a given sampling time the proportion of the fluctuation observed in the samples is less for the shorter periods as given by the following equation:

$$\frac{s^2}{\sigma^2} = \frac{1}{2\pi^2} \left(\frac{t_p}{t_s}\right)^2 \left[1 - \cos\left(\frac{2\pi}{t_p/t_s}\right)\right]$$

where s^2 = variance observed in samples
$\quad\quad\quad \sigma^2$ = true variance
$\quad\quad\quad t_p$ = fluctuation cycle period
$\quad\quad\quad t_s$ = sampling period

Some values of the variance transmittance factor (the fraction on the left of this equation) are given in Table 15.3. They indicate that practically none of the variance is observed for fluctuations shorter than the sampling time and that ratios of cycle period to sampling time of the order of 10 are required before the entire variance can be observed.

In order to select the optimal sampling time it is necessary to know what fluctuation periods are of biological significance. An organism exhibits the same characteristic as a sampling mechanism. There is a resistance to entry of the pollutant, a storage capacity, and a detoxification or excretion rate. These result in attenuation of concentration fluctuations of various periods in the bloodstream or body organs in a similar manner to that observed in sampling for chemical analysis. The exact theoretical relation-

Table 15.3 Averaging Effects of Air Sampling Time

Ratio of Fluctuation Cycle Period to Sampling Time	Variance Transmittance Factor
<1	<0.05
1.5	0.17
2	0.41
2.5	0.57
3	0.68
4	0.81
6	0.91
8	0.95
10	0.96
20	0.99

Source: Saltzman, 1969, 1970.

ship was described by Saltzman (1965, 1969). The conclusion of this study is that the significant parameter for biological effects is the product of the biological half-life of the pollutant in the body and the ratio of entry resistance to the total resistance for the pollutant to pass through the body. When the sampling time is equal to four times this parameter, the fluctuations observed in samples approximate those which occur in the body. When the sampling time is twice this parameter, the sampling program senses all significant fluctuations to a greater extent than would occur in the body. Shorter sampling times would thus give fine detail of little biological significance according to this model.

Similar theoretical studies are in progress to determine responses of monitoring instruments to fluctuating concentrations. Important parameters are the half-life of the chemical reaction utilized and the turbulent flow time of the reagent in the instrument. Again fluctuations of short periods are attenuated, whereas those of longer periods in relation to the instrumental parameters are not.

By means of these studies, the value and significance of the monitoring data can be increased, and the expense and effort minimized.

References

Clemons, C. A., Coleman, A. I., and Saltzman, B. E., 1968. *Environmental Science and Technology, 2:* 551–556.

Larsen, R. I., 1969. *Journal of Air Pollution Control Association, 19:* 24–30.

238   Bernard E. Saltzman

Larsen, R. I., Zimmer, C. E., Lynn, D. A., and Blemel, K. G., 1967. *Journal of Air Pollution Control Association, 17:* 85–93.

Saltzman, B. E., 1965. *Journal of Air Pollution Control Association, 20:* 565–572.

Saltzman, B. E., 1969. Design and interpretation of an air sampling program, Division of Water, Air and Waste Chemistry, 158 National Meeting of the American Chemical Society, New York, N.Y. (September 1969).

Saltzman, B. E., 1970. Simplified methods for statistical interpretation of monitoring data, forthcoming.

Zimmer, C. E., and Larsen, R. I., 1965. *Journal of Air Pollution Control Association, 15:* 565–572.

16

Station Networks for   Commission for Air Chemistry
Worldwide Pollutants   and Radioactivity

Introduction

Air pollution has grown rapidly during the past twenty years in response to the most rapid increase in industrialization and population that the world has ever experienced. Pollution, already known in a few industrialized European cities for over a century, has spread since World War II into the distant countrysides of the industrial nations and unexpectedly appeared in virtually all developing nations of every continent.

Until recently pollutants have been injected into the air without concern that they might accumulate in hazardous amounts at great distances from the sources. Near each polluted zone, upwind the air seemed clean and fresh, downwind the pollutants were expected to disperse by washout or mixing into limitless reservoirs of air, water, and soil. This picture is contradicted by new chemical data from several polar and oceanic air monitoring stations far removed from industry which record alarming increases in certain industrial by-products. Some constituents are altogether new to the air; others are unusual increases in constituents normally present; but all may become global hazards deserving additional study and more careful future monitoring.

The need grows for a worldwide monitoring network to define regional and global trends in atmospheric pollution. Almost all present chemical observing stations record constituents that vary strongly owing to nearby sources. Global trends are difficult or impossible to discern. Stations remote from local influence, on polar ice fields or oceanic islands, are too few and too scattered to define trends accurately. Additional stations at carefully selected locations are needed.

Until now man-made products have been detected on a global scale only years or decades after injection began. By the time pollution can be established, the industry producing it is well established. Also, the evidence of regional or global pollution has almost never been strong enough to convince or justify forcing

Report to the World Meteorological Organization by the Commission for Air Chemistry and Radioactivity (CACR) of the International Association of Meteorology and Atmospheric Physics (IAMAP) of the International Union of Geodesy and Geophysics (IUGG) June 1970.

industries to curtail pollution in order to protect distant environments. In the future, with ever higher pollution capability in a more and more industrialized world, any lag times between occurrence of pollution and detection may be dangerously long, and any hesitation to abate long-distance pollution because of insufficient proof may have serious consequences. We clearly need a set of permanent stations to monitor all presently suspicious global pollutants and to be prepared to add new parameters to the list on short notice.

Atmospheric Constituents to Be Monitored

Any planning on monitoring networks must decide two important questions: What constituents shall be monitored, and how many stations are needed. We will start with a discussion of what to monitor.

Since global air pollution threatens us with long-range consequences felt only gradually, the decision on which chemicals to monitor must be based on very imperfect predictions—on suspicions rather than certain knowledge of effects. The list of incontestably dangerous chemicals is fortunately short, but the list of incontestably benign chemical contaminants is also short, and between these extremes are many substances that we would prefer to see monitored even if their danger is uncertain.

One way in which constituents can be grouped and considered is to consider their lifetime or residence time because this is a factor in determining the density of the monitoring network, as discussed later. Residence times can vary over a wide range. The shortest known is about ten days for tropospheric water vapor. The latter in turn controls the residence time of constituents such as aerosols which are removed by rainout or washout. The residence times of these short-lived constituents are mostly of the order of several weeks. Gases usually have longer residence times. Carbon dioxide, for example, remains in the air several years on the average before it exchanges with the biosphere or ocean water. Although the distinction between short-lived and long-lived chemicals is somewhat arbitrary, we propose that half a year is a reasonable limit between the two categories.

Long-lived chemicals, regardless of the extent of the source region, will spread over an entire hemisphere and beyond. Examples are CO_2, CO, H_2, N_2O, Kr^{85}, hydrocarbons, and all stratospheric constituents. Short-lived chemicals are washed or precipitated out of the air within weeks or months after injection. Since an air parcel can make one complete circuit of the globe in about two weeks, these can still spread over latitude belts hundreds or thousands of kilometers wide and become truly global pollutants if the source regions are widespread. Some may also be spread by freshwater and oceanic currents. Whether a constituent is long-lived or short-lived depends also on the atmospheric reservoir in which it resides. Aerosols and water vapor are short-lived in the troposphere but long-lived in the stratosphere. Most chemicals in this group are present as aerosols, but a few trace gases of high reactivity like SO_2, NO_2, and fluorine also seem to have similar residence times.

Another grouping can be made according to the effects to be expected from the constituents. Some may affect the climate, for example CO_2 and the aerosols. Carbon dioxide has an influence on the climate because it absorbs infrared radiation. Tropospheric and stratospheric aerosols influence the albedo of the earth; tropospheric aerosols may also trigger important changes in the global heat budget by their role as cloud and ice nuclei. For CO_2, although our present knowledge is inadequate to predict the consequences of a given rise in concentration, the suspicion is strong that after A.D. 2000 detectable modification in world climate will occur as the concentrations in air approach double the nineteenth-century preindustrial values.

A few chemicals, like fluorine, lead, certain biocides, and carcinogens, reach toxic levels at already very low concentrations. In this category also fall bomb-produced or industrial radionuclides which, however, are outside the scope of this discussion.

Finally there are urban air pollutants such as CO, SO_2, NO_2, H_2, and lower hydrocarbons which—as far as we know—do not present any direct global hazard but which could serve as global tracers of industrial activity.

For most constituents, very little is known about their history after they are carried away from polluted areas. Most possess natural sources and were present in air before men began contamination. A few are purely industrial in origin. We recommend that all be measured to establish their natural geochemical cycle and thus provide knowledge to predict future levels in the free atmosphere.

It would be tempting to assign priorities to the various constituents proposed for worldwide monitoring, but it soon becomes apparent that our present knowledge is insufficient to do this.

Network Density and Number of Stations

The second important question is on the number or density of the stations. This is not the same for each constituent because each has its own characteristic variability in time and space which in turn depends on the distribution of its sources and sinks and on its atmospheric residence time. The longer the residence time and the more steady and spacially uniform the sources and sinks, the smaller is the variability and the fewer the number of stations needed to monitor it adequately.

The density of a network of monitoring stations is always a compromise between costs and adequate data coverage. Since we have no objective standards for the spacing of stations, suggestions for network densities or the spacing of stations remain somewhat subjective.

A good monitoring network should be dense enough to assume regionally and globally representative data and to establish any natural sources and sinks because these must be known before the degree and significance of the pollution can be reliably assessed. For shorter-lived chemicals the network must be denser because the regional and time variability is more pronounced.

Care should be taken to put stations in the best logistically feasible locations. A good understanding of regional and temporal factors in advance would be helpful, but we cannot afford to wait for this, and in any case, the initial data probably are best obtained from a network of provisional stations. Such a provisional network is, practically speaking, the only kind of network one can set up at this time, and the objective should be to make the best possible use of existing or rapidly attainable new data.

For constituents for which information is very meager, preliminary studies must be made to establish the general features of regional and global distributions before even networks of provisional stations can be set up. Measurements on board aircraft and ships as well as temporary land stations may be useful for such studies.

The network for long-lived species should take advantage of the similarity for the different species of both the anthropogenic and natural sources and sinks. The anthropogenic production of all species occurs predominantly in the highly industrialized temperate zones—mostly in the Northern Hemisphere. Natural sources and sinks, mainly biologic, are concentrated in forests, ocean surface waters, and the stratosphere. CO_2 can be used as an example. There are already several years of continuous data from three remote sites (north coast of Alaska, Hawaii, and Antarctica) supplemented by data from aircraft and oceanic vessels. A comparison of these results with data from less remote stations indicates that land-based measurements of background CO_2 can be made only on sparsely settled or mountainous oceanic islands, major deserts, or polar ice fields. To monitor seasonal and regional variations adequately, the minimum number of stations for CO_2 and constituents of similar character is about 10. Over half might best be placed on a north-south line near the center of the Pacific Ocean (for example, north coast of Alaska, Aleutian Islands, Hawaii, Line Islands, a South Sea island, or New Zealand and the South Pole). Several others would be needed to assess longitudinal differences (for example, North and South Atlantic Ocean, Indian Ocean, and desert stations).

For short-lived pollutants we need to consider the distribution of sources, the vertical mixing of air, the removal mechanisms and their rates, and the general climatological situation in the region of the station. Unfortunately, we have not yet developed any objective standards by which stations can be selected to be sufficiently representative. Sources and sinks in the surroundings of the stations should be avoided as far as possible, and the meteorological and orographical conditions should be normal. The stations suitable to measure long-lived pollutants may also serve to measure short-lived ones, but we will need additional stations, primarily over the continents where the variety of sources and sinks is largest.

Since we cannot expect to be able to establish an unlimited number of stations, however desirable, we propose a spacing of about 1000 km for the short-lived constituents. Over such distances surface-injected material penetrates deeply into the troposphere. The total number of stations to achieve this spacing is of the order of 100 including some oceanic stations, especially weather ships, in the Northern Pacific and Atlantic and a few judiciously located stations in the heavily polluted areas of eastern United States and western Europe.

The criteria for selecting a network suitable to one short-lived pollutant will suffice for most others owing to a similar distribution of anthropogenic sources and natural sinks, the most important being removal by precipitation. But exceptions may be important; for example, DDT is introduced in rural areas often remote from factory pollution.

Some existing World Meteorological Organization (WMO) and other weather stations will be useful, but many of these are too close to locally polluted areas. We expect that some new stations will be required.

In addition to surface monitoring stations, some permanent sampling by aircraft and balloons should be considered, especially in the stratosphere. This region may be particularly advantageous for long-range monitoring because of the considerable damping with respect to tropospheric fluctuations. Periodic stratospheric sampling of CO_2, for instance, could be useful to check the secular increase obtained from ground-level stations where seasonal variations interfere with calculations of long-term trends.

Even tropospheric monitoring by aircraft could be very useful, particularly during the initial period in order to establish the large-scale distribution of new constituents between the hemispheres, for example. Even the setup of a provisional network as mentioned earlier could be supported by such data for constituents for which our knowledge is still too uncertain.

The setup of a global monitoring network is a major undertaking, both with respect to funds and logistics. Because for some pollutants existing measuring techniques work only at the high levels of pollution near cities, new methods have to be developed and tested for reliability. Experience in smaller-scale networks has shown that it is essential to start with this work well in advance at locations with representative clean air condition. To assure accuracy and reliability of measurements so that observed differences between stations are meaningful, we emphasize the necessity to start at the outset with one or more pilot stations where the equipment can be tested under operational conditions. These same stations are later required for the periodic intercomparison of instruments and methods and the preparation of long-lasting reference standards. For certain constituents it is very useful to have only the samples collected at the stations, whereas the analysis is done at one central laboratory. This procedure is, for instance, well established for filter samples. Such analytical laboratories could be combined with the pilot stations.

It is essential that for a given atmospheric constituent the same measuring technique be used at all stations.

One of the major questions addressed by SCEP was that of the
potential heating of the atmosphere as a result of the steady
buildup of CO_2 from the combustion of fossil fuels. The following papers discuss in considerable detail the present state of scientific understanding of this complex problem area, and outline
the kinds of modeling and monitoring programs that must be
undertaken if a definitive answer to this important question is
ever to be obtained.

There are several estimates of the average temperature
change that would result from a given increase in atmospheric
CO_2 (and also of stratospheric water vapor). In the first paper
in this section Dr. Manabe traces historically the development
since 1956 of these estimates, including the most recent which he
and his colleague, Dr. R. T. Wetherald, made in 1967. In addition to describing the assumptions and uncertainties involved
in such calculations, Dr. Manabe also describes the kinds of
models that will be required for climatic change studies. The
ocean circulation model which he and Dr. Kirk Bryan are developing is discussed in more detail in Dr. Bryan's paper in the
second volume of this series, *Man's Impact on Terrestrial and
Oceanic Ecosystems*, edited by Drs. W. H. Matthews, F. E. Smith,
and E. D. Goldberg.

In a brief appendix Dr. Manabe discusses how one could, in
principle, study thermal pollution by numerical models. This
provides a fitting introduction to the following paper by Dr.
Washington, which incidentally was not one of the documents
considered by SCEP but which is included in this volume because
it provides an example of how mathematical models can be used
to "experiment" on the climate. In this paper, Dr. Washington
discusses what would be involved in a simulation of the effects
of massive thermal pollution of the planet—the addition of a
large amount of heat uniformly over the continents. The author
warns that the predicted "climate change" should not be taken
too literally but rather should be considered as a pioneering exercise that shows what can be done with present numerical
models. Such an experiment could also be performed with more
realistic inputs when we know what they are. This paper pro-

vides a concrete application of the modeling concepts discussed in Part III.

Another approach to the investigation of the influence of CO_2 buildup on the atmospheric temperature structure is outlined in the paper by Professor R. E. Newell and Captain T. G. Dopplick. In this short note the authors apply previous computations of radiative heating rates to the CO_2 problem. This piece attempts to place the CO_2 contribution to temperature change in perspective, especially with respect to the influence of water vapor.

The fourth paper in this series by Professor Roger Revelle was prepared during SCEP, and its contents are significantly reflected in the final recommendations of the SCEP Work Group on Monitoring (see Part I). In it Professor Revelle provides an estimate of the magnitude of increase of atmospheric CO_2 which man may add by the end of the century, and then he discusses the reasons why the estimate may be low. Four specific kinds of monitoring are outlined which, if undertaken now, can provide the information to reduce the uncertainty in these estimates.

The final paper in the series is a description by Dr. Bischof of air collection and analysis techniques that are used for monitoring CO_2 from aircraft. In making cost estimates of global monitoring systems, the SCEP Work Group on Monitoring assumed the type of use of commercial aircraft which is analyzed in this paper.

17

**Estimates of Future Change   Syukuro Manabe
of Climate Due to the Increase
of Carbon Dioxide
Concentration in the Air**

Introduction

According to the estimate by the U.N. Department of Social and Economic Affairs (1956), the concentration of carbon dioxide in the atmosphere may increase as much as 25 percent during this century as a result of fossil fuel combustion. It has been speculated that an increase in the carbon dioxide content in the atmosphere may result in the gradual rise of the atmospheric temperature (Callender, 1949; Plass, 1956). The magnitude of temperature increase resulting from a given increase in the carbon dioxide content has been estimated by many authors, for example, Plass (1956), Kaplan (1960), Kondratiev and Niilisk (1960), Möller (1963), and Manabe and Wetherald (1967). First, we shall briefly describe the basic principles used for their estimates and some of the results from their studies. Then, we shall discuss how we can improve the estimate in the future.

Carbon Dioxide Content and the Heat Balance of the Atmosphere

Plass (1956), Kaplan (1960), and Kondratiev and Niilisk (1960) estimated the increase of the temperature of the earth's surface required for compensating the increase in the downward terrestrial radiation due to the increase in the carbon dioxide content in the atmosphere. Their basic assumption may be written as follows:

$$dE(C, T_s) = 0 \tag{17.1}$$

or,

$$\frac{\partial E}{\partial C} dC + \frac{\partial E}{\partial T_s} dT_s = 0 \tag{17.2}$$

where E is net upward terrestrial radiation and is equal to the difference between the upward terrestrial radiation and downward terrestrial radiation. Here C is carbon dioxide content in the atmosphere and T_s is the temperature of the earth's surface. Obviously, E should depend not only on T_s, but also on the distribution of temperature at higher levels. The reason E is only a

Prepared for SCEP.

function of C and T_s is because it is assumed that the vertical distributions of static stability and absolute humidity are not affected by the change of carbon dioxide content in the atmosphere. From equation (17.2), one can get the following equation for computing the temperature rise dT_s due to the increase in the carbon dioxide content dC in the atmosphere:

$$
dT_s = -\left(\frac{\dfrac{\partial E}{\partial C}}{\dfrac{\partial E}{\partial T_s}} \right) \cdot dC
\tag{17.3}
$$

Plass (1956) adopted this approach for estimating the influence of the change in carbon dioxide content upon the temperature of the earth's surface. His study indicates that the doubling or halving of the carbon dioxide content in the atmosphere results in a temperature change of $+3.8°C$ or $-3.6°C$, respectively. (Strictly speaking, the relation Plass used is slightly different from equation [17.3]. Equation [17.3] is originally obtained by Möller [1963].) Kaplan (1960) attempted to revise Plass's estimate by taking into consideration the effect of cloudiness, as well as by improving the computation scheme of radiative transfer. Kondratiev and Niilisk (1960) considered the effect of overlapping between the 15-micron band of carbon dioxide and the rotation band of water vapor. The magnitude of temperature changes estimated by both Kaplan and Kondratiev and Niilisk are significantly less than that estimated by Plass.

Möller (1963) reviewed these studies critically and tried to improve these estimates by making a more realistic assumption. According to Möller the atmosphere tends to restore a certain climatological distribution of relative humidity responding to the change of temperature. The change in surface temperature is accompanied by the change in air temperature, which in turn causes the changes in absolute humidity of air and in the downward terrestrial radiation. The change in absolute humidity also affects absorption of solar radiation and, accordingly, the amount of solar radiation reaching the earth's surface. Taking into consideration all these factors, Möller reevaluated the effect of the increase of carbon dioxide content upon the temperature of the earth's surface. His basic assumption may be expressed by the following equation:

$$d \{S [W(T_s)] - E[T_s, W(T_s), C]\} = 0 \qquad (17.4)$$

where S is net downward solar radiation at the earth's surface and W is the total liquid equivalent of water vapor. The reason that E is a function of only T_s and W is because the vertical distribution of relative humidity and static stability is assumed to be invariant. The total differentiation of equation (17.4) yields:

$$\frac{\partial S}{\partial W}\frac{dW}{dT_s} dT_s - \frac{\partial E}{\partial T_s} dT_s - \frac{\partial E}{\partial W}\frac{dW}{dT_s} dT_s - \frac{\partial E}{\partial C} dC = 0 \qquad (17.5)$$

From this equation, one gets the following equation, which gives the temperature rise dT_s, resulting from the increase of carbon dioxide content dC:

$$dT_s = \frac{\dfrac{\partial E}{\partial C}}{\dfrac{dS}{dW}\dfrac{dW}{dT_s} - \dfrac{\partial E}{\partial T_s} - \dfrac{\partial E}{\partial W}\dfrac{dW}{dT_s}} \cdot dC \qquad (17.6)$$

Using this formula, Möller obtained rather surprising results. According to his estimate, an increase in the water vapor content of the atmosphere with rising temperature causes a self-amplification effect which results in an almost arbitrary temperature change. When the air temperature is around 15°C, the doubling of carbon dioxide content results in an increase of temperature by as much as 10°C. For other air temperatures, the result may be completely different. The reason he gets such arbitrary results is because the denominator on the right-hand side of equation (17.6) changes signs and magnitude, depending upon the surface temperature T_s. In other words, net upward radiation at the earth's surface does not necessarily increase with increasing surface temperature because of the temperature dependence of water vapor content in the atmosphere.

Radiative, Convective Equilibrium
Examining Möller's method, Manabe and Wetherald (1967) felt that it is necessary to take into consideration another physical factor in order to obtain reasonable results. They maintain that

the change in carbon dioxide content not only affects the flux of net radiation at the earth's surface but also the turbulent flux of energy from the earth's surface to the atmosphere. In other words, equation (17.4) does not hold. Instead, the heat balance equation of the earth's surface without any heat capacity should have the following form:

$$S - E - H = 0 \tag{17.7}$$

accordingly,

$$d(S - E - H) = 0 \tag{17.8}$$

where H is the sum of sensible and latent heat fluxes from the earth's surface to the atmosphere. In order to find out how H depends upon the carbon dioxide content it is necessary to take into consideration the heat balance of the atmosphere as well as that of the earth's surface. Therefore, Manabe and Wetherald proposed to use a model of radiative, convective equilibrium to investigate this problem. By comparing the state of radiative convective equilibrium which was obtained for various carbon dioxide concentrations, they estimated the dependence of atmospheric temperature upon carbon dioxide concentration. A brief description of this study will be made here.

In the atmosphere, three gaseous absorbers, that is, water vapor, carbon dioxide, and ozone have strong absorption bands and affect the field of both solar and terrestrial radiation. In this computation, the radiative effects of these gaseous absorbers as well as that of clouds are calculated by using the equation of radiative transfer.

The state of radiative, convective equilibrium was approached asymptotically by the numerical time integration of the model from the initial condition of an isothermal atmosphere. In order to simulate the macroscopic behavior of moist convection, they introduced a very simple concept of a so-called "convective adjustment." Whenever the vertical temperature gradient exceeds the neutral gradient for moist convection, it was assumed that the neutral lapse rate is restored instantaneously by the effect of the free moist convection. For this study the neutral lapse rate is assumed to be 6.5°C/km. This value is chosen based upon the observation of the static stability in the actual troposphere. It should be pointed out, however, that the static stability of the troposphere is influenced not only by the moist convection but also by the stabilizing effect of cyclone waves.

Figure 17.1 shows the approach toward the state of equilibrium. Toward the end of this time integration, the magnitude of the net downward solar radiation is almost exactly equal to that of net upward terrestrial radiation at the top of the atmosphere, that is, the atmosphere is in complete thermal equilibrium as a whole. In Figure 17.2 the state of radiative, convective equilibrium for the hemispheric mean insolation is compared with U.S. standard atmosphere. The agreement between the two distributions is excellent.

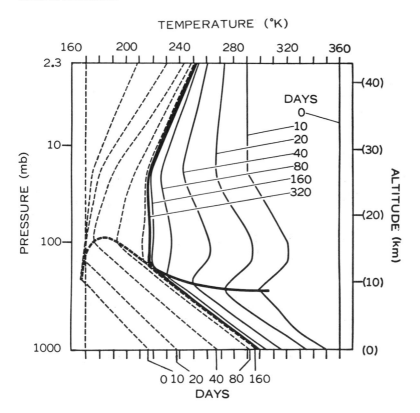

Figure 17.1 Approach toward the state of radiative, convective equilibrium. The solid and dashed show the approach from a warm and cold isothermal atmosphere.

Source: Manabe and Strickler, 1964.

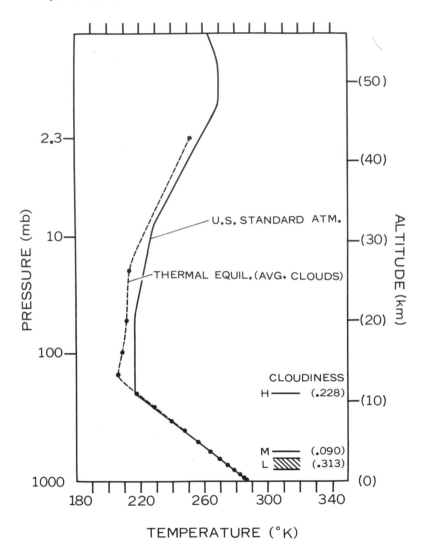

Figure 17.2 Dashed lines show the radiative, convective equilibrium of the atmosphere with cloudiness indicated in the right hand side of the figure. The solid line shows the U.S. Standard Atmosphere.

Source: Manabe and Strickler, 1964.

Encouraged by this agreement, they decided to evaluate the dependence of the equilibrium temperature upon the concentration of carbon dioxide by using this model. In the equilibrium computation described so far, the state of radiative equilibrium has been obtained for the given distribution of absolute humidity.

It is well known, nevertheless, that the warmer the atmosphere, the more moisture it usually contains. As Möller (1963) pointed out, the atmosphere tends to preserve the general level of relative humidity rather than that of absolute humidity through the process of condensation and evaporation. Owing to the dependence of the so-called greenhouse effect upon the concentration of water vapor, the equilibrium temperature of the atmosphere with a given distribution of relative humidity is almost twice as sensitive to the change of CO_2 concentration as that of the atmosphere with a given distribution of absolute humidity. Table 17.1 shows the results of our computation.

This table indicates that doubling or halving the CO_2 concentration increases or decreases the surface temperature of the atmosphere with given relative humidity by about 2.3°C. Assuming the concentration of CO_2 increases by about 25 percent from A.D. 1900 to 2000, as the U.N. Department of Social and Economic Affairs predicts (1956), the resulting increase of temperature would be about 0.75°C, which may have significant effect upon the climate of the earth's surface. Figure 17.3 shows how the vertical distribution of temperature depends upon the CO_2 content. It is interesting that in the stratosphere, the larger the CO_2 concentration, the colder the temperature is.

(Recently, Manabe revised the computation described above by using the scheme of radiative transfer which is proposed by Rodgers and Walshaw [1966]. The scheme is slightly modified by Stone and Manabe [1968]. In addition, the range of wave number, where 15-micron band of carbon dioxide is overlapped with the rotation band of water vapor, is subdivided into four intervals in order to evaluate the effect of the overlapping between bands accurately. The preliminary results indicate that the doubling of carbon dioxide contents raises the equilibrium temperature of the

Table 17.1 Change of Equilibrium Temperature of the Earth's Surface in °C Corresponding to Various Changes of CO_2 Content of the Atmosphere

Change of CO_2 Content (ppm)	Fixed Absolute Humidity		Fixed Relative Humidity	
	Average Cloudiness	Clear	Average Cloudiness	Clear
300 → 150	−1.25	−1.30	−2.28	−2.80
300 → 600	+1.33	+1.36	+2.36	+2.92

Source: Manabe and Wetherald, 1967.

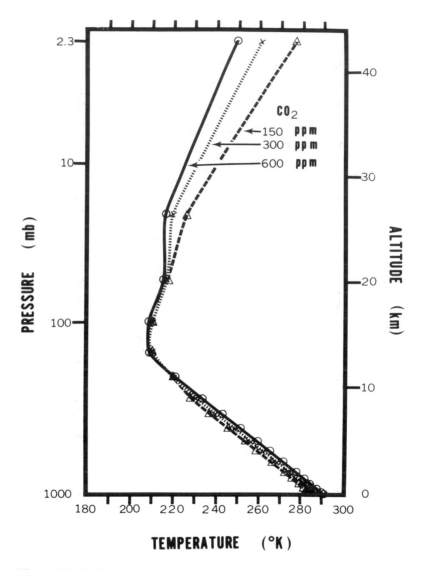

Figure 17.3 Vertical distribution of temperature in radiative, convective equilibrium for various values of CO_2 content

Source: Manabe and Wetherald, 1967.

earth's surface by about 1.9°C in case of average cloudiness. In short, the new estimate is slightly less than the previous one.)

One of the important assumptions made here is that the change in the carbon dioxide content in the atmosphere does not affect cloudiness. In their study, Manabe and Wetherald found that the state of radiative, convective equilibrium is highly sensitive to the distribution of clouds. It is possible that the change in convective activity, resulting from the change in the carbon dioxide concentration, significantly affects the distribution of clouds. Because of this and other reasons discussed in the following section, the estimate obtained here may be altered.

Budyko's Model

The model of radiative, convective equilibrium, described in the previous section, is one dimensional. However, the climate of the planet Earth is maintained by the nonlinear coupling among various processes such as the poleward heat transfer by the atmospheric and oceanic circulation, as well as the vertical heat transfer by radiative transfer and convection. Thus, it is clear that one cannot obtain a definitive conclusion by the use of a one dimensional model, concerning the effect of the increase of carbon dioxide concentration upon climate.

Recently, Budyko (1969) constructed a very simple two-dimensional model for the atmosphere in which the effects of poleward heat transport and radiative transfer are parameterized. (See also the study by Sellers, 1969.) In his model, he incorporated the following two positive feedback mechanisms that enhance the sensitivity of climate:

1.   Water vapor—Greenhouse coupling
2.   Snow cover—Albedo coupling

The first coupling was already considered by Möller (1963), and Manabe and Wetherald (1967). It has the following positive feedback mechanism:

(increase (decrease) of atmospheric temperature)
$\longrightarrow$ (larger (smaller) water vapor content of the atmosphere)
$\longrightarrow$ (more (less) greenhouse effect)
$\longrightarrow$ (more (less) intense convection)
$\longrightarrow$ (increase (decrease) of atmospheric temperature).

The second coupling has the following positive feedback mechanism:

(decrease (increase) of atmospheric temperature)
$\longrightarrow$ (wider (narrower) area of snow cover or ice pack)
$\longrightarrow$ (larger (smaller) albedo of earth's surface)
$\longrightarrow$ (decrease (increase) of atmospheric temperature).

By using his model, Budyko (1969) suggested that the thermal regime of the model atmosphere is highly sensitive to a small change in various parameters such as the solar constant, when both of these couplings are considered simultaneously. According to his estimate, a 1 percent change in the solar constant results in a change in the mean temperature of the earth's surface by as much as 5°C because of these positive feedback mechanisms. For this reason he believes that the present arctic ice cover is in rather delicate balance and could be removed by artificial means. Although Budyko did not discuss the problem of carbon dioxide, his study indicates that the simultaneous action of the two positive feedback mechanisms discussed may be important. Since the effect of the heat advection is parameterized and many other drastic simplifications are made in Budyko's model, it is obvious that we need a better model for the quantitative study of climatic changes.

(One of the important simplifications is the elimination of the seasonal variation. Obviously, the ice cap cannot grow equatorwards unless it survives the melting due to intense summer heating. Colder temperatures during the winter may result in a more extensive ice cover by pushing the boundary of the snow toward the equator, but may not necessarily create a deeper and more stable snow cover. Therefore, it is not certain that the more extensive snow cover formed during winter has a better chance of surviving the summer heating. The speculation described here indicates that it is necessary to take into consideration the effect of seasonal variation of solar radiation for the discussion of the stability of the ice cap.)

Joint Ocean-Atmosphere Model
Following the works of Phillips (1956) and Smagorinsky (1963), intensive effort has been made for constructing the dynamical model of atmospheric and oceanic circulation (Bryan and Cox, 1967; Leith, 1965; Manabe, Smagorinsky, and Strickler, 1965; Mintz, 1965; Smagorinsky, Manabe, and Holloway, 1965; and Washington and Kasahara, 1970). Recently, Manabe and Bryan (1969) constructed a joint ocean atmosphere model in which the oceanic and the atmospheric part of the model interact with each other thermally as well as dynamically. Since a large-scale change

of climate usually accompanies a change in sea surface temperature and vice versa, it is necessary to construct such a joint model in order to investigate the possibility of climatic change. Here, we shall briefly describe the joint ocean-atmosphere model, constructed by Bryan and Manabe.

Figure 17.4 shows how the various components of the atmospheric part of the model interact with those of the oceanic part of the model. The distribution of the zonal mean temperature of the joint system, which is obtained from the time integration of the joint model, is shown in the left-hand side of Figure 17.5. This distribution can be compared with the observed temperatures shown in the right-hand side of Figure 17.5. There are some differences between the two distributions. For example, the temperature of both the troposphere and the stratosphere of the model is too low in higher latitudes. The depth of the thermocline in the tropics of the model ocean is too great. General features of the two distributions, however, are very similar.

There are many important factors which have not been incorporated into the joint model just described. For example, the effects of seasonal variation of solar radiation are not taken into consideration. The distribution of clouds is not computed by the model. Instead, the climatological distribution of clouds is assumed for the computation of radiative transfer. Also, the water

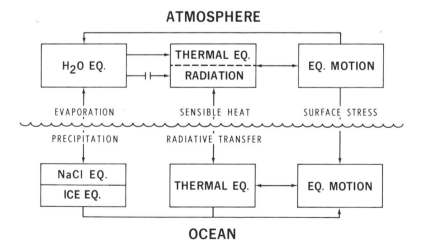

Figure 17.4 Box diagram of the joint model structure

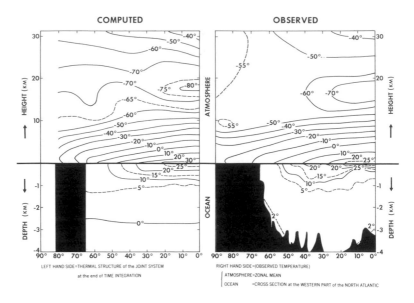

Figure 17.5 Zonal mean temperature of the joint ocean-atmosphere system, left-hand side
This distribution is obtained by taking the average of two hemispheres. The right-hand side shows the observed distribution in the Northern Hemisphere. The atmospheric part represents the zonally averaged, annual mean temperature. (The oceanic part is based on a cross section for the western North Atlantic from Sverdrup, Johnson, and Fleming, 1942.)

vapor greenhouse coupling, discussed in the previous section, is not incorporated into the model. Furthermore, the prognostic system of snow cover and ice pack, which have already been incorporated into the model, have not yet been tested against observation. Obviously, the incorporation of various positive feedback mechanisms increases markedly the degrees of freedom of the model and makes the simulation of climate much more difficult.

Once completed, the joint model has many potential applications. It would be useful not only for the study of the climatic changes in the past but also for the study of future change of climate due to various human activities such as fossil fuel combustion. Given the future output of CO_2 in the atmosphere, one may be able to evaluate future increases of carbon dioxide concentration in the atmosphere and the ocean and its effect upon the climate by performing the long-term integration of the joint model. The results of Manabe and Bryan (1969) indicate that the time constant of the thermal adjustment of the joint ocean-atmosphere system is more than 100 years. Therefore, to predict the future evolution of climate for such a long period of time, it seems to be essential to use an extremely high-speed computer. We understand that such a computer will be available in the very near future thanks to the remarkable advance of computer technology.

As we know, the intensive efforts for improving the atmospheric part of the model has been undertaken by the Global Atmospheric Research Program (GARP). We believe, however, that we cannot make very long-range predictions of the change of climate unless we have a realistic model of the oceans and a satisfactory understanding of the mechanism of ocean-atmosphere interaction. Thus, it would be very desirable to initiate a joint ocean-atmosphere research program in collaboration with physical oceanographers for the improvement of the joint model such as the one described.

Appendix: Thermal Pollution and Climate
Recently, Budyko (1969) made a simple computation concerning the future energy production by man's activities. Assuming that the present rate of the growth of the energy production continues, he estimated the rate of energy output to become comparable with the energy coming from the sun in less than 200 years.

In order to evaluate how an increase of the output of heat energy affects the atmospheric temperature, the states of radiative, convective equilibrium of the atmosphere with a fixed distribution of *relative humidity* are computed for various values of solar constant. The terrestrial radiation model, which was proposed by Rodgers and Walshaw (1966) and modified by Manabe and Stone (1968), is used for this computation. The formulation of convective adjustment is identical with that used by Manabe and Wetherald (1967). Figure 17.6 shows the dependence of surface temperature in radiative, convective equilibrium upon the solar constant. This figure indicates that the surface temperature of the model earth increases about 1.2°C, resulting from the 1 percent increase of the solar constant.

In computing the states of radiative, convective equilibrium, it is assumed that the cloudiness is not affected by the change in convective activity resulting from the change in the solar con-

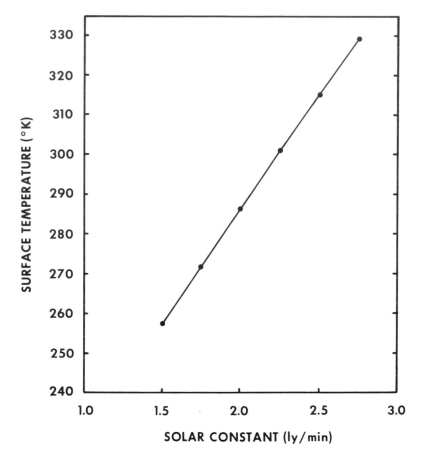

Figure 17.6 Abscissa: Solar Constant; Ordinate: The equilibrium temperature of the earth's surface

stant. Since the surface temperature of the atmosphere in radiative, convective equilibrium is very sensitive to the cloud distribution, the results of Figure 17.6 may be altered significantly if the foregoing assumption does not hold.

The model of radiative, convective equilibrium is a one-dimensional model. Obviously, in order to discuss this sort of problem, it is necessary to construct a three-dimensional model because of the three-dimensional nature of energy source and other considerations discussed in this paper. Particularly, note the last section for the description of the joint ocean-atmosphere model which may be useful for solving this sort of problem in the future.

References

Bryan, K., and Cox, M. D., 1967. A numerical investigation of the oceanic general circulation, *Tellus, 19:* 54–80.

Budyko, M. I., 1969. The effect of solar radiation variations on the climate of the earth, *Tellus, 21:* 611–619.

Callender, G. S., 1949. Can carbon dioxide influence climate?, *Weather, 4:* 310.

Kaplan, L. D., 1960. The influence of carbon dioxide variations on the atmospheric heat balance, *Tellus, 12:* 204–208.

Kondratiev, K. Y., and Niilisk, H. I., 1960. On the question of carbon dioxide heat radiation in the atmosphere, *Geofisica Pura e Applicata, 46:* 216–230.

Leith, C. E., 1965. Numerical simulation of the earth's atmosphere, *Methods in Computational Physics,* Vol. 4 (New York: Academic Press), pp. 1–28.

Manabe, S., and Bryan, K., 1969. Climate calculations with a combined ocean-atmosphere model, *Journal of Atmospheric Sciences, 26:* 786–789.

Manabe, S., Smagorinsky, J., and Strickler, R. F., 1965. Simulated climatology of a general circulation model with a hydrologic cycle, *Monthly Weather Review, 93:* 769–798.

Manabe, S., and Strickler, R. F., 1964. Thermal equilibrium of the atmosphere with a convective adjustment, *Journal of Atmospheric Sciences, 21:* 361–385. Figures 17.1 and 17.2 reprinted with permission of American Meteorological Society.

Manabe, S., and Wetherald, R. T., 1967. Thermal equilibrium of the atmosphere with a given distribution of relative humidity, *Journal of Atmospheric Sciences, 24:* 241–259. Table 17.1 and Figure 17.3 reprinted with permission of the American Meteorological Society.

Mintz, Y., 1965. Very long-term global integration of the primitive equation of the atmospheric motion, *WMO Technical Note No. 66,* WMO-IUGG Symposium on Research and Development Aspect of Long-range Forecasting, Boulder, Colorado, 1964, pp. 141–167.

Möller, F., 1963. On the influence of changes in the CO₂ concentration in the air on the radiation balance of the earth's surface and on climate, *Journal of Geophysical Research, 68:* 3877–3886.

Phillips, N. A., 1956. The general circulation of the atmosphere: a numerical experiment, *Quarterly Journal of the Royal Meteorological Society, 82:* 123–164.

Plass, G. N., 1956. The influence of the 15μ carbon-dioxide band on the atmospheric infra-red cooling rate, *Quarterly Journal of the Royal Meteorological Society, 82:* 310–324.

Rodgers, C. D., and Walshaw, C. D., 1966. The computation of infra-red cooling rate in planetary atmospheres, *Quarterly Journal of the Royal Meteorological Society, 92:* 67–92.

Sellers, W. D., 1969. A global climatic model based on the energy balance of the earth-atmosphere system, *Journal of Applied Meteorology, 8:* 392–400.

Smagorinsky, J., 1963. General circulation experiments with the primitive equations: I. The basic experiment, *Monthly Weather Review, 91:* 99–164.

Smagorinsky, J., Manabe, S., and Holloway, J. L., Jr., 1965. Numerical results from a nine-level general circulation model of the atmosphere, *Monthly Weather Review, 93:* 727–768.

Stone, H. M., and Manabe, S., 1968. Comparison among various numerical models designed for computing infra-red cooling, *Monthly Weather Review*, *96:* 735–741.

Sverdrup, H. U., Johnson, M. W., and Fleming, R. H., 1942. *The Oceans* (Englewood Cliffs, New Jersey: Prentice-Hall), pp. 1087. Figure 17.5 reprinted with permission of Prentice-Hall. Copyright, 1942, renewed 1970.

United Nations Department of Economic and Social Affairs, 1956. World energy requirement in 1975 and 2000, *Proceedings of the International Conference on the Peaceful Use of Atomic Energy*, pp. 3–33.

Washington, W. M., and Kasahara, A., 1970. A January simulation experiment with the two-layer version of the NCAR Global Circulation Model, *Monthly Weather Review*, *98:* 559–580.

18

On the Possible Uses of Global   Warren M. Washington
Atmospheric Models for the
Study of Air and Thermal
Pollution

Introduction

Man is becoming increasingly aware that he may be inadvertently changing weather and climate by polluting the atmosphere. We already know that under certain weather conditions some of our cities become almost uninhabitable because of the accumulation of pollutants. We have had a striking example of this the latter part of July 1970 along the East Coast. Little, however, is known about global aspects of air pollution and its possible effects on weather and climate. One of the ways to study this problem is to construct a computer model of the earth's atmosphere which includes many of the essential features of the real atmosphere. If such a model duplicates all of the important aspects of the real atmosphere, then in principle we can use such a model for studies in testing man's influence on his environment.

In this paper we will give a brief outline of the construction of a model and an example of how it can be used for future air pollution experiments. The example is one of thermal pollution caused by industrialization. We will show from model results that if enough energy is put into the atmosphere by power plants and cities, the surface temperatures over our continents will warm up. We do not wish to draw any definitive conclusions with the model since the model does not describe all of the features of the real atmosphere. We only want to indicate how we could use such a model in the future for pollution experiments.

Mathematical Basis

We shall not attempt to go into detail about various equations that are used in the construction of a computer model. They are essentially Newton's second law of motion, the law of conservation of mass, and the first law of thermodynamics. These equations are the well-known hydrodynamic and thermodynamic laws for a gas or fluid system. The solution of these equations yields the time

Prepared for a monograph on *Workshops on Research Problems in Air and Water Pollution*, University of Colorado, Boulder, August 3–15, 1970.

evolution of the meteorological variables such as temperature and pressure as a function of space.

Basically, the idea is that if we know all of the variables at one instant in time, we can solve the equations for the time change of the meteorological variables. This advances the variables forward in time. The variables at the new time provide new "data" for another advance in time. Because of numerical constraints, the time interval for the time change is of the order of a few minutes for the models now in use. To give some idea of the space structure of the model, we divide the earth's atmosphere into subvolumes.

Figure 18.1 shows the vertical grid of the National Center for Atmospheric Research (NCAR) model where we have subdivided the lower part of the atmosphere into six layers of 3 km each. The placement of the meteorological variables is also shown. Here p is the atmospheric pressure; w is the vertical component

SIX-LAYER NCAR MODEL WITH OROGRAPHY

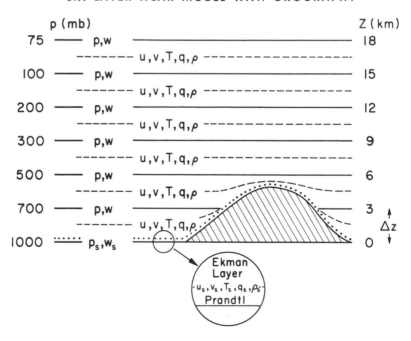

Figure 18.1 The vertical grid structure of the NCAR six-layer model showing the placement of meteorological variables and the mountains

of velocity; u and v are the east-west and north-south components of velocity, respectively; T is temperature; q is moisture; and ρ is density. We also show the boundary layer adjacent to the ground which is called the Prandtl layer, and the layer above it is called the Ekman layer. Many of the important energy, momentum, and moisture exchanges between the atmosphere and the earth's surface take place within the Prandtl layer so that it requires special treatment within the model.

Atmospheric Heating/Cooling Sources Included in the Model
Essentially all atmospheric energy comes from the sun. The flux of energy received at the top of the atmosphere from the sun is approximately 2.0 ly min^{-1}. (One cal cm^{-2} is frequently called a langley or abbreviated as ly.) This amount of energy flux is assumed constant and is often referred to as the *solar constant*. Because of the reflection by clouds and the earth's surface, approximately 35 percent of this incoming solar radiation is reflected back to space. About 20 percent is absorbed within the atmosphere itself, and the remaining 45 percent is absorbed at the earth's surface. If the earth is assumed to be a disc perpendicular to the solar flux, then all of the 2.0 ly min^{-1} would be available to the earth; however, we must take into account that the earth is a sphere. (The solar flux is reduced by a factor of approximately 4.) Therefore, the amount of solar radiation available for absorption within the atmospheric system is approximately 0.5 ly min^{-1}.

We know that the mean temperature of the earth's climate has not changed much in recent years, so that the absorbed solar energy must be balanced by an equal amount of energy being re-radiated back to space. Since the mean temperature of the earth is much lower than the sun's temperature, we would expect most of this energy to be radiated at lower frequency on the electromagnetic spectrum. It happens that most terrestrial radiation is in the infrared region of the spectrum. We show on Figure 18.2 the balance of absorbed incoming solar radiation with the outgoing infrared radiation. This figure also shows that we have a net radiational heating from the equator to 38° latitude and a net radiational cooling from 38° to the poles. (Note that 1 ly min^{-1} = 1440 ly day^{-1}.) In the absence of any circulation of the atmosphere and oceans we would expect that the tropics would be heat-

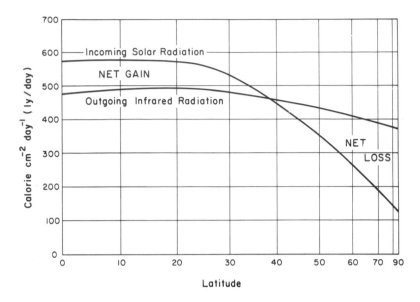

Figure 18.2 The heating/cooling balance between the incoming solar radiation and the outgoing infrared radiation as a function of latitude

ing and the poles would be cooling continually. If we consider atmospheric and oceanic motions, then there is heat transport from the tropics to the poles. It has been estimated that the oceans transport approximately 10 percent of the heat, while atmospheric motions take care of the remainder.

We can look at the heating and cooling of the atmosphere in greater detail if we look at individual processes involved. Figure 18.3 shows a schematic diagram of the various major physical processes. Let Q represent the heating/cooling which can be broken down into three major areas: (1) radiation, Q_a; (2) condensation of water vapor, Q_c; and (3) diffusion, Q_d. We can subdivide further the radiation into solar (short wave–solar, Q_{as}) and infrared (long wave–terrestrial, Q_{al}). The heating/cooling by diffusion can be divided into vertical, Q_{dV}, and horizontal, Q_{dH}.

Also shown on Figure 18.3 is the earth's boundary surface where heat, momentum, and water are transferred to the atmosphere and vice versa. For the present model calculations we assume that the ocean temperatures are constant for a season; however, over land or snow-ice surfaces we consider a surface heat flux balance taking into account solar radiation received at the earth's surface, infrared radiation interchange between the earth's surface and the atmosphere, sensible heat transfer, evaporation, and heat conduction below the earth's surface.

Some Possible Methods in which Air Pollution Can Affect the Heating/Cooling Processes

We will not try to list all possible ways that air pollution can interact with the heat balance of the atmosphere, but we will list some that we feel may have a major effect.

The radiation calculations must include the distribution of clouds, water vapor, dust and aerosols, and gases such as carbon dioxide. If we change any or all of these quantities, then the balance of radiation could be changed sufficiently to affect the climate. We should point out that many of the radiative processes are poorly understood (such as solar absorption within clouds); therefore, we cannot improve the treatment within the model until more basic research is carried out. To illustrate the complex

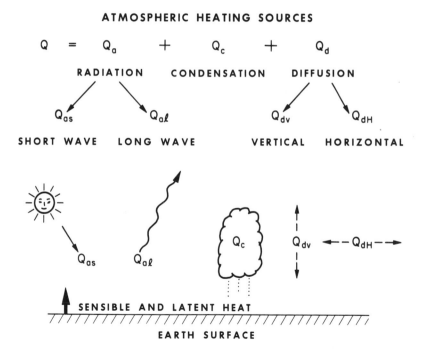

ATMOSPHERIC HEATING SOURCES

$$Q = Q_a + Q_c + Q_d$$

RADIATION   CONDENSATION   DIFFUSION

Q_{as} Q_{al} Q_{dv} Q_{dH}

SHORT WAVE   LONG WAVE   VERTICAL   HORIZONTAL

Q_{as} Q_{al} Q_c Q_{dv} ← − Q_{dH} − →

SENSIBLE AND LATENT HEAT

EARTH SURFACE

Figure 18.3 The various components of the heating/cooling processes which are included in the model

role that aerosols, for example, play in the heat balance of the atmosphere, we will give the possible feedback. The presence of aerosols such as natural haze from vegetation, smoke, dust, and smog tends to increase the albedo; thus, there is less absorption of solar radiation within the atmosphere and on the ground. We can conclude that the aerosols would tend to decrease atmospheric temperatures. On the other hand, the same aerosols would trap terrestrial radiation which would increase atmospheric temperature. Obviously, we cannot settle the question of what really happens by qualitative arguments; we must resort to some kind of model which gives quantitative results.

It has been observed that the percentage of carbon dioxide within the atmosphere has increased from 10 to 15 percent since the turn of the century. Over the same period it has also been observed that the earth's average surface temperature has increased approximately 0.2°C. The question is whether this temperature increase is because of a carbon dioxide increase. We cannot easily resolve this question by models because the effect of a small increase in cloudiness is much more important than the small increase of carbon dioxide as far as the heat balance is concerned. Since presently there is uncertainty of how to model accurately some of the important physical processes in the atmosphere, we prefer to do only "extreme" types of experiments and leave the more subtle experiments for the future.

A Thermal Pollution Experiment

We know that in urban areas the air temperatures are usually warmer than in the surrounding rural areas. This effect, called "urban heat island," is in part a result of the generation of heat energy by human activities (see Myrup, 1969 and 1970). Morrison and Readling (1968) have estimated for the year 1965 that there is 53.8×10^{15} Btu of heat energy released over the entire United States. If we assume the total area of the United States is approximately 9.4×10^{16} cm^2, then the energy flux becomes 2.8×10^{-4} ly min^{-1} or 0.4 ly day^{-1}. As we mentioned previously in reference to Figure 18.2, the flux balance in the tropics or high latitude, for example, is approximately 100 ly day^{-1}. Therefore, present-day industrialization accounts for less than 1 percent, especially if we consider that almost all of the land areas are not as industrialized as the United States.

We at NCAR asked the question what would happen to the climatology if we added to continental areas an additional heat source at the ground. We made two major assumptions: (1) that the additional heating is uniform, which does not take into account that most metropolitan areas are clustered; and (2) that the heat energy flux is 50 ly day^{-1} which is more than 100 times larger than the present-day activity level. Since the models are not refined, we would prefer extreme experiments to see at least some quantitative effect.

The model chosen for this experiment is described in a report by Oliger et al. (1970). We used a version of the model which simulates the month of January. A complete hydrological cycle is included as well as mountains. The amount of clouds is determined by an empirical relationship based upon relative humidity and vertical velocity. We started the experiment by using an atmosphere at rest at day zero. By approximately day 20 the flow patterns became similar to those appropriate for January. We ran the experiment to day 57. We will call this the control experiment. The experiment was repeated after putting 50 ly day^{-1} into the surface temperature calculation over continental areas at day 35. We will refer to this as the thermal pollution experiment. The weather patterns of the two experiments began to diverge quite rapidly. Figure 18.4 shows the global Root Mean Square (RMS) difference of the surface temperature between the control and thermal pollution experiments. The solid circles are at local noon

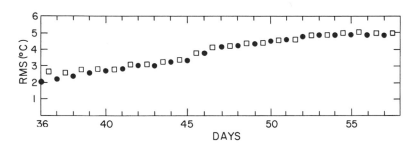

Figure 18.4 The global Root Mean Square (RMS) difference of surface temperature between the control and thermal pollution experiments

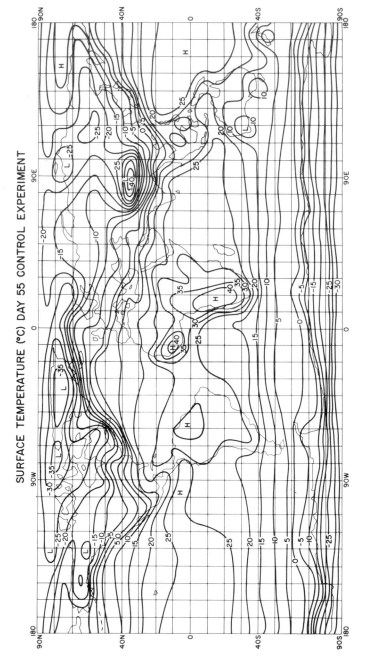

Figure 18.5 The surface temperature distribution at day 55 of the control experiment

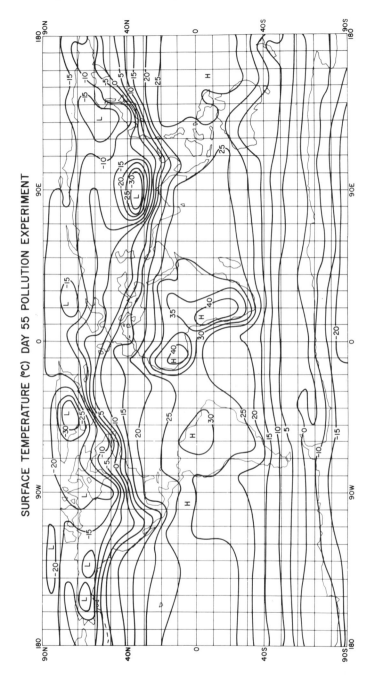

Figure 18.6 The surface temperature distribution at day 55 of the thermal pollution experiment

Greenwich Mean Time (GMT), and the open squares are at 12 midnight GMT. We see that at day 36 the surface heat balance gives an RMS temperature difference which is approximately 2°C. This RMS increases to 5°C by day 57. The weather patterns which are not shown look somewhat different by day 57. The RMS difference does not increase after day 55 indicating that the climatology difference between the two experiments does not continue to increase.

The distribution of surface temperature for the control and pollution experiments is shown on Figures 18.5 and 18.6. The contour interval is 5°C. The additional 50 ly day^{-1} is a small energy flux at the surface at local noon so that the solar flux dominates. The time chosen for these figures is such that local noon is exactly over Greenwich. Therefore, the surface temperatures over Africa show a difference of 1°–2°C maximum. However, since it

ZONAL AVERAGE TEMPERATURE (°C) DAY 55
CONTROL EXPERIMENT

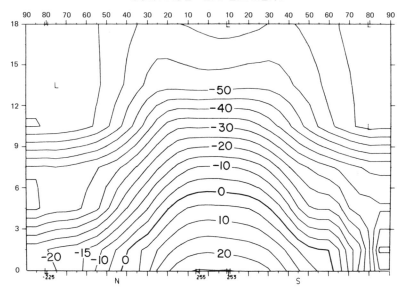

Figure 18.7 The zonal distribution of temperature at day 55 of the control experiment as a function of latitude and height
Contour from −7.000+01 to 2.500E+01, contour interval of 5.000E+00, scaled by 1.000E+01.

is January and the solar flux does not reach beyond 68°N, the 50 ly day^{-1} contributes a great deal to the heat balance in, for example, Siberia. The difference there is about 8°C.

The zonal distribution of temperature as a function of latitude and height is shown on Figures 18.7 and 18.8 for the two experiments. We note that the temperatures in the tropical regions are not changed much whereas in the polar regions we see significant changes, particularly near the North Pole.

Finally, as already mentioned, the experiment we have conducted is quite extreme. We do not want to speculate at this time what the present levels of thermal pollution are doing to weather and climate because of its minor role in the heating and cooling processes within the atmosphere.

Acknowledgments

The author wishes to thank Akira Kasahara for many fruitful discussions on this experiment and John Firor for the initial sug-

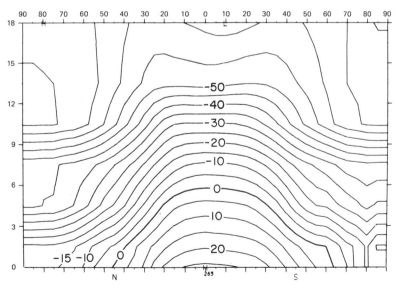

ZONAL AVERAGE TEMPERATURE (°C) DAY 55
POLLUTION EXPERIMENT

Figure 18.8 The zonal distribution of temperature at day 55 of the thermal pollution experiment as a function of latitude and height

gestion. The processing of the output was capably handled by Gloria De Santo.

References

Kasahara, Akira, and Washington, Warren M., 1967. NCAR global general circulation model of the atmosphere, *Monthly Weather Review, 95* (7) (July), pp. 389–402.

Kasahara, Akira, and Washington, Warren M., 1969. Thermal and dynamical effects of orography on the general circulation of the atmosphere, *Proceedings of the WMO/IUGG Symposium on Numerical Weather Prediction, Tokyo, Japan, November 26–December 4, 1968,* Meteorological Society of Japan (Tokyo, March 1969), pp. IV–47–56.

Morrison, W. E., and Readling, C. L., 1968. An energy model for the United States featuring energy balances for the years 1947 to 1965 and projections and forecasts to the years 1980 and 2000, U.S. Department of the Interior, Bureau of Mines Information Circular 8384.

Myrup, L. O., 1969. A Numerical Model of the Urban Heat Island, *Journal of Applied Meteorology, 8* (6): 908–918.

Myrup, L. O., 1970. Corrigendum to a numerical model of the urban heat island, *Journal of Applied Meteorology, 9* (3): 541.

Oliger, Joseph E., Wellck, Robert E., Kasahara, Akira, and Washington, Warren M., 1970. *Description of NCAR Global Circulation Model* (Boulder, Colorado: National Center for Atmospheric Research), p. 94.

Washington, Warren M., and Kasahara, Akira, 1970. A January simulation experiment with the two-layer version of the NCAR global circulation model, *Monthly Weather Review, 98* (August 8, 1970), pp. 559–581.

19

The Effect of Changing Carbon Dioxide Concentration on Radiative Heating Rates

Reginald E. Newell and T. G. Dopplick

Carbon dioxide is the largest global pollutant by mass and atmospheric concentrations are increasing at about 0.7 parts per million by volume (ppmv) per year. The influence of this change on the atmospheric temperature structure is unknown; it can be investigated by three progressively more complicated techniques. In the first, one seeks the radiative equilibrium temperature profile (variation with height) for different levels of CO_2 concentration, implying that there is no net radiative flux divergence at any point in the atmosphere. The approach has been used by a number of workers, Möller (1963) and Manabe and Wetherald (1967) being the most recent. The latter authors also incorporated a convective adjustment so that one aspect of atmospheric motion was included. In fact, there are no extensive regions in the atmosphere where the local change of temperature by radiative processes is zero (e.g., see Newell et al., 1970). In the second approach, illustrated here, one computes the radiative heating rate as the local imbalance of radiative fluxes, for each altitude and for different CO_2 mixing ratios, keeping other factors constant. The third approach would be to include the other factors governing local temperature such as latent heat liberation, cloudiness, air motion, change in water vapor content, change in CO_2 content with change in oceanic temperature, and so on. Only the ESSA Geophysical Fluid Dynamics Laboratory group (e.g., Manabe, 1969; Bryan, 1969) could consider this possibility.

Radiative heating rates have been computed using Rodgers' (1967) work; the details are presented elsewhere (Dopplick, 1970). The present note is a simple application of the work. The local rate of temperature change produced by infrared radiative divergence due to CO_2, H_2O and O_3 appears in Table 19.1. It can be seen that at lower levels the net cooling rate is slightly diminished while at upper levels it is significantly increased by a change in CO_2 from 320 to 400 ppmv. The physical reasons for the two changes differ. Consider a layer in the troposphere centered at 800 mb. The pressure thickness which gives the layer approxi-

Reprinted with permission of the American Meteorological Society from the *Journal of Applied Meteorology, 9* (1970), pp. 958–959.

Table 19.1 Cooling Rates (°C day^{-1}) by Net Thermal Radiation for Different CO_2 Levels

Layer (mb)	January (40N)			July (40N)		
	320 ppmv	400 ppmv	Differ-ence	320 ppmv	400 ppmv	Differ-ence
5–10	−2.696	−2.901	−0.205	−3.344	−3.597	−0.253
10–15	−1.998	−2.214	−0.216	−2.470	−2.748	−0.278
15–20	−1.860	−1.989	−0.129	−2.035	−2.194	−0.159
20–30	−1.309	−1.379	−0.070	−1.220	−1.299	−0.079
30–50	−0.692	−0.725	−0.033	−0.502	−0.537	−0.035
50–70	−0.386	−0.402	−0.016	−0.157	−0.172	−0.015
70–100	−0.301	−0.310	−0.009	−0.008	−0.012	−0.004
100–150	−0.268	−0.272	−0.004	−0.118	−0.121	−0.003
150–200	−0.470	−0.471	−0.001	−0.575	−0.575	0.000
200–300	−1.216	−1.216	0.000	−2.113	−2.113	0.000
300–500	−1.438	−1.436	0.002	−1.991	−1.985	0.006
500–700	−1.357	−1.349	0.003	−1.694	−1.679	0.015
700–850	−1.258	−1.246	0.012	−1.924	−1.906	0.018
850–1000	−1.269	−1.255	0.014	−2.038	−2.019	0.019

Table 19.2 Components of Radiative Heating (°C day^{-1}) at 40N for July with a CO_2 Content of 320 ppmv

Layer (mb)	Thermal radiation			Solar radiation		Total
	H_2O	CO_2	O_3	O_3	$(H_2O + CO_2 + O_2)$*	
5–10	−0.37	−2.26	−0.71	2.97	0.37	0.00
850–1000	−1.79	−0.25	0.00	0.01	0.54	−1.50

* Near infrared.

mately unit optical depth for radiation absorbed by CO_2 can easily be obtained from the curves of absorption vs. "pressure × path" in the literature. Using Brooks' (1958) curve for example and a factor of 1.66 for the diffuse character of the radiation gives unit optical depth for a slab as ~150 mb if the CO_2 mixing ratio is 320 ppmv and about 120 mb for 400 ppmv. At 800 mb part of the flux from below originates from the surface and is attenuated en route, while part comes from the layer in between. The first part will decrease as the mixing ratio and therefore attenuation increases; the second part likewise decreases because it is governed by the effective temperature and, as it is the CO_2 close to the level that counts most, this effective temperature is slightly lower (provided temperature decreases with altitude as it does in the selected cases). The infrared flux arriving at 800 mb from above is governed again by the effective temperature which is a little higher for the larger mixing ratio as the important radiation originates from slightly lower altitudes and this flux value therefore increases. With less radiation coming up from below and more radiation coming down from above, the local cooling rate is diminished, but note that the effect is small, i.e., about a hundredth of a degree per day. In the stratosphere the pressure broadening effect dominates over the temperature increase with altitude, and the net effect of increasing CO_2 is to increase the cooling rate by several tenths of a degree per day.

The total column divergence of thermal radiation (net upward radiation at 5 mb minus that at 1000 mb) is 140,012 and 139,908 ergs cm^{-2} sec^{-1} for 320 and 400 ppmv, respectively, for January, while in July the corresponding figures are 196,009 and 195,487 ergs cm^{-2} sec^{-1}. Between 100 and 500 ergs cm^{-2} sec^{-1} additional radiation are therefore available to the atmosphere for the higher CO_2 concentration, corresponding to columnar heating rates of $1-5 \times 10^{-30}C$ day^{-1}.

The greenhouse theory as usually discussed puts such a "heating" interpretation on the CO_2 changes even though the actual effect of a CO_2 increase is to diminish the cooling rate. It is well to stress that the conditions here are such that all other items are unchanged. The term greenhouse is of dubious applicability because the greenhouse glass leads to higher temperatures by reducing turbulent eddy heat losses, rather than by a radiative influence (Kondratyev, 1965).

To place the CO_2 contribution to temperature change in perspective it is compared with other radiative components at two levels in Table 19.2. Clear skies are assumed. Carbon dioxide is secondary to water vapor in the troposphere as noted by others (e.g., Rodgers and Walshaw, 1966) and dominant in the lower stratosphere under the conditions assumed here. When looking for a potential influence of global pollution on the tropospheric temperature it would therefore be wise to pay careful attention to the water cycle and its possible modification, particularly as it enters also through the effect of latent heat.

Acknowledgments
Capt. Dopplick was assigned to M.I.T. by the Air Force Institute of Technology. Our work was supported by the U.S. Atomic Energy Commission under Contract AT(30-1)2241.

References

Brooks, D. L., 1958. The distribution of carbon dioxide cooling in the lower stratosphere, *Journal of Meteorology, 15,* 210–219.

Bryan, K., 1969. Climate and the ocean circulation III. The ocean model. *Monthly Weather Review, 97,* 806–827.

Dopplick, T. G., 1970. Global radiative heating of the earth's atmosphere. Planetary Circulation Project, Rept. 24, Department of Meteorology, M.I.T.

Gebhart, R., 1967. On the significance of the shortwave CO_2 absorption in investigations concerning the CO_2 theory of climatic change. *Archiv für Meteorologie, Geophysik und Bioklimatologie, B15,* 52–61.

Kondratyev, K. Y., 1965. *Radiative Heat Exchange in the Atmosphere,* Oxford, Pergamon Press, 411 pp.

Manabe, S., 1969. Climate and the ocean circulation I. The atmospheric circulation and the hydrology of the earth's surface. *Monthly Weather Review, 97,* 739–774.

Manabe, S., and R. T. Wetherald, 1967. Thermal equilibrium of the atmosphere with a given distribution of relative humidity. *Journal of Atmospheric Science, 24,* 241–259.

Möller, F., 1963. On the influence of changes in the CO_2 concentration in air on the radiation balance at the earth's surface and on the climate. *Journal of Geophysical Research, 68,* 3877–3886.

Newell, R. E., D. G. Vincent, T. G. Dopplick, D. Ferruza, and J. W. Kidson, 1970. The energy balance of the global atmosphere. *Proceedings of the Conference on the Global Circulation of the Atmosphere,* Royal Meteorological Society (in press).

Rodgers, C. D., 1967. The radiative heat budget of the troposphere and lower stratosphere. Planetary Circulations Project, Rept. A2, Department of Meteorology, M.I.T., 99 pp.

Rodgers, C. D., and M. Walshaw, 1966. The computation of infrared cooling rate in planetary atmospheres. *Quarterly Journal of the Royal Meteorological Society, 92,* 67–92.

Note: Professor Julius London has drawn our attention to a paper by Gebhart (1967) which emphasizes the absorption of solar near infrared radiation by CO_2 and the tendency to compensate thermal radiation changes. Numerical comparisons will be made in a future paper.

20

The Probable Future Increase   Roger Revelle
in Atmospheric Carbon
Dioxide—Kinds of Monitoring
for Prediction

It has been estimated that by the year 2000 world energy consumption will have increased nearly fourfold and the consumption of fossil fuels more than threefold—from approximately 5.4 billion tons at present to 17.8 billion tons by the end of the century. The total fossil fuel consumption over the next thirty years will then be nearly 350 billion tons. This is about 12 percent of Hubbert's estimate of the total recoverable remaining reserves of fossil fuels (Hubbert, 1962), and about 2 percent of Cambel's estimate of total reserves (Energy Study Group, 1965). In other words, at the beginning of the twenty-first century mankind will be using up the earth's fossil fuels at a rate between 0.1 percent and 0.6 percent per year, depending upon which value one chooses for the total recoverable reserves. There are of course many uncertainties in these figures, the two principal ones being the rate of increase in world energy consumption as a function of the rate of increase in the Gross World Product and the rate of development of nuclear energy.

Consumption of fossil fuels at the estimated rates would produce 1,080 billion tons of carbon dioxide by the year 2000 or approximately half of the present atmospheric content of CO_2. If the proportion of this total production remaining in the atmosphere stays the same as that computed from Keeling's measurements at the Mauna Loa Observatory between 1958 and 1967, about 44 percent of the fossil fuel CO_2, or 475 billion tons will remain in the atmosphere. This is 21 percent of the present atmospheric CO_2 content. Adding this amount to the 9 percent increase over the past hundred years, we arrive at an increase of 30 percent of atmospheric carbon dioxide by 2000, resulting from fossil fuel combustion since the middle of the nineteenth century. This is probably a minimum figure, for several reasons:
1.   The carbon locked in the biosphere, chiefly in forests, may be diminished by forest clearing to increase the area of agricultural land, or by greater forest exploitation. CO_2 released to the

Prepared during SCEP

atmosphere from these processes will probably be small during the next thirty years. One of the results of the "green revolution" is that the area of cultivated land is likely to increase more slowly than in the past.

2.   Between 20 and 40 percent of the fossil fuel CO_2 that does not remain in the atmosphere is taken up in the biosphere. An increase of atmospheric CO_2 increases the rate of photosynthesis and this can be balanced by increased oxidation of plant materials only if the total mass of the biosphere increases. But the rate of oxidation of plant materials may increase somewhat more rapidly than the rate of increase in the mass of the biosphere. If this is so, the mean residence time of carbon in the biosphere may become shorter in the future, and the proportion in the biosphere of the carbon dioxide added by fossil fuel combustion may diminish, resulting in a larger proportion remaining in the atmosphere.

3.   If the carbon dioxide added to the atmosphere results in a rise of average air temperatures, the surface layers of the ocean would also be warmed, and the solubility of carbon dioxide in these layers would correspondingly decrease, driving some of the oceanic CO_2 into the atmosphere. For example a rise of 1°C in the temperature of the upper ocean waters would cause about a 6 percent increase in atmospheric CO_2.

4.   The partial pressure of carbon dioxide in the deep ocean waters is much higher than that of the waters near the surface. When these deep waters come to the surface at high latitudes, they release large quantities of carbon dioxide to the atmosphere. It is not know whether this exchange occurs at a relatively constant rate from year to year or whether there are occasional large overturns at high latitudes that result in a sudden sharp increase of atmospheric carbon dioxide, separated by intervals, which may be several decades long, during which there is little exchange between the atmosphere and the deep ocean waters. A release of the excess CO_2 from 5 percent of the deep ocean water would result in a 25 percent increase in atmospheric CO_2. This would be larger than the entire injection of carbon dioxide from fossil fuel combustion over the next thirty years.

These uncertainties in the rate of addition of carbon dioxide to the atmosphere strongly suggest that several kinds of monitoring and measurement should be undertaken:

1.   The total content of atmospheric CO$_2$ should be continuously measured at stations where the atmosphere is thoroughly mixed, and where local effects from vegetation, human activities, and vulcanism can be kept to a minimum. Some four to twelve such observing stations are required. If only four were to be selected, we suggest Point Barrow, the Mauna Loa Observatory, Kerguelen Island, and a station on the high Antarctic plateau, such as the South Pole station. The accuracy of measurement should be about 0.1 part per million of the atmosphere or plus or minus 0.03 percent of the atmospheric CO$_2$ content.

2.   The average temperature of the top 400 meters of the ocean waters should be monitored with sufficient accuracy to permit observation of changes from year to year of about 0.5°C and continuing increases or decreases over several years. A variety of observations, including satellite, ship, and buoy measurements should be combined to provide data for annual maps of ocean temperature.

3.   Deep ocean water comes to the surface and releases CO$_2$ to the atmosphere in three areas: the northern North Atlantic, the far northwestern Pacific, and the Weddell Sea. These should be observed annually by oceanographic ships, which would look for intermittent large-scale upwelling of the deep waters that could release very large quantities of carbon dioxide to the atmosphere.

4.   The partition of added carbon dioxide between the ocean and the biosphere is uncertain by about a factor of 2. This uncertainty could be resolved if better information were obtained about the rate of downward mixing of the ocean waters between the surface and approximately 600 meters. A good possibility exists for obtaining this information by following the carbon 14 released from nuclear tests during the past fifteen years. Present indications are that this added carbon 14 will be partitioned over the next several decades between the atmosphere and the upper 600 meters of the ocean waters, in such a way as to result in about a 10 percent increase in the carbon 14 content of both fluids. But the rate of mixing is uncertain. The necessary data could be obtained by three meridional transects, one in each of the three major oceans, repeated at intervals of about ten years, in which careful carbon 14 measurements were taken at different latitudes down to a depth of 1,000 meters.

These four kinds of monitoring and measurement would permit better prediction of the total change in atmospheric carbon dioxide over the next several decades, perhaps by a factor of 2. The actual changes in atmospheric carbon dioxide content from year to year would of course be given by the atmospheric monitoring stations. The possible effects of these measured changes on climate could be followed in part by the mapping of sea surface temperatures proposed here. The possible effects on the biosphere would become much better known if the proportion of carbon dioxide entering the ocean were established more accurately by the proposed investigations of carbon 14 in the mixing layer.

It should be emphasized that the effects on weather and climate of increasing atmospheric carbon dioxide may be more far-reaching than the predicted increase in the average atmospheric temperature near the earth's surface and decrease in the average temperature of the stratosphere. This is because of the rather complex interrelationships between atmospheric water vapor and atmospheric carbon dioxide. Both water vapor and carbon dioxide absorb and reradiate in the infrared. In the moist warm air of low latitudes, the effect of water vapor predominates, while carbon dioxide is of greater importance in the cold dry air of higher latitudes. Similarly, the relative effect of carbon dioxide increases in the wintertime, when atmospheric cooling reduces the absolute humidity, and becomes smaller in the summer when the water vapor content of the lower atmosphere rises. One would expect, therefore, that an important result of the addition of carbon dioxide to the atmosphere would be a reduction in the global temperature gradient between low and high latitudes and between summer and winter. This could bring about a slowing down of the general atmospheric circulation and of the heat exchange between low latitudes and high latitudes. Such possible changes should be introduced into theoretical and computer models of the general atmospheric circulation.

References

Energy Study Group, 1965. *Energy R&D and National Progress*, prepared for the Interdepartmental Energy Study, Executive Office of the President (Washington, D.C.: U.S. Government Printing Office).

Hubbert, M. K., 1962. *Energy Resources*, NRC Publication 1000-D (Washington, D.C.: National Academy of Sciences-National Research Council).

21

Carbon Dioxide Measurements from Aircraft   Walter Bischof

Abstract

During 1962–68, 1,482 air samples were collected from aircraft, partly from jet airliners, at both tropospheric and stratospheric flight levels. The sampling and analysis technique is described and the accuracy of measurement discussed.

Introduction

Measurements of CO_2 concentrations in air samples collected from aircraft were begun in 1957 at the Institute of Meteorology in Stockholm as a complement to the Scandinavian ground station network. Data obtained during 1957–1961 from flight levels above 1,000 m showed that the irregular fluctuations in the CO_2 content of the air decrease with elevation. The seasonal variations therefore are less obscured (Bischof, 1960, 1962). These variations and the annual mean were found to be close to those reported by Keeling (1960). Unfortunately, an attempt to calculate an annual increase for this period failed because of insufficient accuracy in the analysis method used at that time.

In 1962, a project was started to collect air samples from commercial jet aircraft on flights over longer distances and at higher flight levels. A new analysis technique was developed to improve the accuracy and gas standards were established in cooperation with Scripps Institution of Oceanography, Univ. of California.

The results obtained during 1963 at 5,000 m were in good agreement with data reported by Bolin and Keeling (1963) from flights at the 500 mb level during 1958–61 over the Pacific, confirming a rise in tropospheric CO_2 of 0.6–0.7 ppm/yr as calculated by Bolin and Keeling. The 1963 data also show how release and absorption of CO_2 at the earth's surface are reflected in the seasonal CO_2 variations at all tropospheric levels up to the tropopause. In contrast, no or only small variations were found in the lower stratosphere (Bischof and Bolin, 1966). This confirmed that the vertical exchange of CO_2 is damped by the tropopause, as already tentatively concluded by Bischof (1965) from data collected during flights between Copenhagen and Los Angeles in 1963.

Reprinted from *Tellus, 22* (1970), pp. 545–549.

Between 1962 and 1968, 837 samples were collected from flights with DC-8 aircraft mainly over Scandinavia, the North Atlantic, and the North American continent, but also on flights to Tokyo, Rio de Janeiro and Bombay.

Furthermore, in cooperation with other institutions, such as Environmental Science Services Administration (ESSA) and Woods Hole Oceanographic Institution (WHOI) in the USA, air samples were collected on flight expeditions over the Caribbean Sea, the east coast of North America, and the Indian Ocean. Samples have also been obtained from aircraft belonging to the Swedish Air Force and from small private planes. A total number of 1,482 samples were collected during 1962–68.

A data separation was made in order to obtain a mean annual CO_2 cycle for the troposphere as well as for the stratosphere, and the average rate of annual CO_2 increase for this period of time. The variability of the data has also been studied. An extensive presentation of these studies is given in a separate article (Bolin and Bischof, 1970).

In this article a description of the air collection and analysis technique is presented and the accuracy of the measurements is discussed.

Air Sampling and Analysis

Air Sampling Technique

Although many kinds of aircraft were used in this project, air samples were always collected in the same type of 500 ml Pyrex glass flask having two stopcocks and 14 mm standard taper joint connections, see Figure 21.1. In general samples were obtained by connecting the flasks to an air ventilation system, for example, using air from a pitot tube. The air was drawn through the sample flask for some time, occasionally a sucking pump was connected to the outgoing stopcock, or a venturi tube was used, in order to increase the air exchange in the sample flask. Usually, also the flask was evacuated before sampling, in order to ensure complete air exchange in the flask.

To collect air samples from a DC-8, a surprisingly easy way was found by connecting the flasks to the fresh air ventilation system in the cabin. Unlike other jet airliners, the DC-8 is equipped

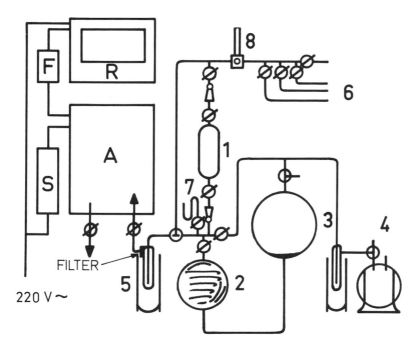

Figure 21.1 Analysis equipment
A, Analyzer; S, voltage stabilizer; R, recorder; F, recorder filter; 1, sample flask; 2, 1 l Hg reservoir; 3, 2 l Hg reservoir; 4, vacuum pump; 5, water vapour trap; 6, standard gas connections; 7, vacuum meter; 8, flow meter and regulator.

with a separate air intake and a turbo compressor for cabin air conditioning. Pressurized and heated air is brought to the cabin, while fresh air via a bypass is brought to the fresh air outlets at the passenger's seat. Tests were made to check the contamination risk by comparing the results from DC-8 flights during descent with results obtained from a Piper plane shortly afterwards. In addition, samples from different DC-8 aircraft and from different outlets on the same aircraft were compared. The results were found to be identical.

Wherever air samples are taken, the sampling time must be long enough to ensure a complete air exchange in the sample flask. Tests with flasks not evacuated before sampling show that a sampling time of 3 minutes was required when using an air flow of 0.6 l/min, which is equal to about three times the volume of the sample flask.

Analysis Equipment

A special flask sample analysis equipment was developed at the Institute in Stockholm (Figure 21.1). By avoiding high vacuum techniques and pressure regulation it has been possible to construct a portable equipment which is easy to operate. The accuracy of measurement was found to be about 0.2 ppm, i.e. about the same as reported by Keeling (1968) (cf. below).

Similar equipment has been installed at the Istituto Universitario Navale, Naples, Italy (Bischof and De Maio, 1964) and at ESSA, APCL, Boulder, Colo, USA. A detailed description is available (Bischof, 1970).

In general, a somewhat modified infrared gas analyzer of type URAS, manufactured by Hartmann and Braun, Germany, was used, but other types (IRGA I and II, by Grubb Parsons, London) also have been in service. Since 1966, a UNOR of MAIHAK, Hamburg, replaced the earlier analyzers. This instrument is now used also for continuous measurements on aircraft.

Standard gas was prepared by AGA, Stockholm, and stored in 50 l high-pressure gas tanks. Standard gas calibration was made annually against the Scripps standards. We thereby obtained our primary standards.

Description of the Analysis Procedure (Figure 21.1)

Before entering the *analyzer (A)*, all gases must pass a *water vapor trap (5)*. Then two different operations are performed: (*a*) *Standard gas calibration,* by flowing standard gas with a known CO_2 concentration through the analyzer; (*b*) *Flask sample analysis,* by bringing sample air into the analyzer.

Standard calibration is made before starting any flask analysis to obtain a calibration factor, i.e. a recorder-scale factor, and to check the linearity of the calibration curve. Ten standard gas tanks are available, containing gas with CO_2 concentrations from 300 to 340 ppm. Standard gas is brought into the analyzer sample cell with a flow rate of 0.3 l/min controlled by a *flow meter and regulator (8)*. Calibration is made either with a constant flow rate or after flushing the cell with the resting air in the sample cell (static air analysis).

A complete calibration includes the use of at least three different standards, each one used ten times. A mean value is obtained and a comparison made with two of our primary standards. With the equipment in good condition, no deviation larger than ± 0.2 ppm is accepted. Otherwise the calibration has to be rechecked. Some of the errors are due to hysteresis and non-linearity in the membrane condensor. By following the procedure described above we ensure that these errors are less than ±0.2 ppm.

In performing the *flask sample analysis* the flask is attached to the equipment by a 14 mm standard taper joint connection. The sample air is brought into the sample cell by using a mercury pump system containing two *mercury reservoirs (2), (3)*, which are connected to a *vacuum pump (4)*. The vacuum pump is also used to evacuate all remaining air in stopcocks and connection lines between the sample flask and the 1 l Hg reservoir. The vacuum is checked by a *vacuum meter (7)*.

After evacuating the connection lines, the flask stopcock is opened and the sample air is brought into the 1 l Hg reservoir by pumping the mercury into the 2 l Hg reservoir, using the vacuum pump. In reverse order, the mercury is used to push the sample air from the 1 l reservoir into the analyzer and through the sample cell. To have complete gas exchange in the sample cell, about 150 ml of sample air at room pressure is required (static air analysis). The analysis result is noted on the *recorder (R)* and is compared with one of the standard gases. While preparing a new sample for analysis, a comparison between the standard just used and one of the other standards is performed. This procedure allows an analysis capacity of about ten samples per hour, including standard gas comparison for each sample. In order to check the analysis reliability flask samples containing standard gas usually are analysed at the beginning and occasionally also during routine analysis work.

Accuracy of the Measurements

Investigations were made to check the reproducibility of the analyses, i.e. to determine the sum of possible errors caused by the infrared gas analyzer itself and the rest of the analysis equipment, respectively.

Check of Analyzer Calibration

An analyzer characteristic was found by flowing standard gases of known concentration through the sample cell as described under "Air sampling technique." A maximum deviation of ±0.2 ppm between individual measurements and a standard deviation of ±0.1 ppm was found, which is equal to the accuracy of the recorder. Part of this error was due to the hysteresis of the membrane condensor and lack of linearity of the calibration. The results are summarized in Table 21.1.

By analysing samples of standard gas filled in flasks, we can assess the reproducibility of the measurements. From Table 21.2 it can be seen that almost the same accuracy is attained in this way.

Since standard gas is very dry, this test is, however, not adequate to judge the accuracy of the analysis of samples from an atmosphere which is moist. It can be shown by bringing a few water droplets into the sample flask, that water vapour disturbs the measurements considerably. The quality of the analyses, therefore, is very much dependent on the efficiency of the water vapour trap. In other words, it is not enough to freeze out the water, because ice crystals may pass into the sample cell and evaporate. To avoid this, a water vapour trap has been given a special shape as shown in Figure 21.1. In practice, samples received from DC-8 flights are, however, very dry.

Field Work

Laboratory conditions still may be too idealized to judge the accuracy of the measurement. Therefore flask samples were taken during a period with high atmospheric moisture. Sample air was pumped through the sample flask and to the analyzer. A check of the flask analysis compared with results obtained by having air

Table 21.1 Analyzer Calibration

Deviation from average	±0	0.1	0.2	0.3 ppm	
No. of cases	63	39	5	0	107
% of total	59	36	5	0	100
S.D.	0.08 ppm				

Table 21.2 Analysis of Standard Gas Filled in Flasks

Deviation from average	0	0.1	0.2	0.3 ppm	
No. of cases	3	6	1	1	11
% of total	27	55	9	9	100
S.D.	0.14 ppm				

flowing directly into the analyzer showed that moist air flask analysis can be made with an accuracy better than ±0.2 ppm.

Comparison of Data

FLIGHT SAMPLES

On a flight on December 14, 1962, two samples were obtained in 1 l flasks and analysed at Scripps Institution and in Stockholm respectively. The result was: Stockholm, 314.48 ppm; Scripps, 314.20 and 314.58 ppm. During a RFF (ESSA) flight over the Caribbean Sea, a number of samples were collected in our 1/2 l flasks and also in 2 l high vacuum flasks belonging to Scripps Institution. Analyses were made in Stockholm and at Scripps Institution. The result is shown in Figure 21.2.

DOUBLE ANALYSIS

On a DC-8 flight, both 1/2 l and 2 l flasks were used and analysed in Stockholm. Due to the larger volume, double analysis could be made of the 2 l samples. From the result shown in Table 21.3 it

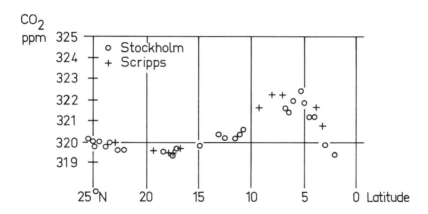

Figure 21.2 Comparison between flask samples analysed in Stockholm and at Scripps Institution respectively
(RFF flight (ESSA) of April 27–29, 1966, over the Caribbean Sea and south of Panama, the samples were collected at 1,600 m.)

Table 21.3 Double Analysis from 21 Sample Flasks

Deviation from average	0	0.01–0.05	0.06–0.10	0.11–0.15	0.15 ppm	
No. of cases	15	29	17	9	0	70
% of total	21	42	24	13	0	100
S.D.	0.04					

can be seen that no deviation larger than 0.15 ppm occurred and that 87% of all deviations were found within a 0.1 ppm range.

Storing Time

Samples were consistently analysed within a few days of collection because it had been found (by analysing standard gas samples) that a contamination risk exists after a few weeks. The risk is different for different flasks. This is probably due to diffusion through stopcocks in the presence of an external excess pressure and concentration difference between sample and room air, and depends of course also on the quality of the individual flask.

References

Bischof, W. 1960. Periodical variations of the atmospheric CO_2 content in Scandinavia. *Tellus 12* (2) 216–226.

——— 1962. Variations in concentration of carbon dioxide in the free atmosphere. *Tellus 14* (1) 67–90.

——— 1965. Carbon dioxide concentration in the upper troposphere and lower stratosphere, I. *Tellus 17* (3) 395–402.

Bischof, W. and De Maio, A. 1964. Über die Installation einer Apparatur für die Analyse des atmosphärischen Kohlensäuregehalts. *Pure and appl. Geophys. 58* (II) 204–207.

Bischof, W. and Bolin, B. 1966. Space and time variations of the CO_2 content of the troposphere and lower stratosphere. *Tellus 18* (2) 155–159.

Bischof, W. 1970. A flask analysis equipment for atmospheric CO_2 AC-11 report, Inst. of Met. Univ. of Stockholm.

Bolin, B. and Bischof, W. 1970. Variations of carbon dioxide content of the atmosphere in the Northern hemisphere. *Tellus 22*, no. 4, pp. 431–442.

Bolin, B. and Keeling, C. D. 1963. Large-scale atmospheric mixing as deduced from seasonal and meridional variations of carbon dioxide. *J. Geoph. Res. 68* (13) 3899–3920.

Keeling, C. D. 1960. The concentration and isotopic abundances of carbon dioxide in the atmosphere. *Tellus 12* (2) 200–203.

Keeling, C. D., Harris, T. B., and Wilkins, E. M. 1968. The concentration of atmospheric carbon dioxide at 500 and 700 millibars. *J. Geoph. Res. 73* (14) 4511–4528.

Predictions that the ice caps will melt as a result of the buildup of CO_2 are often countered with the equally apocalyptical forecast that because of particles in the atmosphere the major climate modification brought on by man will instead be an ice age. Such a forecast is based on the argument that particles will change the albedo (reflectivity) of the planet and essentially shield the earth from the sun, resulting in a significant cooling effect.

During the Study, Dr. J. M. Mitchell prepared a paper in which he stressed that without further study it is impossible to determine whether the overall effect of particles will be a cooling *or* a warming of the earth's surface. This finding was based on new calculations using a modified model of particle scattering and absorption. Following SCEP, he expanded that paper, and a final version is published here.

One of the standard references on the subject of particles and aerosols in the atmosphere is by Dr. C. E. Junge. At SCEP Dr. Junge prepared a summary of the relevant topics in this complex area. His discussion of tropospheric aerosols, monitoring, trends, and residence times provides an overview for the papers that follow.

Because particles enter the atmosphere from many sources and are also produced through a variety of physical and chemical processes, it is extremely difficult to document the sources and quantities of the atmospheric particulate load. The interaction of specialists from several fields at SCEP generated some insights into this problem and the resulting paper by Drs. Peterson and Junge summarizes what is presently known or can be reasonably conjectured about this complex, but critically important, area.

Recently Drs. Ludwig, Morgan, and McMullen of the National Air Pollution Control Administration (NAPCA) published a paper assessing the present state of air quality in our cities and anticipating some of the future consequences of trends in human activity. This paper is included here because it provides documentation of the progress that has been made in reducing air pollution—or at least holding it to a standstill—in most of our larger cities in spite of growing population and growing demands for energy. The paper also points out the inexorable increase in the amount of suspended matter at nonurban stations at a rate of about 4 percent per year. This trend may have profound implications on a regional or global basis.

The complex origins and reactions of the gases and particles of a major contaminant of the urban atmosphere—smog—are treated in the paper by Dr. R. D. Cadle. In it Dr. Cadle explains that smog is not an easy substance to define, that it varies from place to place depending on what is introduced into the atmosphere, and that it can be hazardous to man, animals, and plants. Through the kind of understanding that has been gained in the past few decades, summarized in part in this paper, it may eventually be possible to institute intelligent controls on polluting sources.

The last four papers in this series focus on techniques of monitoring particles and turbidity. (Additional discussions of some of these methods appear in the papers in Part X of this volume.) Dr. Robinson's paper is a short piece prepared during the Study to provide an outline of the methods that have been used to measure turbidity. This paper also critically comments on the quantity—and the quality—of much of the data now available for analysis.

While a consultant at SCEP, Dr. Schaefer presented a paper that summarized the experience of his research group with simple and effective methods of particle counting—using portable condensation nuclei counters and coated sedimentation slides. The paper also emphasizes the possibilities for rapid implementation of global monitoring with this equipment. The following paper, prepared after SCEP by Dr. A. J. Drummond, describes much more elaborate instrumentation which has also been proved in the field and which monitors solar radiation in a manner that provides data for drawing conclusions about the nature of atmospheric aerosols.

The final paper in this series is a lightly edited collection of working papers prepared during SCEP by Drs. Altshuller and Cadle. Both authors have had wide experience in making and evaluating atmospheric chemical measurements, and the paper is a distillation of many of their thoughts on reasons and techniques for monitoring a wide variety of trace constituents in the atmosphere. Many of the recommendations and cost estimates produced by the SCEP Work Group on Monitoring were based on these documents.

22

The Effect of Atmospheric Particles on Radiation and Temperature   J. Murray Mitchell, Jr.

General Statement of the Problem

The potential climatic effects of atmospheric particles, or aerosols, can be divided into two principal categories. On the one hand, there are effects that involve the role of particles as either cloud condensation nuclei or freezing nuclei, which bear on the physical structure and precipitability of clouds. On the other hand, there are effects that involve the role of particles as scatterers and absorbers of radiation passing through the atmosphere, which bear on the radiative flux divergence field and the temperature distribution in the atmosphere. This note is concerned only with the latter category of effects, which we define as the heat-budget effects of aerosols.

Atmospheric particles in the size range 0.1 to 1 μ are of special interest to the earth's heat budget since they are relatively abundant in the atmosphere, and they are well recognized as having the greatest effect in scattering, absorbing, and attenuating solar radiation. Particles in this size range have a relatively small impact on terrestrial long-wave radiation, which is therefore to be ignored in first approximation (for example, Robinson, 1970). Larger sized particles that might interfere more strongly with long-wave (infrared) radiation are less abundant in most natural aerosols, but may become important locally around heavy industrial pollution sources.

It has commonly been assumed in the past that the attenuation of direct solar radiation by particles is attributable mainly to scattering rather than to absorption. For particles in the 0.1 to 1 μ range, the scattering is predominantly in forward (down-beam) directions, with only a relatively small fraction of the total ($\sim$10 percent) in backward (up-beam) directions. Any extent of backscatter tends to increase the albedo of the atmospheric layer containing the particles. Whether it increases or decreases the albedo of the earth-atmosphere system as a whole, however, depends upon the surface albedo.

Recently the neglect of *absorption* in atmospheric particulate layers has been called into question. There are indications that

Prepared during SCEP and expanded after SCEP.

in certain anthropogenic aerosols, especially those of industrial origin containing carbon, iron oxides, and other materials, the amount of absorption is of the same order as the amount of back-scattering (for example, Charlson and Pilat, 1969). Backscattering of solar radiation reduces the total heating within and below the scattering medium. Absorption, on the other hand, increases the heating within the absorbing medium. The combined effect of the two processes on the *net* heating within and below an aerosol layer depends on their relative magnitudes. To determine this net heating in actual cases, it is considered necessary to measure the absorption and backscattering coefficients by airborne instruments. Very few measurements of this nature have in fact been made in the past.

The Relative Importance of Scattering and Absorption

The combined influence of backscattering and absorption by aerosol has been analyzed in relatively crude but highly illuminating terms by Charlson and Pilat (1969). Those authors considered a homogeneous aerosol layer in which both backscattering and absorption of solar radiation (but neglecting long-wave radiation) could be described by a simple Beer-Lambert relationship (not wavelength dependent). They considered the total heating, H, of *both* the aerosol layer and the underlying surface to be given (in my notation) by

$$H = \underbrace{S_o(1 - A)e^{-(a+b)}}_{\substack{\textit{total surface} \\ \textit{heating}}} + \underbrace{S_o(1 - e^{-a})}_{\substack{\textit{aerosol} \\ \textit{heating}}}$$

where S_o is the insolational heating available at the top of the aerosol layer, A is the reflectivity (albedo) of the underlying surface, a is the total absorption of solar radiation by the aerosol, and b is the total backscatter of solar radiation by the aerosol.

Charlson and Pilat noted that the basic effect of backscattering ($b > 0$) is to decrease the total heating H, and thus to cool the earth-atmosphere system. They also noted that the effect of absorption ($a > 0$) is twofold:

1.   Since $A > 0$, the heating H increases with increasing absorption (other things being equal); and

2.   Part of the heating is removed from the surface to a level above the surface, which tends to stabilize the atmosphere.

These authors point out that the net effect of the presence of an aerosol may be either to heat or to cool the earth-atmosphere system, depending on the relative magnitudes of absorption and backscattering. In terms of their model, however, they did not indicate quantitatively what combination of values of a and b would result in a net heating, as opposed to a net cooling.

Refinement of Charlson-Pilat Model

Two modifications of the Charlson and Pilat model seem appropriate in order to make it more realistic for quantitative application to the problem. First, one should allow the solar radiation reflected from the surface (when $A > 0$) to be made available for absorption in the aerosol along with the incident radiation and also be made subject to downward backscatter by the aerosol. Second, one should allow for the fact that a part of the solar radiation reaching the surface is used for evaporation or evapotranspiration to which extent the heating of the surface is realized as sensible heat *not* at the surface but at higher altitudes where the evaporated water finally condenses as clouds and precipitation.

After incorporation of these two modifications, Charlson and Pilat's model becomes

$$H = \underbrace{S_o C(1 - A)e^{-(a+b)}[1 + A(1 - e^{-b})]}_{\text{sensible surface heating}}$$

$$+ \underbrace{S_o(1 - e^{-a})[1 + Ae^{-b}e^{-(a+b)}]}_{\text{aerosol heating}} \quad (22.1)$$

where $C = B/(B + 1)$, with B equal to the ratio of sensible heating to latent (evaporative) heating more familiarly known as the Bowen ratio.

Solutions of equation (22.1) are shown graphically in Figure 22.1 for six combinations of surface parameters A and C that approximate to the natural surface types indicated (snowfields, oceans, forests, prairies, deserts, and urban areas). In the figure, the total sensible heating H is expressed as a ratio to the total incident solar energy S_o, less its value under conditions of no aerosol ($a = b = 0$), and plotted as a function of aerosol absorption a for an assumed value of backscatter ($b = 0.01$).

Figure 22.1 indicates that for relatively dry surfaces (for example, deserts and urban areas), the net sensible heating effect of

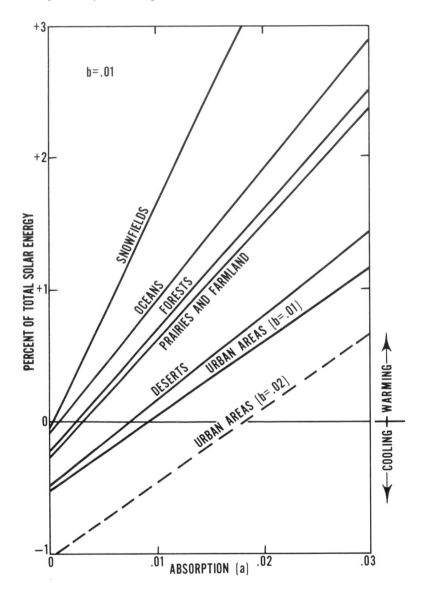

Figure 22.1 Net sensible heating of the lower atmosphere attributable to aerosol, as a function of aerosol absorption a above various types of surface All curves are based on equation (22.1) with a constant aerosol backscatter, $b = 0.01$ (except $b = 0.02$ for bottom curve).

aerosol is positive (warming) only if the absorption a by the aerosol is equal to or larger than the backscatter b by the aerosol. On the other hand, it indicates that for relatively moist surfaces (especially oceans and snowfields), the net effect of aerosol is positive (warming) even when the absorption a is a very small fraction of the backscatter b. This result is of considerable interest because according to various authors the efficiencies of real tropospheric aerosols (natural or man-made) in absorbing and in backscattering solar radiation appear to be of the same order of magnitude (Charlson and Pilat, 1969; Lettau and Lettau, 1969; Robinson, 1970). If this is the case of tropospheric aerosols generally, then it appears that the prevailing effect of such aerosols is more likely to warm the lower atmosphere (below cloud level) than to cool it. In such an event, increasing tropospheric aerosol produced by human activities is likely to supplement increasing CO_2 and thermal pollution of the atmosphere in producing a general warming of the climate of the earth. This, of course, is diametrically opposed to the cooling effect of aerosol postulated by many previous investigators of the problem and tends to contradict the idea that the global cooling trend of the past quarter century is traceable to increasing background levels of atmospheric turbidity from human activities.

Criterion for Distinguishing Cooling from Warming Effect of Aerosol

As a diagnostic aid in judging the direction of the atmospheric temperature response to the presence of particles in the lower atmosphere we may set equation (22.1) equal to zero and solve for the critical ratio of aerosol absorption to backscatter, $(a/b)_o$, for which no net sensible heating or cooling of the lower atmosphere results. After neglect of second order effects in a and b, this leads us to the criterion

$$\left(\frac{a}{b}\right)_o = \frac{C(1-A)^2}{[1+A-C(1-A)]} \tag{22.2}$$

solutions of which are plotted in Figure 22.2. If the observed ratio a/b of real aerosols is greater than $(a/b)_o$ for a given type of underlying surface shown in the figure, the effect of the aerosol is to warm the lower atmosphere. If the observed ratio a/b is less

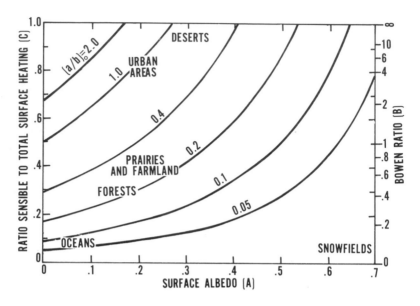

Figure 22.2 Isolines of the critical aerosol absorption-to-backscatter ratio, $(a/b)_o$, for which the net heating of the lower atmosphere is zero, mapped in coordinates (A, C) that define relevant properties of the underlying surface The various surface types shown in Figure 22.1 are located appropriately in this figure.

than $(a/b)_o$ for a given surface type, the effect is to cool the lower atmosphere.

Implications for Cloudiness and Precipitation

Inasmuch as the extinction of solar radiation by aerosols is very likely to decrease evaporation (assuming a constant C or Bowen ratio), the net effect of aerosols over moist surfaces (for example, oceans and vegetated areas) may extend further than a climatic warming to a slight decrease in the global average rates of evaporation and precipitation. Such an effect on precipitation might be further encouraged also by the role of aerosols in increasing the atmospheric static stability, cited by Charlson and Pilat.

Stratospheric Aerosol Case

It should be stressed that the situation with regard to stratospheric aerosols, or to high-tropospheric aerosol layers, is different in one very important respect. The heating by absorption within the aerosol layer then occurs well above the earth's surface. In such a case, *surface climate* is likely to be cooled regardless of the magnitude of the absorption (and associated warming) in the aerosol layer itself. Direct observation of changes of atmospheric temperature following volcanic eruptions tends to confirm this in the case of volcanic aerosols in the stratosphere (compare Mitchell, 1961, and Newell, 1970).

Recommendations

As a first and essential step in evaluating the thermal effect of atmospheric aerosol layers, both natural and man-made, it is of the utmost importance to measure the backscattering, absorption, and other optical properties of each type (each combination of chemical constituents) of aerosol layer commonly encountered in the atmosphere.

With this basic information more refined analyses should be made of the influence of each aerosol type on the atmospheric radiation and heat-balance regime, to advance our badly needed understanding of the climatic impact of changing levels of atmospheric particulate concentrations which are attributable to both human and natural agencies.

For a more comprehensive and detailed analysis of aerosol effects along the lines presented in this note, refer to Mitchell (1971).

References

Charlson, R. J., and Pilat, M. J., 1969. Climate, the influence of aerosols, *Journal of Applied Meteorology, 8:* 1001–1002.

Lettau, H., and Lettau, K., 1969. Shortwave radiation climatonomy, *Tellus, 21:* 208–222.

Mitchell, J. M., Jr., 1961. Recent secular changes of global temperature, *Annals of the New York Academy of Science, 95:* 235–250.

Mitchell, J. M., Jr., 1971. The effect of atmospheric aerosol on climate with special reference to temperature near the earth's surface, *Journal of Applied Meteorology, 10* (in press).

Newell, R. E., 1970. Stratospheric temperature change from the Mt. Agung volcanic eruption of 1963, *Journal of the Atmospheric Sciences, 27:* 977–978.

Robinson, G. D., 1970. *Long-Term Effects of Air Pollution—A Survey* (Hartford: Center for the Environment and Man, Inc.), pp. 21–24.

23

The Nature and Residence Times of Tropospheric Aerosols

Christian E. Junge

General Description of Tropospheric Aerosols

So far as the relatively short-lived atmospheric aerosols are concerned, we can divide the troposphere very roughly into two reservoirs:

1. The lower half of the troposphere over continents, which shows usually high concentrations of particles because most sources, including man-made sources, are on continents; and

2. The upper half of the troposphere above continents and most of the total troposphere over the oceans. This part of the troposphere can be considered "clean" in the sense that it is not yet affected very much by anthropogenic activities.

Very roughly, Reservoir 1 comprises about 20 percent and Reservoir 2 about 80 percent of the mass of the troposphere. Since Reservoir 2 is now becoming influenced by man, any influence on cloud formation and precipitation by global pollution may have far-reaching effects on the global climate.

The main natural sources of tropospheric aerosols are (see Chaper 24 for a more detailed discussion):

1. Sea spray particles over the oceans: size range is above about $0.5\text{-}\mu$ diameter;

2. Dust storms from deserts and dry areas: practically nothing is known about these important sources. Size range is apparently mostly above about $0.5\text{-}\mu$ diameter;

3. Photochemical and homogeneous gas reactions between hydrocarbons released from plants and ozone, NO, NO_2, NH_3, H_2S_2, and SO_2 released from the earth's surface, ocean or land; and

4. Volcanic eruptions: sizes may cover the whole range from $10^{-3}\mu$ to above $10^2\text{-}\mu$ diameter.

The population of atmospheric aerosols has a size distribution that is continuous from about $10^{-2}\text{-}\mu$ diameter or less to above $10^2\text{-}\mu$ radius. Maximum concentrations usually lie between 0.01 to 0.1 μ. Very little is known about the distributions below 0.01 μ because they are difficult to collect and measure. Since they are subject to Brownian motion, coagulation would rapidly remove

Prepared during SCEP and expanded after SCEP.

these particles but—at least for clean air over the North Atlantic —there seem to be particles present as small as 10^{-3} μ, suggesting constant production even within the "clean" air. If this can be shown to be correct for most of our tropospherical aerosols, their total concentration as well as the small end of the size distribution must be considered a dynamical and not a static system, with small particles being constantly added to the collection.

Although the particles smaller than 0.1 μ control the total *number* concentration, most of the aerosol mass is carried by particles greater than 0.1-μ diameter. Chemical analysis of filter samples of aerosols gives practically no information on the chemical composition of the smaller Aitken particles of less than 0.1 μ. The chemical composition of the larger particles can vary with the region and the particle size, depending on what sources are predominant.

Atmospheric aerosols show a high degree of "internal mixing"; this means that the large variety of inorganic and organic compounds formed in natural aerosols is found within each individual particle. Particularly under clean air conditions, all particles will have soluble fractions of material, which enables them to act as condensation or cloud nuclei. The difference between condensation and cloud nuclei is the required supersaturation: cloud nuclei require supersaturations of not more than a few percent of relative humidity and are measured by diffusion chambers; condensation nuclei are counted with an expansion cloud chamber using much higher supersaturations. The distinction is an artificial one and is not based on real cloud physics principles.

Some Thoughts about Monitoring Aerosols

[This question of monitoring aerosols, including many of Junge's ideas—since he played an active role in the monitoring Work Group, is treated at length elsewhere in this volume. We have therefore shortened this section, deleting many of the details of how the measurements should be made. We include the discussions of why various aerosol parameters are important to know. Eds.]

Monitoring the tropospheric aerosols from ground-based stations is not as simple as it sounds, because to describe an aerosol properly, one needs several parameters—not just one or two. Complete knowledge of aerosols would be obtained by measuring the complete size distribution plus the chemical composition as a function of particle size. This is not possible at the present time, and certainly much too complicated for routine monitoring even if it could be done. Therefore, it is necessary to determine the most important parameters that can be monitored using the techniques which are presently available. These parameters are discussed in this paper in terms of their role in meteorology.

Total Number Concentration

In clean tropospheric air this ranges between about 200 and 600/ cm^3. Of all the aerosol parameters the total number is the most sensitive to air pollution. The reason for this is that pollution always affects the small end of the size spectrum primarily, that is, the range smaller than 0.1-μ diameter. Since these particles represent a small mass fraction of the total aerosol mass, although they represent practically the whole number concentration, there is a large effect on the total number for a very small mass effect. Therefore, the total number concentration is an excellent parameter to monitor with respect to pollution.

The total number concentration is important in meteorology because in clean air it controls to a considerable degree the number of droplets formed in clouds, and by this the stability of the cloud and the probability of rainfall (especially when the warm cloud rain formation process dominates). An increase in particle concentration will result in larger number of cloud droplets of smaller size, which do not coalesce as readily as bigger cloud droplets. This is, however, only true for clean air. If the initial aerosol concentration is higher than 2,000/cm^3 or so, an increase will have little effect, because the cloud droplet concentration cannot be much higher than about 500/cm^3, leaving most of the aerosol particles unused for condensation.

Optical Effects of Aerosols

Short wave radiation is primarily influenced by aerosol particles in the size range of 0.1 to 1.0 μ. A convenient means of measuring the optical effect of the total atmospheric aerosol is by determining the "turbidity." (See chapters by Robinson and Mitchell for a more complete discussion.) Any change of turbidity is an integral effect, depending on the vertical distribution of the aerosols in the troposphere and stratosphere, on the concentration and size distribution within the indicated size range, and on the optical qualities of the aerosol particles. Turbidity is therefore difficult to interpret in terms of aerosol properties, but it has the advantage that it gives some measure of the overall influence on the radiation balance of the atmosphere as a whole if measured together with the total sky radiation, and that it can easily be measured if the sun is shining. A disadvantage is that it cannot separate tropospheric and stratospheric effects, which can be annoying after volcanic eruptions. However, there is hardly an easier way to obtain a representative value for the optical effects than by measuring turbidity.

Because the effects of air pollution are usually not so pronounced in the optically important size range of about 0.1- to 1.0-μ diameter than for the total number concentration, turbidity most likely will not respond so sensitively to global air pollution as the total number concentration.

Cloud Nuclei Concentration

In recent years it has become fashionable to measure the cloud nuclei concentration. This is the concentration of those particles of the atmospheric aerosol population that require supersaturation of less than a few percent relative humidity to form droplets, depending on the type of diffusion chamber used. It is claimed that this number gives the approximate cloud droplet concentration if a cloud would form under natural conditions, but this is not firmly established and cannot be entirely true since the number of cloud droplets depends not only on the characteristics of the aerosol but also on the cooling rate of the cloud air and on other cloud parameters.

Since the cloud nuclei are those particles larger than the size determined by the supersaturation of the cloud chamber (mostly between 0.1 and 0.01-μ diameter) their number is not so sensitive to pollution as the total number of particles, and therefore not so well suited for monitoring global pollution. On the other hand, the cloud nuclei concentration may be a better indicator for the overall effect of pollution on cloud behavior than the total number, but this has still to be convincingly shown. In the present situation, therefore, it may be advisable to monitor both the total number concentration and the cloud nuclei concentration. As with the turbidity, the cloud number concentration is difficult to interpret in terms of simple aerosol parameters because it is not only influenced by the size distribution of the aerosols but also by the amount of soluble and insoluble material as a function of particle size and other chemical factors.

Ice Nuclei Concentration

Another parameter of the aerosol population which is of great importance for effects on clouds is the ice nuclei concentration. Apparently the largest particles are those that act primarily as ice nuclei (soil mineral particles), but industrial sources are likely to produce ice nuclei of every size. The ice nucleating ability of an aerosol particle is a very complex integral quality, depending on a variety of physical chemical properties (not only on the crystal structure). It is recommended that ice nuclei concentrations should be considered for inclusion in any monitoring program.

Total Mass Concentration

So far we have discussed only one clearly defined aerosol parameter —namely, the total aerosol number concentration—whereas the three others were concerned with effects of the aerosol on radiation and on clouds, the two important ways by which the aerosol may act upon the atmosphere. But total number alone is not enough to characterize an aerosol sufficiently, even if one measures the optical effects, the cloud nuclei concentration, and ice nuclei concentration. The *total mass concentration* of aerosols is another important parameter that should be monitored, for instance, by taking filter samples. This gives the total weight of all particle sizes if the filter is good. These filter samples can also be analysed for specific chemical constituents, and the analysis can be further refined by using impactors to subdivide the aerosols into several size ranges.

Trends in Aerosols

We have discussed in the previous section various aerosol parameters which we feel are important for defining a complex atmospheric aerosol. The question arises: What trends are already observed for these parameters?

Total Number Concentration

There is no study that has critically surveyed all the nuclei counts made thus far. (Aitken nuclei counts are taken as a good measure of total number of particles.) There are a large number of counts of Aitken particles reported in the literature. My impression is that the counts have gone up in the Atlantic. For instance, Lettau observed in 1935 about $300/cm^3$, whereas on the Meteor voyage through the middle of the North Atlantic (1969) we found around $500/cm^3$. To my recollection the data from the Pacific did not show such an increase. However, they were mostly obtained by the GE Gardner counter, and this instrument has a lower limit around $200/cm^3$, therefore the accuracy of the data is not as good. The data in the Atlantic were obtained with absolute counters and are much more reliable. It may, however, be that a critical compilation of all Aitken nuclei data will produce more clear-cut results.

The best data on trends were obtained from air conductivity measurements, which were made as early as 1907 over the North Atlantic and Pacific Oceans and again as recently as 1967. They show quite clearly that in the North Atlantic the total concentration has apparently increased by a factor of about 2, whereas in the South Pacific there was no change between 1907 and today. No data are given from the North Pacific.

Optical Effects

Measurements in Davos, Switzerland, and Washington, D.C., indicate increases, but this is primarily local. The only really interesting data for clean air are those from Mauna Loa, but they seem to be not reliable enough. So essentially there is nothing available on turbidity for long-term trends in unpolluted air.

Cloud Nuclei Concentrations

Since these measurements were only started about ten years ago, any trends can hardly be expected, particularly since the method was further developed and changed during this decade.

Ice Nuclei

The longest available and reasonably consistent series of measurements is that from Mount Washington, New Hampshire, by Vincent Schaefer. As far as I know no trends have shown up. I do not know of other series of measurements covering a number of years.

Total Weight Concentration

The North American network on high volume sampling included a number of rural stations. The measurements are now available over a period of about ten to fifteen years. (See Chapter 25 summarizing these measurements.) But these stations would not indicate a global effect—if anything can be considered as representative of regional changes. I do not know of any set of data which could give information on global trends in clean air.

Residence Times of Aerosols

The residence (or life) time of tropospheric aerosols is defined as the average time an individual aerosol particle stays in the troposphere after formation before it is removed from the troposphere by rainout and washout (removal processes inside and below raining clouds), sedimentation, or impaction on the earth surface. Because of the continuous modification processes of atmospheric aerosols, the individual particles may undergo changes in size and composition during this time. For this reason the foregoing definition refers to the material of which the aerosol is made up and not to individual particles. Since natural aerosol particles cover a broad size range from less than 0.01 to greater than 100-μ diameter, the residence time of any constituent of the aerosol material may depend on the original distribution of such a constituent over this size range. However, we have no knowledge on this dependence; all we know at the present time are some figures on average values derived from natural and artificial radioactive species which became attached to the natural aerosols.

Unfortunately the values derived for the average lifetimes differ considerably within the range of a few days up to more than three weeks, depending on the method used. (It also depends on the altitude of the aerosol.) The first determination of the lifetime was by Stewart and his associates who followed the decrease of total β-activity following the early Nevada test, the debris of which had supposedly not entered the stratosphere. They could follow the radioactive cloud during many circulations around the Northern Hemisphere. After an initial period of lateral mixing and uniform spreading within the Northern Hemisphere the decrease was supposed to be due to the removal processes mentioned earlier. The observations indicated a lifetime of three to four weeks, and it was this figure which was up to very recent times considered the most reliable.

Other work by Indian scientists using radioisotopes produced by cosmic radiation resulted in similar figures. Other groups using radon and its decay products indicated shorter lifetimes, but the method cannot be considered reliable.

Very recently Martell and his group measured the Pb^{210}/Bi^{210} ratio in natural aerosols. These late daughters of radon should provide relatively reliable values, and it turned out that for most of the lower and middle troposphere an average of about six days was obtained which increased to two to three weeks within the upper troposphere. It is not yet clear how this discrepancy can be explained, but it is not unlikely that the different injection mechanisms into the atmosphere (fission products injected into the upper troposphere, and radon diffusing from the continental earth's surface) play a role.

The aerosol lifetime has important implications with respect to the transport and distribution of all substances associated with atmospheric aerosols. Since the air in middle latitudes circles the globe in about one to two weeks, all aerosol substances can spread fairly widely in longitude. Even in the tropics Sahara dust has been carried within one week westward across the Atlantic. But in all these cases the concentration drops markedly with increasing distance from the source.

Radioactive data from the 80th meridian network shows that the equator is practically a barrier to the spread of low level natural aerosols. The two hemispherical tropospheres, so far as aerosols are concerned, are apparently completely independent from each other. This is an important fact that should be kept in mind for all global considerations.

24

Sources of Particulate Matter   James T. Peterson and
in the Atmosphere   Christian E. Junge

Introduction
When considering the injection of particulate matter into the
atmosphere on a global basis, it is worthwhile to discuss several
facts about atmospheric aerosols which are sometimes overlooked.

The atmospheric aerosol, particularly that observed in remote places, is the product of a variety of processes which result
in a mixture of substances, about 50 percent of which are water
soluble. Even dry dust from deserts injected into the atmosphere
will accumulate such substances as $(NH_4)_2SO_4$ and other soluble
or insoluble material. Such mixing of substances results from
several processes: (1) coagulation with other particles of different
origin, (2) modification of particle composition in water clouds
due to repeated cycles of condensation and evaporation of cloud
droplets, and (3) formation within the atmosphere of available
compounds due to homogeneous or inhomogeneous (on the surface of aerosol or cloud droplets) gas reactions, such as the formation of $(NH_4)_2SO_4$ from SO_2 and NH_3 or the formation of organic
compounds produced by oxidation and other reactions from natural and man-made hydrocarbons.

Because of the considerable fraction of water soluble organic
materials generally present in atmospheric particles due to this
mixing, they cannot properly be called "dust," which is normally
understood to be entirely water insoluble, and are, therefore, appropriately referred to as "aerosols." Atmospheric aerosols are
produced by: (1) *direct injections* of particulate material into the
atmosphere, such as by dust storms or industrial processes, and
(2) *formation of particulate material within the atmosphere* due
to the formation of nonvolatile compounds by gas reactions, some
of which are photochemical in nature.

The particles injected by source (1) will stay within the atmosphere until removed by such processes as rainout, washout,
sedimentation, and so on. However, until removal, their chemical
and physical properties may be considerably modified by the processes outlined above. Original "dust" particles will thereby be
converted into atmospheric aerosols. The composition of these

Prepared during SCEP and expanded after SCEP.

aerosol particles will depend, of course, on the nature of other aerosols with which they come in contact and the physical processes to which they are subjected and, therefore, will vary with time and space within the atmosphere. In addition, *new material will be added* to the diversely injected particles, as a result of gas reactions.

Gas reactions accounting for source (2) can occur on the surface of aerosol particles, within cloud droplets (which after reevaporation form a bigger aerosol particle), or homogeneously to form new nonvolatile molecules which become readily attached to the aerosol particles present. Under special circumstances, homogeneous gas reactions can also form new populations of aerosol particles (an example of this being photochemical "smog").

On the basis of this discussion, it is clear that both sources of particulate material listed above must be considered for any inventory of the origins of atmospheric aerosols. For the second source, the assessment is incomplete, because only a few processes by which particulate matter is formed by gas reactions are known —such as the reaction of SO_2 and of hydrocarbons. In addition, present knowledge indicates that in both cases the larger fraction of these gases are removed only *after* conversion to particulate matter in the atmosphere. Therefore, these reactions will be discussed later to show the substantive contributions by source (2).

Calculations
This section documents the various sources of atmospheric particulate matter on a worldwide basis, with emphasis on a comparison of anthropogenic versus natural emission sources. These sources include both direct injections and formation within the atmosphere of particulate material. The data used for this study were taken from previously published material and from reports compiled during the Study of Critical Environmental Problems (SCEP). Most of the numbers presented lack precise accuracy and are rounded off, but the estimates are believed to be accurate within a factor of 2. Estimates of natural and man-made particulate emissions are made for 1968, along with rough projections for 1980 and 2000 based on current emission controls.

The data are presented in terms of the mass of all particles emitted into the atmosphere which do not directly fall out, and also of those with diameter less than 5 μ. The 5-μ cutoff was used since that is the smallest size class given for industrial emission data commonly published by the National Air Pollution Control Administration (NAPCA). Ideally, the various source emissions should be categorized by more complete size distribution curves or number concentrations. These detailed data would be more widely applicable to specific geophysical problems, but they were not available for inclusion in this report.

The results of this study for 1968 are shown in Table 24.1 where the contributions from the individual sources are listed for the two size classifications in units of 10^6 metric tons (all units are metric tons unless otherwise stated). Before discussing the significance of these numbers, the assumptions and techniques used to derive the basic data will be set forth.

Eriksson (1959) has estimated the annual emission of sea salt into the atmosphere to be 1000×10^6 tons. With a Beaufort wind of 3 to 5, approximately half of these particles are less than 5 μ.

In a first-week interim report prepared during SCEP (unpublished), G. Arnason calculated the amount of windblown dust derived from the continents, which originates primarily from arid regions, particularly the Sahara. His world estimate, about 500×10^6 tons/year, was extrapolated from a calculation for the

Table 24.1 Estimated Mass of Global Emissions of Atmospheric Particulate Material, 1968, in 10^6 Tons

Source	$d < 5\ \mu$	All Sizes
Sea salt	500	1000
Converted natural sulfate	335	420
Natural windblown dust	250	500
Converted man-made sulfate	200	220
Converted natural hydrocarbons	75	75
Converted natural nitrates	60	75
Converted man-made nitrates	35	40
Man-made particles	30	135
Volcanoes	25	?
Converted man-made hydrocarbons	15	15
Forest fires	5	35
Meteoric debris	0	10
Total	1530	2525+

United States (Wadleigh, 1968) which included both natural and agricultural sources of windblown dust. If the particle size distribution of this material is assumed to follow a standard Junge distribution, $dn/dr \, \alpha \, r^{-4}$, between the sizes of 0.1 and 100 μ, and a concave distribution at smaller sizes, then roughly one-half of the total mass is associated with particles less than 5 μ in diameter.

Two rough, independent checks of this estimate of 250×10^6 tons/year for small particles are available. The SCEP Report (1970) estimated that the annual global small particle dustfall is approximately 70×10^6 tons, as determined from glacial sediments. Since the glaciers sampled were 2000 to 3000 meters above sea level this figure is certainly a lower limit. Goldberg (1970) also made a rough estimate of the natural nonsoluble, nonbiological dustfall by considering sea sediments. The average deep ocean sediment rate is about 10^{-4} cm/year with a density of 2 g/cm^3 and 50 percent water composition. If this rate is assumed to totally result from dustfall (no water transport of river sediments into the deep ocean) and thus represent an upper limit of dustfall, and it is assumed further that there is a similar fall of fine particles over all 5×10^{18} cm^2 of the earth, then the natural global dustfall is 500×10^6 tons/year, the majority of which is certainly less than 5 μ in diameter.

These methods for determining the global rate of removal of water insoluble material from the atmosphere suggest an interesting application of the results of this study. If the accuracy of these estimates of the removal of insoluble particles were improved along with those for the corresponding emissions presented in Table 24.1, then their relative difference could be used to estimate whether these particles are accumulating in the atmosphere or being removed at the same rate as their input.

The direct emissions of particles into the atmosphere by man's activities are based on a report by the Division of Air Quality and Emission Data (DAQED) (NAPCA, 1970), which lists atmospheric particulate emissions for the United States, and on other unpublished calculations from the Study. Table 24.2 shows the U.S. and world atmospheric emissions by source for 1968. Since statistics of global particulate emissions and emission controls have not been published, the world estimates presented in Table 24.2 are for the most part "educated guesses" based on the more accurate

Table 24.2 United States and World Direct Atmospheric Emissions of Particles for Various Anthropogenic Sources, 1968, in 10^6 Tons/Year

Source	United States		World	
	Total	$d < 5\,\mu$	Total	$d < 5\,\mu$
Transportation	1.1	0.9	2.2	1.8
Stationary source fuel combustion	8.0	2.5	43.4	9.6
Industrial process	6.7	2.9	56.4	12.4
Solid waste disposal	1.0	0.2	2.4	0.4
Miscellaneous	2.6	0.4	28.8	5.4
	19.4	6.9	133.2	29.6

United States data. The transportation emissions are primarily the result of automobile exhaust. Estimates of emissions from fuel combustion from stationary sources were provided by DAQED of NAPCA. The industrial process data were based on the SCEP Report (1970). The solid waste and miscellaneous emissions were based also on the SCEP Report (1970).

Estimates for the mass of particles emitted into the atmosphere from forest fires and meteoric debris were taken directly from estimates made during SCEP and Rosen (1969), respectively.

A rough estimate of the emission of fine particles from volcanoes was obtained from a report by Mitchell (1970). Mitchell estimated that over the last 120 years the Northern Hemisphere's stratospheric loading has been 4.2×10^6 short tons. If a similar loading is assumed for the Southern Hemisphere along with Mitchell's fourteen-month residence time, a global annual injection rate of 3.3×10^6 tons/year can be calculated. A SCEP working paper (unpublished) also made a rough estimate (4×10^6 tons) for the average tropospheric loading from volcanic activity. If one assumes a residence time of 0.1 year, an injection rate of 36×10^6 tons/year results for the troposphere or a total injection rate of about 40×10^6 tons/year. An estimate of 25×10^6 tons/year is presented in Table 24.1 which is more representative of those years with no large eruptions than the long term average, which is biased by several big eruptions. In the last 120 years, there have been 20 years when large eruptions occurred and when emissions were about 2 to 10 times greater than the long term average.

The importance of gaseous SO_2 emissions and their conversion to sulfate in determining the total burden of atmospheric particles has been discussed in the introduction. The fraction of total sulfur emissions resulting from human activity can be calculated fairly well since the global consumption of fuels and smelting of copper are documented. A historical record of pollutant SO_2 emissions is shown in Figure 24.1. Such accuracy, however, cannot be carried over to natural emissions since, for example, scientists still debate whether the oceans are sources or sinks of atmospheric sulfur. In this section two estimates of the global sulfur balance will be presented. The first is taken directly from a recent report by Robinson and Robbins (1969), and the second was compiled from industrial process emissions data presented in the SCEP Report (1970) and unpublished data on emissions from stationary sources estimated by DAQED-NAPCA at SCEP.

The data on Figure 24.1 were linearly extrapolated to 1968 to yield the first estimate of global pollutant SO_2 emissions, 142×10^6 tons/year. The second estimate, 179×10^6 tons/year,

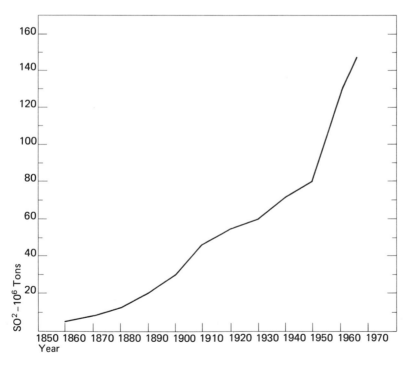

Figure 24.1 Estimated historical global SO_2 emissions
Source: Robinson and Robbins, 1969.

is based on a calculation of 143×10^6 tons/year for stationary sources compiled by DAQED-NAPCA, 27×10^6 tons/year for industrial process emissions, calculated at SCEP (1970), and an additional 5 percent estimated for all other sources. The main difference between the calculations of these two sources lies in their values for the sulfur content of coal, 2.0 and 2.7 percent for the first and second estimates, respectively. The average of these two estimates of the global emission of SO_2, 160×10^6 tons/year, was subsequently used herein to determine the amount of particulate matter produced from the gaseous SO_2. Two-thirds of this SO_2 was assumed to be converted to ammonium sulfate aerosol $(NH_4)_2 SO_4$, and the remainder of the gas was not converted to aerosol but was either washed out by precipitation or absorbed by the earth's surface. By applying the ratio of molecular weights, 132/64, the result of this exercise is that

$$160 \left(\frac{2}{3}\right) \left(\frac{132}{64}\right) = 220 \times 10^6 \text{ tons/year}$$

of atmospheric sulfate particles result from the emission of pollutant SO_2. Ninety percent of this mass was assumed to be smaller than 5μ in diameter.

Table 24.3, taken directly from Robinson and Robbins (1969), lists their estimates of hemispheric sulfur emissions for both pollutant (man-made) and natural sources in units of 10^6 short tons of sulfur. The estimates in Table 24.1 for natural emissions were calculated from these data as follows. In the same way as for pollut-

Table 24.3 Total Annual Worldwide and Hemispheric Sulfur Emissions, 10^6 Short Tons Sulfur

Source	Total	Northern Hemisphere	Southern Hemisphere
Pollutant SO_2 sources	73	68	5
Biological H_2S (land)	68	49[a]	19[a]
Biological H_2S (marine)	30	13[b]	17[b]
Sea spray	44	19[c]	25[c]
Total	215	149 (69%)	66 (31%)

[a] Based on ratio of land area between 0 and 65° N. and S.
[b] Based on ratio of ocean areas between 0 and 65° N. and S.
[c] Based on ratio of ocean areas in both hemispheres.

ant SO_2, two-thirds of the H_2S emissions were assumed to be converted to ammonium sulfate,

$$98 \left(\frac{2}{3}\right) \left(\frac{132}{32}\right) (0.907) = 244 \times 10^6 \text{ tons/year},$$

and all the atmospheric sea salt was assumed converted to sodium sulfate,

$$44 \left(\frac{142}{32}\right) (0.907) = 177 \times 10^6 \text{ tons/year}.$$

Sea spray sulfate is generally larger than the other sulfates so 80 percent, or 335×10^6 tons/year, were assumed to be particles smaller than 5 μ in diameter.

Large quantities of nitrogen, which are subsequently converted to nitrate aerosol, are annually emitted into the atmosphere both from pollution sources, primarily NO and NO_2, and from natural sources, primarily NO. Some estimates of the nitrogen cycle have been made, for example see Robinson and Robbins (1969), but because of uncertainties in the estimates an indirect method was used here. From an analysis of more than 17,000 actual urban and nonurban 24-hour high volume samples an average sulfate to nitrate ratio of 5.5 to 0.1 was determined (Stern, 1968). Thus, the values for natural and man-made sulfur emissions listed in Table 24.1 were divided by 5.5 to yield a rough estimate of the nitrogen emission converted to aerosol.

The final source categories of Table 24.1 to be considered are those for man-made and natural hydrocarbons. As in the case of sulfur and nitrogen, these gaseous hydrocarbons are converted within the atmosphere to aerosols. United States emissions of pollutant hydrocarbons in 1968 were 29×10^6 tons (DAQED-NAPCA, 1970). If it is assumed that three-fourths of these are nonmethane (methane is a very stable compound), that world emissions are twice that of the United States, and that one-third of the mass is converted to aerosol, then about 15×10^6 tons/year of pollutant hydrocarbons are converted to aerosols. Went (1960) estimated that the global emissions of natural hydrocarbons released by plants and soil are 154×10^6 tons/year. One-half of these emissions were taken to be converted to aerosols, and all of the hydrocarbon aerosols were assumed to be less than 5 μ in diameter.

Discussion

Table 24.1 shows that on a global scale man is now emitting 280×10^6 tons/year of small particulate matter into the atmosphere compared to a natural rate of 1250×10^6 tons/year, or a ratio of about 4.5 to 1. Moreover, man-made emissions are now about 18 percent of the total emissions of small particles.

These figures for particles smaller than 5 μ in diameter can be put in a different perspective by considering emissions from northern hemispheric land areas only—those areas of the globe on which most of the people reside. Robinson and Robbins (Table 24.3) determined that 93 percent of pollutant SO_2 is emitted in the Northern Hemisphere. Based on this number, 90 percent, or 250×10^6 tons/year, of all man-made pollutants were assumed to originate from the continents of the Northern Hemisphere.

Emissions from natural sources over the northern hemispheric continents were estimated in the following ways. Using the data from Table 24.3, natural sulfate emissions converted to particles smaller than 5 μ were calculated to be

$$49 \left(\frac{2}{3}\right) \left(\frac{132}{32}\right) (0.907)(0.9) = 110 \times 10^6 \text{ tons/year.}$$

Based on the fraction of land area between 0° and 65° N. and S. in the Northern Hemisphere (72 percent) and the fact that the Sahara is the largest source of natural dust, 80 percent or 200×10^6 tons/year of the natural wind blown dust was calculated to originate from this area. Seventy-two percent and 60 percent of the natural hydrocarbons and nitrates, respectively (a total of 90×10^6 tons/year), were assumed to be emitted from the land areas, whereas volcanoes and forest fires annually contributed an additional 15×10^6 tons. Thus, for the continents in the Northern Hemisphere particulate emissions from human activity are 250/415, or 60 percent, of those from natural sources.

Some rough figures for future global particulate emissions have been calculated, based on the 1968 level of control; these are presented in Table 24.4. These data are based on worldwide tenuous projections of industrial growth as well as current levels of emission controls and fuel desulfurization. They should be used only as a guide to what may actually occur since there probably will be technological advances in air pollution control. The future

Table 24.4 Future Estimated Mass of Particulate Emissions Annually from Man-made Sources for the World for Particles less than 5 μ Diameter, in 10^6 Tons/Year

Source	1968	1980	2000
Converted sulfate	200	260	450
Converted nitrate	35	45	80
Directly emitted particles	30	45	100
Converted hydrocarbons	15	25	50
Total	280	375	680

Based on 1968 rate of emission control.

estimates for sulfate and nitrate are based on data from the same sources and determined in a manner identical to that used for the 1968 global emission estimates. Projections for emissions of particulate matter were based on calculations used for the SCEP Report (1970) estimate of particulate emissions and unpublished data from DAQED-NAPCA; a similar rate of increase was applied to the hydrocarbon data.

The data from Table 24.4 show that unless more stringent emission controls are applied the mass of man-made fine particles emitted into the atmosphere annually will increase nearly 250 percent by 2000. By this time the amount of particles resulting from the conversion of man-made SO_2 emissions will have surpassed that due to natural sulfur emissions. For the Northern Hemisphere's land areas only, man-made particles will outweigh those from natural sources soon after 1980.

In summary, man today is contributing nearly one-fifth of the total atmospheric burden of small particles. This ratio will probably increase slightly during the next thirty years even if all nations apply moderate to stringent emission controls. The ratio can be reduced only if strict emission controls are widely implemented throughout the world.

References

Division of Air Quality and Emission Data—NAPCA, 1970. *Nationwide Inventory of Air Pollutant Emissions, 1968,* Publication No. AP-73 (Raleigh, North Carolina: National Air Pollution Control Administration).

Eriksson, E., 1959. The yearly circulation of chloride and sulfur in nature; meteorological, geochemical and pedological implications, Part I, *Tellus, 11,* (4): 375.

Goldberg, E., 1970. Personal communication at SCEP.

Mitchell, J. M., 1970. A preliminary evaluation of atmospheric pollution as a cause of the global temperature fluctuation of the past century, *Global Effects of Environmental Pollution,* edited by F. Singer (Dordrecht: D. Reidel Publishing Company).

Robinson, E. and Robbins, R. C., 1969. Sources, abundance and fate of gaseous atmospheric pollutants, supplement, SRI project PR-6755.

Rosen, J. M., 1969. Stratospheric dust and its relationship to the meteoric influx, *Space Science Reviews, 9* (1): 58.

Study of Critical Environmental Problems (SCEP), 1970. *Man's Impact on the Global Environment* (Cambridge, Massachusetts: The M.I.T. Press).

Stern, A. C., 1968. *Air Pollution,* Vol. I (New York: Academic Press).

Wadleigh, C. H., 1968. Wastes in relation to agriculture and forestry, Miscellaneous publication No. 1065 (Washington, D.C.: U.S. Department of Agriculture).

Went, F. W., 1960. Organic matter in the atmosphere and its possible relation to petroleum formation, *Proceedings of the National Academy of Sciences of the United States of America, 46* (2): 212–221.

25
Trends in Urban Air Quality   John H. Ludwig,
                                      George B. Morgan, and
                                      Thomas B. McMullen

An evaluation of trends in air quality is a means to assess the net result of numerous interrelated factors that affect the quality of our air resource. Factors such as population growth, industrial activity, energy consumption, rural-urban distribution of our population, the shape and size of our metropolitan areas, various social and economic changes, and the effectiveness of our efforts to control pollution at its source all play an important role in determining the resulting air quality. It must be noted that the trends in ambient levels of pollutants at any one place is a complex function of some or all of these factors and not merely the direct assessment of the degree of control applied to the various sources of pollution.

We assess trends presumably because we would like to anticipate what is happening to us now, and we inject into this picture projections of potential pollution increases related to our nation's goals in population growth and standard of living, in order to anticipate some of the future consequences of trends in human activity. This also allows us to assess the need to control pollution and to undertake now the research and development required to bring such control about. It is to be noted that the subject of this article is by no means merely of scientific curiosity; rather, it is the basic building block on which our nation's need and program, to restore and protect our environment, is founded.

First, let's look briefly at some of the current effects of urban centers on air quality and the trends associated with population and geographical location, since these considerations are important in assessing temporal trends. Samples of total suspended particulates collected at various urban and nonurban areas have been analyzed for certain metal and nonmetal substances that exhibit biological activity. These substances include sulfates, nitrates, fluorides, benzene soluble organics, benzo(a)pyrene, copper, chro-

This paper was presented at the National Fall Meeting of the American Geophysical Union, San Francisco, California, December 17, 1969.
Reprinted from *EOS, Transactions of the American Geophysical Union, 51* (1970), pp. 468–475.

mium, manganese, nickel, tin, titanium, vanadium, iron, zinc, lead, etc.

The relative influence of urban activity on the ambient levels of ten metals is shown in Figure 25.1. The ambient concentrations of these suspended particulate fractions are expressed in terms of micrograms per cubic meter of air. The urban values are drawn from a group of 31 stations covering 11 years (1957–1967); the nonurban values are from 30 stations for 1966–1967—the first two years that new laboratory techniques made possible the analyses of such low level constituents. The relative concentrations of these

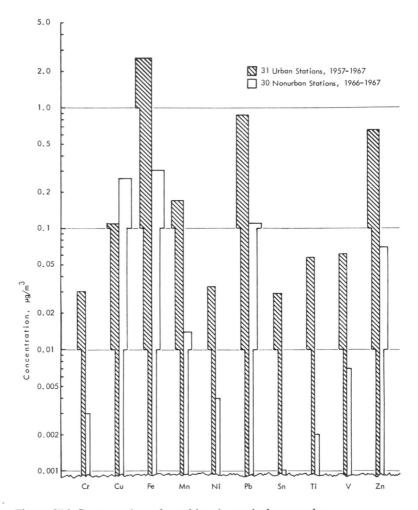

Figure 25.1 Concentrations of metal ions in particulate samples

metals in the atmosphere are quite different, with iron, lead, and zinc being the most abundant. For all but copper the urban fractions are typically twice those found at nonurban locations, indicating that urban activities probably contribute significantly to the nonurban background levels of these pollutants. For copper the opposite is true; the nonurban samples are found to be higher than in urban areas. This points to the mining and smelting of ores, typically located in rural areas, and reentrainment of soil as the major sources of atmospheric copper. Also, it is interesting to note that iron, lead, and zinc are prevalent since these are common in natural or in high use by man.

That the size of urban area is a determining factor in the atmospheric levels of some metals is illustrated in Figure 25.2, where the ambient concentrations of copper, iron, manganese, nickel, lead, and vanadium found in the various cities are grouped by population. Of the six metals, three—lead, vanadium, and nickel —show increased levels in the larger cities. This is quite understandable since the major source of lead is from additives to automotive fuels and most of the vanadium comes from the burning of Caribbean fuel oils that are being used in the large eastern seaboard and Great Lakes cities. Nickel concentrations reflect the

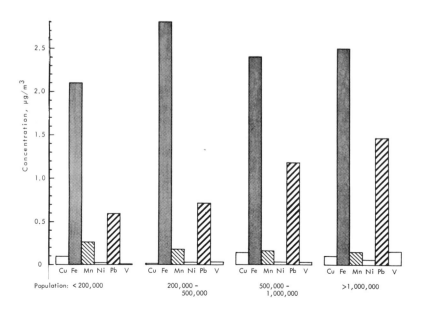

Figure 25.2 Concentrations of six particulate metal ions by city size

metal industries as well as emissions from automotive exhaust that result from normal wear of engine parts and, to a lesser extent, from fuel additives. The relative uniformity of iron reflects the large natural abundance of this metal.

There are also geographical differences in ambient levels of these metals. For purposes of illustrating these differences, the country, as shown in Figure 25.3, was divided into four broad geographical areas. The Great Lakes area has been included with the Northeast and Atlantic Coast areas because of similarities in industrial activity and fuel usage. The sampling results were grouped according to these regions, and the urban concentrations of the six metals are shown in Figure 25.4. All except copper show some geographical variation. Iron is more prevalent in the Great Lakes-Mid Coastal and Appalachian regions than elsewhere, reflecting the manufacture of iron and steel. The high manganese in the Appalachian area results from the influence of the unusually high values found in the Kanawha Valley on the regional average. Vanadium is higher in the eastern Great Lakes-Mid Coastal Region because of the use of Caribbean fuel oils. Lead and nickel, previously shown to be population dependent, depict a very minor geographical variation.

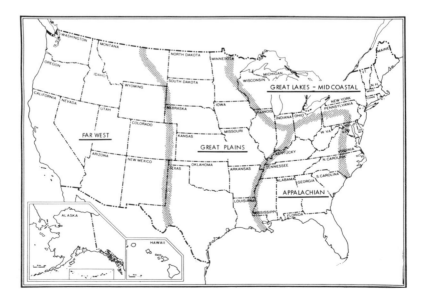

Figure 25.3 Major geographical areas

The significant differences in metal levels in the atmospheres of urban and nonurban locations can further be illustrated by comparing the distributions of individual observations made at both urban and nonurban locations. This type of comparison for manganese, nickel, and vanadium is shown in Figure 25.5. The typical concentration in urban locations is up to 10 times higher than the typical nonurban concentrations. The range of concentrations at the nonurban sites overlap the concentrations ranges of the urban locations. The degree of overlap varies and reflects the complex interrelationships, not well understood, between man-made and natural sources, transport of pollutants, and removal mechanisms in the atmosphere. Both the urban and nonurban profiles for vanadium exhibit two bulges. This bimodal distribution is directly related to use of imported residual fuel oil, as mentioned previously, since most of the urban sites averaging above 0.1 $\mu g/m^3$ and most of the nonurban sites averaging above 0.01 $\mu g/m^3$ are from the northeastern section of the U.S. Figure 25.6 shows that copper is present in nonurban samples in approximately the same relative magnitude as in urban areas. On the other hand, lead shows significantly higher urban concentrations.

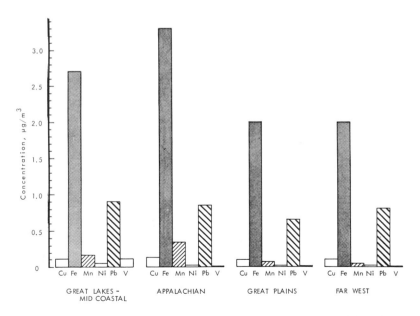

Figure 25.4 Concentrations of six particulate metal ions by geographic region

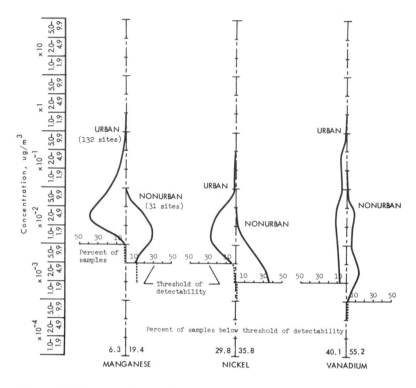

Figure 25.5 Concentration profiles of particulate constituents, 1966–1967

Iron is between copper and lead in this regard. It is interesting to note that the profile for urban and nonurban SO_4, Figure 25.7, is distributed over urban and nonurban areas in approximately equal concentration, which is to be expected because the average size of the SO_4 particle is in the submicron range.

The manner in which pollution levels diminish as we move away from the core of an urban area is illustrated subjectively in Table 25.1. Here the nonurban stations are grouped according to their nearness to major urban centers and are classified as proximate (urban background), intermediate (reflecting agricultural activity), and remote (isolated from man's normal activity). Urban stations show average ambient pollutant concentrations by a factor of 3 to 50 higher than the remote background stations. For example, lead concentrations within the core of urban areas average about 1.11 $\mu g/m^3$, decrease to 0.21 $\mu g/m^3$ in the proximate

stations, and further to 0.096 $\mu g/m^3$ and 0.022 $\mu g/m^3$ at the intermediate and remote stations respectively; and suspended particulates decrease from 102 $\mu g/m^3$ to 21 $\mu g/m^3$.

This demonstrates the wide distances over which urban produced pollutants are dispersed. This evidence of widespreading
influence from population centers supports the decision to establish multi-county air quality control regions around major urban
centers to deal with our more severe pollution problems, and history may judge even this as only the first feasible step toward national and worldwide environmental husbandry.

The combined temporal-spatial trends in suspended particulates are depicted in Figure 25.8. The jagged plots are the averaged
values at each sampling interval for 20 nonurban and 58 urban
sites. The trend curve was calculated and plotted by computer
using a technique called the Whittaker-Henderson smoothing
formula based on the optimization of a mathematical expression

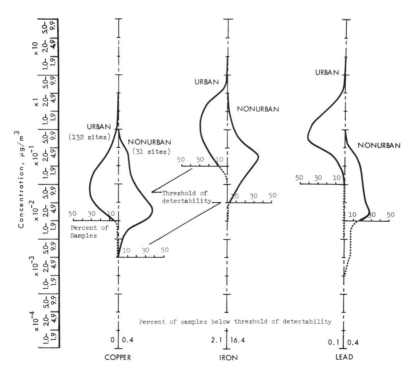

Figure 25.6 Concentration profiles of particulate constituents, 1966–1967

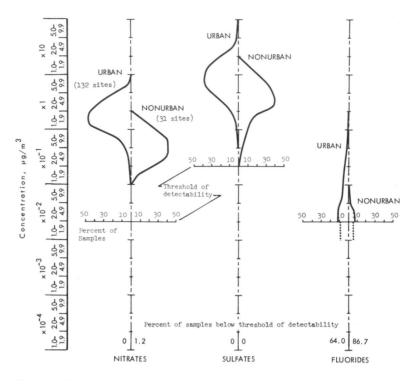

Figure 25.7 Concentration profiles of particulate constituents, 1966–1967

Table 25.1 Selected Particulate Constituents as Percentages of Gross Suspended Particulates (1966–1967)

	Urban (217 Stations)		Proximate (5)		Nonurban Intermed. (15)		Remote (10)	
	$\mu g/m^3$	%	$\mu g/m^3$	%	$\mu g/m^3$	%	$\mu g/m^3$	%
Suspended particulates	102.0		45.0		40.0		21.0	
Benzene-soluble organics	6.7	6.6	2.5	5.6	2.2	5.4	1.1	5.1
Ammonium ion	0.9	0.9	1.22	2.7	0.28	0.7	0.15	0.7
Nitrate ion	2.4	2.4	1.40	3.1	0.85	2.1	0.46	2.2
Sulfate ion	10.1	9.9	10.0	22.2	5.29	13.1	2.51	11.8
Copper	0.16	0.15	0.16	0.36	0.078	0.19	0.060	0.28
Iron	1.43	1.38	0.56	1.24	0.27	0.67	0.15	0.71
Manganese	0.073	0.07	0.026	0.06	0.012	0.03	0.005	0.02
Nickel	0.017	0.02	0.008	0.02	0.004	0.01	0.002	0.01
Lead	1.11	1.07	0.21	0.47	0.096	0.24	0.022	0.10

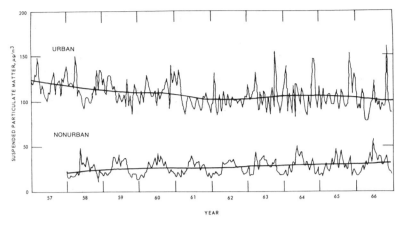

Figure 25.8 Long-term trends at 58 urban sites and 20 nonurban sites

combining weighted measurements of curve fit and smoothness. The intrusion of human detritus on nonurban air quality indicates a small but statistically significant increase with time based on averages for 20 nonurban sites operated since 1958. The average level of 25.4 $\mu g/m^3$ for the period 1958–1961 rose by about 12% to 28.5 $\mu g/m^3$ in the 1962–1966 period. The smoothed plot of all the data for all of these 20 nonurban stations over the 9-year period shows an unrelenting upward slope.

In contrast, a similar analysis of data for 58 urban sites for 10 years (1957–1966) shows a statistically significant decrease of about 7% from 111.5 $\mu g/m^3$ (1957–1961) to 103.2 $\mu g/m^3$ (1962–1966). This decrease, indicating trends at single city-center sites, is attributed mainly to control measures being applied to this most conspicuous form of air pollution, changes in use of cleaner fuel associated with increased standard of living, and to a certain amount of decentralization of sources.

It is important to realize that the above trend is only for particulate matter. Conspicuous as the particulate matter is, it is well to remember that by weight it represents only 1% of the total weight of the 6 major gaseous pollutants typically found in urban air—sulfur dioxide, carbon monoxide, nitric oxide, nitrogen dioxide, hydrocarbons, and oxidants.

Another measure of particulate pollution is settleable dust loading—the large particulate matter that is an annoyance to us by soiling our clothes, settling on our window sills and porches

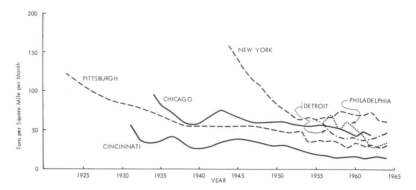

Figure 25.9 Trends in settleable dust in 6 cities

and blowing into our eyes. As shown in Figure 25.9 there has been a very noticeable decrease in all our major urban areas. The decrease with time reflects both the application of control techniques on particulate pollution sources and the change in domestic fuel from coal to oil and gas, principally the changes in fuel usage. We are indeed fortunate as a nation that our affluence and our desires to eliminate the inconvenience of coal for home heating justified the additional costs associated with switching to more expensive fuels, for along with the ridding of this chore came a significant decrease in not only settleable dust and fly ash, but many other pollutants as well.

Unfortunately, the sampling for gaseous pollutants is not as complete and widespread as that for particulate matter. The routine monitoring for gases in a fashion amenable to trend analyses was started by the federal government in 1962 and then in only six urban areas, two of which were relocated in 1964–1965. The annual average concentrations of sulfur dioxide for the six cities are shown individually in Figure 25.10. Data available from local agencies, where gaseous air monitoring data have been obtained, exhibit similar characteristics. On the basis of these data, no definite general pattern or trend can be established. There are indications that the levels are decreasing somewhat in the core areas of these cities. Many of these areas are now enacting rules limiting the allowable sulfur in coal as well as fuel oil. This limiting of sulfur, together with urban renewal that replaces buildings that in many cases used heavy fuels for space heating, is the primary cause of the slight downward trend in sulfur dioxide concentrations noted. This trend is further illustrated in Figure 25.11, where the annual average concentrations for 22 other urban sites are grouped and plotted. Here, the downward trend is more clearly evident. It is important to note that these changes have been observed in the downtown areas where the effect of fuel switching or urban renewal would be most pronounced.

With respect to other gaseous pollutants such as carbon monoxide and oxidant (Figures 25.12 and 25.13), very little, if anything, can be said about the changes in ambient levels. The limited number of sampling locations coupled with the short sampling duration restrict our ability to establish conclusive trends and associated rates of change. The variability in meteorology from year to year further confuses the issue by influencing annual average concentrations in a sporadic and unpredictable manner.

In the case of carbon monoxide (Figure 25.13), which is derived principally from motor vehicles, we may be seeing a levelling associated with near saturation of our downtown streets by automobiles. Unfortunately, we have no data on nonurban gaseous pollutant measurements to explore the spatial trends as de-

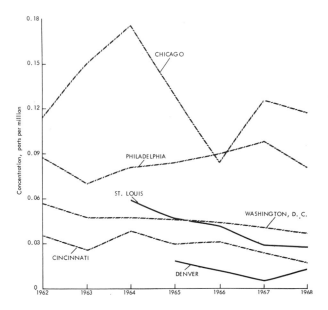

Figure 25.10 Sulfur dioxide, CAMP annual averages

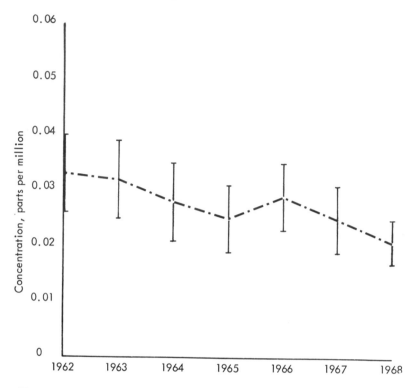

Figure 25.11 Annual average sulfur dioxide concentrations at 22 urban NASN sites

picted by particulates, but there is no reason to conclude that such dispersion is not occurring. This implies that the principal problem may not be one of great intensification of levels in city cores, but more an extension of pollution so that concentrations over larger and larger areas may be raised above minimum critical effects levels. A visual evaluation of trends over the years as viewed by air travellers approaching cities gives subjective credence to such a supposition.

Another area of need in trend analysis, which to date has been almost entirely neglected, concerns the evaluation of high values that are important in relation to short-term effects. The so called "stagnation conditions" associated with high air pollution levels occur only on 5% to 10% of the days. It would be interesting to assess the effects of the numerous physical factors previously men-

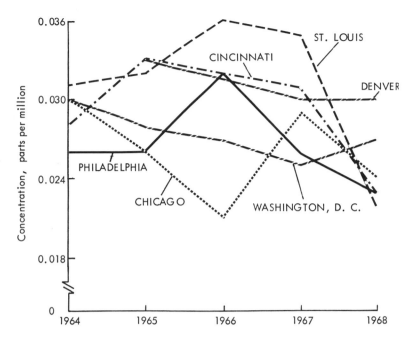

Figure 25.12 Total oxidants, CAMP annual averages

tioned on the occurrence and severity of these high pollution episodes; such data would provide an additional insight into the
effects of our changing social patterns on environmental air quality.

   This lack of sufficient amounts of data from which to establish
trends in air quality reflects the only recent attention of our society
to the impact of air pollution on our environment and the costs
associated with collecting data of quality and quantity sufficient
to assess trends. Furthermore, the availability of funds for such
investigations, including methods and instrumentation development, is limited by competing demands for the air pollution program dollar, let alone the general competition by other social
problems for the tax dollar. The question is: How do we optimize
our present resources in this area and program additional future
resources that may become available so as to meet our most critical
needs in an optimum manner? Is it possible to utilize other technology in the light of the multitude of pollutants and locations
to be assessed?

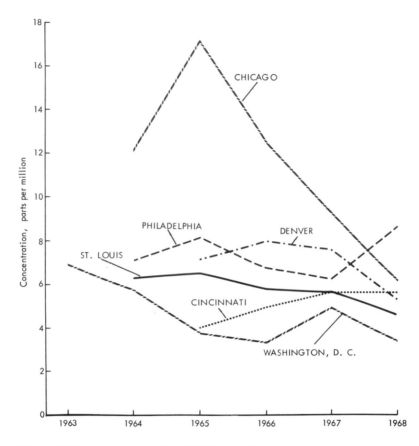

Figure 25.13 Carbon monoxide, CAMP annual averages

The answer to these questions lies in the utilization of two techniques: (1) projecting potential to pollute commensurate with our expectations of population growth and standard of living, reflecting items such as future energy use, industrial growth and social advancement. Such projections must be worldwide as well as nationwide if we are to evaluate long-term climatic effects as well as the immediate effect on man's health and welfare; and (2) use of mathematical models to assist in the evaluation of source-receptor relationships, both to interpret existing data and to plan the networks for collecting additional data. This latter item—mathematical modeling—should be utilized more extensively to enable a more meaningful assessment of the effects of projected increases in total source emissions in the light of the changing nature of the areal distribution as well as the quantitative aspects of individual source emissions as controls are applied.

What are the prospects for upward trends in our abilities to produce pollutants? A very simple model we would propose is that our abilities to produce pollution is a function of the product of population and standard of living raised to some exponent. We have no specific data on what this exponent might be, but we suspect it is not less than 1. But the precise value of this exponent is not of great importance at the moment, in that the problem posed is generally the same.

Our national expectations are to double the population and to increase the standard of living by a factor of 3 or 4 by the year 2025 or so. Energy needs of the nation are at present doubling every 10 years, and this rate of increase is expected to continue for the next few decades. This, then, tells us that our potential to pollute, using the simple model proposed, will increase by at least a factor of 6 or 8. These increases will not be the same for all pollutants, of course. Facetiously, there is a limit to the number of vehicles each of us can drive at one time. But most assuredly we will be witness to great increases in our abilities to produce pollutants such as carbon monoxide, sulfur oxides, nitrogen oxides, hydrocarbons, and other organic compounds, fine particulates, odors, and others. In spite of the fact that all of these will be increasingly controlled at the same time as we satisfy our inclinations to avail ourselves of the benefits of activities that produce them, there will be an ever-increasing need to improve our abilities to assess trends.

As an example, projected trends in power generation and total sulfur dioxide emissions to the atmosphere are shown in Figures 25.14 and 25.15. These trends indicate an increase over present levels of SO_2 emissions of from 2 to 5 times, depending on the degree of controls applied to fossil fuel use and recognizing that nuclear power generation will increase as shown in Figure 25.14, so that by the 1990s energy derived from atomic energy (assuming development of breeder reactors) would supply about 50% of the total needs. It is indeed interesting to note that about the best we can do is to hold total emissions resulting from power generation to somewhat above 2 times current emissions.

Although we have only alluded to the need to measure trends

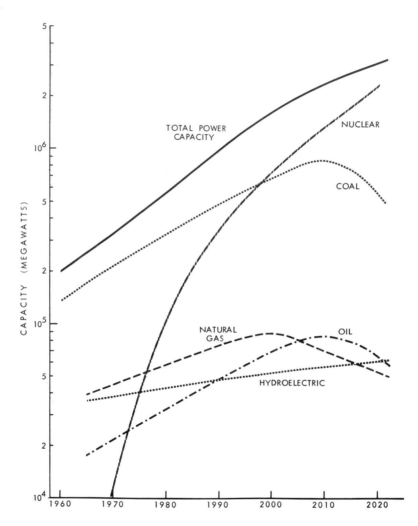

Figure 25.14 Projected power generating capacity of electric utilities in the U.S.

in pollutants of geophysical significance—those involved with possible long-term climate modification—others have addressed themselves to this problem. It is mentioned here only with respect to the fact that such pollutants of manmade origin have their principal sources in our urban centers, and that control of such sources will, if need be, reflect back to man's activities in these complexes. The fact that the full technological development of the underde-

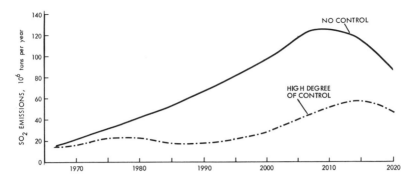

Figure 25.15 Effect of fuel substitution and control processes on annual SO₂ emissions

veloped nations in the world would probably result in a 6-or-so fold additional increase in the world's potential to pollute gives heed to serious thought as to how to stem this tide, particularly with regard to possible worldwide climatic changes.

In the future, development of resources on both a national and international basis will challenge our ingenuity to develop ways to minimize adverse effects and to preserve environmental quality so that man can live in health and happiness in his environment. But whatever these effects, the alarming promise of the future underlines the importance of improving our nation's commitment to assessing trends in our environment.

Acknowledgment

Data presented in Figures 25.14 and 25.15 was provided by the Division of Process Control Engineering, National Air Pollution Control Administration, based on projected electric utility capacity data published by FPC and AEC.

Selected Literature

Air Pollution—A National Sample, PHS Publ. No. 1562, for National Conference on Air Pollution, Wash., D.C., Dec. 12–14, U.S. Public Health Serv., 1966.

Air Pollution Measurements of the National Air Sampling Network, 1963, U.S. Public Health Serv., 1965.

Air Quality Data, 1966 edition, No. APTD 68–9, National Air Pollution Control Admin. Publ., U.S. Public Health Serv., 1968.

Air Quality Data from the National Air Sampling Networks, 1964–1965, U.S. Public Health Serv., 1966.

Air Quality Data of the National Air Sampling Network, 1962 (out of print).

McHale, J., *The Ecological Context: Energy and Materials, World Design Science Decade, 1965–1975, Document 6,* World Resources Inventory, Southern Ill. Univ., Carbondale, Ill., 1967.

McMullen, T. B., R. B. Faoro, and H. B. Morgan, Profile of pollutant fractions in nonurban suspended particulate matter, *APCA Paper No. 96–165,* presented at Ann. Meeting of Air Pollution Control Ass., New York, June 22–26, 1969.

Spirtus, R., and H. F. Levin, Patterns and trends in levels of suspended particulate matter at 78 NASN sites from 1957 through 1966, *APCA Paper No. 6908,* presented at Ann. Meeting of Air Pollution Control Ass., New York, June 22–26, 1969.

1962–1969 Summary of Monthly Means and Maximums, Continuous Air Monitoring Projects, APTD No. 69–1, National Air Pollution Control Admin. Publ., April 1969.

26

The Chemistry of Smog   Richard D. Cadle

Smog is only one of many types of air pollution, and it is useful to differentiate between smog and these other types. Smog may be loosely defined as the widespread type of air pollution found in the atmospheres of many cities. It is a personal nuisance, often causes considerable financial loss, is injurious to health, and has often caused death. There is usually a multitude of sources, although they may all be of a similar type. For example, the smog may come largely from automobiles, but of course there is a large number of automobiles. Several other classes of air pollution can be defined; for example, industrial air pollution can be defined as a single-source or nearly single-source pollution which is emitted to the air from a factory or other industrial operation. Radioactive fallout, of course, is another type of air pollution and has the potential for destroying a large part of the earth's population.

Just as there are several types of air pollution there are also several types of smog. That over most communities can be considered to be one of two kinds, although most smog consists of both kinds, one predominating. The type which has plagued mankind the longest is usually a combination of coal smoke and fog or even just coal smoke. When fog is present, the individual particles of the smoke act as condensation nuclei on which the fog droplets condense so the fog is extremely dirty. The term "smog" is a contraction of the words "smoke" and "fog." The London Fog of December 1952 was a particularly severe example, and about 3,500 deaths were attributed to its effects. A number of years ago the author observed such a smog in Cleveland that was so intense in the middle of one afternoon that street lights were turned on automatically and shopkeepers turned on their neon signs. The well-known "before and after" photographs demonstrating the effectiveness of smog control in cities, such as Pittsburgh, generally relate to this type of smog.

The other predominant type has been named photochemical smog, although it has little relationship to smoke and fog. It is the type that has plagued many cities in the western United States, particularly Los Angeles, and actually is now worldwide. The name originates from the fact that most of the unpleasant proper-

Prepared for SCEP.

ties of such smog result from the products of chemical reactions induced by sunlight.

There are, however, still other types of smog. Ice fog is an example that is found in polar regions, especially the Arctic, and is a combination of smoke particles and ice crystals. It occurs only when temperatures are extremely low. The smoke particles and the water responsible for the formation of the ice crystals are both produced by various combustion processes that are part of man's operations. For example, a single airplane, it has been stated, taking off from an airport under certain meteorological conditions, will produce sufficient ice fog to close down an airport for hours.

History of Smog Control

An understanding of the history of smog control is helpful to understanding what smog is, where it comes from, and the differences between the different types of smog. Smog and air pollution in general have been with the human race for a long time (at least as long as man has had fire), and the answers as to how to clean up our air have been found to be anything but simple. Perhaps the first serious attempt to control it was the prohibition of coal burning in London in 1273, which was followed many years later by a Royal Proclamation by Edward I in 1306 that prohibited manufacturers from burning coal in their furnaces. The records show that after one man was beheaded enforcement lapsed. Elizabeth I prohibited coal burning during sessions of Parliament. The first air pollution ordinance in the United States became law in 1876 in Saint Louis, but it required merely that factory chimneys be at least twenty feet higher than adjacent buildings and did not restrict in any way what came from them.

Photochemical smog in the Los Angeles area seems to be particularly intense, and most field investigations have been undertaken there. Actually the Los Angeles basin has a long history of smog. About 300 years ago, Spanish explorers described the Los Angeles region as a "land of smoke and fires." The fires and the accumulated smoke were almost certainly the results of activities of Indians. It is also of historical interest that a photograph taken in 1922 from a hill above Hollywood shows an intense haze lying over the basin. Of course, there is no way of knowing whether this haze had a natural origin, such as dust, pollen, and sea salt, or whether it was photochemical smog. But it does show the tendency for materials of this sort to accumulate in the air of that area. Smog as we know it in Los Angeles today first appeared about 1943. At first it was thought to be similar to that experienced in London and other cities having a coal-burning economy, but it was soon observed that its properties differed markedly. Photochemical smog for one thing seems to be cleaner (if any smog can be said to be cleaner), in the sense that it does not contain the very large particles of soot that are so characteristic of smog from a coal-burning economy. It also has at least two other rather unique properties. One is the eye irritation it produces which can actually lead to lachrymation, and the other is a type of plant damage that it produces. Typically, the damaged surfaces of the leaves acquire a metallic sheen that has been described as silvering or as bronzing.

The first clue relating to the origin of smog in Los Angeles and other cities in the western United States was obtained by Nellan and coworkers (1950), and by Bartel and Temple (1952), both groups of the United States Rubber Company, in the late 1940s. They observed that rubber goods in western cities seemed to undergo severe cracking during periods of smog, a type of cracking that laboratory tests suggested was caused only by ozone. They analyzed the air of a number of western cities and found that the air often contained many times the amount of an oxidant, which was probably ozone, found in uncontaminated air near the earth's surface. The next major finding was by Fritz Went who at that time was a professor at the California Institute of Technology. He observed that when ozone was allowed to react with olefins, such as those emitted to the air in automobile exhaust gases, and plants were subjected to the diluted products, damage to the plants occurred which was similar to that produced by Los Angeles smog. Haagen-Smit and his coworkers (1953), also at Cal Tech, reported that a substance was produced which cracked rubber when olefins and nitrogen dioxide in air at concentrations comparable to those found in smog were irradiated with sunlight. Haagen-Smit at first believed that the substances he had produced were organic peroxides, but later studies showed that the substance responsible for the rubber cracking was indeed ozone. A group at Stanford Research Institute demonstrated by field tests with automobiles that automobile exhaust was the main source of olefins in the Los Angeles atmosphere and a major source of the nitrogen oxides (Magill, Hutchison, and Stormes, 1952). All of this was in the early 1950s. These results furnish the basis for practically all of the research and development on photochemical smog which has been done since that time.

Gaseous Components of Smog

Typical concentrations of trace constituents in photochemical smog are shown in Table 26.1. These same gases are found in smogs in the cities having primarily a coal-burning economy, but the relative concentrations tend to be different. For example, there are lower concentrations of the oxides of nitrogen and much higher concentrations of sulfur dioxide than in photochemical smog. The high sulfur dioxide concentrations reflect the high sulfur content of many types of coal. Furthermore, the organic compounds given off by coal burning are more aromatic in character than those produced by the combustion of petroleum, and this has some important implications. For example, smog in primarily coal-burning cities has a much higher concentration of carcinogenic compounds than those in cities whose smog is pri-

Table 26.1 Typical Concentrations of Trace Constituents in Photochemical Smog (parts of constituent per hundred million parts of air by volume, pphm)

Constituent	Concentration (pphm)
Oxides of nitrogen	20
NH_3	2
H_2	50
H_2O	2×10^6
CO	4×10^3
CO_2	4×10^4
O_3	50
CH_4	250
Higher paraffins	25
C_2H_4	50
Higher olefins	25
C_2H_2	25
C_6H_6	10
Aldehydes	60
SO_2	20

Source: Cadle and Allen, 1970.

marily of a photochemical origin, and this reflects the differences in the structure between petroleum products and coal.

The relative concentrations shown in Table 26.1 are becoming increasingly like those in most cities in the United States, and the reason for this is that there is a tendency away from coal burning and toward the combustion of petroleum products throughout the country and for that matter throughout the world. Table 26.2 demonstrates this trend as it occurred during the 1950s and is reflected in the decreasing sale of coal-burning home heating equipment and the increasing sale of equipment for burning gas. Some of the effects of switching from coal to gas in a power plant are shown in Table 26.3. (See also Chapter 25.)

It is well established that the main contaminants in photochemical smog originate from automobile exhaust, although certainly there is a contribution from industrial activity and the incineration of wastes. The exhaust gases consist primarily of nitro-

Table 26.2 Sales of Heating Equipment (units)

	1950	1955	1959
Central heating equipment			
Gas-fired			
Warm air furnaces	599,800	874,400	1,053,400
Conversion burners°	345,300	209,100	156,200
Boilers	79,600	90,100	147,662
Floor and wall furnaces	613,000°	554,800	546,800
Total	1,637,700	1,728,400	1,904,062
Oil-fired			
Furnaces	156,700	371,200	344,000
Conversion burners	608,000	241,000	152,900
Boilers	82,100	196,500	140,200
Floor and wall furnaces	73,400	59,700	40,000
Total	920,200	868,400	677,100
Coal			
Stokers (domestic and commercial class 1 and 2)	19,600	13,300	12,200
Direct heating equipment			
Gas-fired	2,023,300	1,729,100	1,446,300
Oil-fired	1,320,600	634,500	473,900
Coal-fired	888,500	654,200	727,900

° Estimated.
Source: Gas Appliance Manufacturers Association, 1960.

Table 26.3 Coal Emits More Contaminants than most other Fuels (typical 175-Mw. steam power plant)

Fuel	Contaminant (tons per day)		
	Sulfur dioxide	Nitrogen oxides	Particulate matter
Natural gas	<0.5	100	<0.05
California residual fuel oil	450	210	10
Coal	1915	335	3000
Low-sulfur fuel oil	<150	<105	<3

Source: Chemical and Engineering News, and the Los Angeles Air Pollution Control District.

gen oxides, nitrogen, uncombusted and partially combusted hydrocarbons from the fuel, oxides of carbon, and water vapor. Carbon oxides and water vapor are relatively inert to solar radiation at the ground and therefore are not involved in the primary photochemical processes, but they may play a role in secondary reactions.

Particulate Composition

As mentioned earlier, smog of the coal-burning type tends to contain larger particles, especially flakes of soot, than the photochemical smog, and when this type of smog is truly smoke and fog the size distribution is of course that of the droplets, which for the most part are very much larger than the particles that are found either in smoke or in smog of the photochemical variety. Particles in smog of the coal-burning type vary more in nature from city to city and time to time than in the case of photochemical smog. In most of the coal-burning cities various dusts, smokes, and fumes that have little or nothing to do with coal are also emitted. Many of these are industrial. Backyard incineration and burning at city dumps of course may also contribute. Most of the studies of particulate material in the air of highly industrialized cities have been concerned with total particle loading, that is, with mass concentrations. Extensive studies of the concentrations of atmospheric pollutants in urban and nonurban environments have been made by the National Air Sampling Network of the U.S. Public Health Service. This network was established in 1953 and consists of 185 urban and 51 nonurban stations in all 50 states, Puerto Rico, and the District of Columbia. The urban stations are located in central business or commercial places, and nonurban stations are located as remotely as possible. The latter have been established on ocean and lake shores, in the desert, in forests, and in mountain and farmland areas. The locations and sampling times obviously involve a great assortment of smoggy and nonsmoggy conditions. The particulate material is collected by filtering 70,000 to 80,000 cubic feet of air over a twenty-four-hour time through glass fiber filters using so-called high-volume samplers. All of the samples are analyzed for the weight concentration of particles, for the benzene-soluble fraction of the particles, and also for gross radioactivity. Some of the samples are also analyzed for nitrate, sulfate, and a large number of metals. Table 26.4 summarizes the results for the concentrations of suspended particulate material at urban stations.

It is especially interesting to compare Pacific Coast data with the Midwest and Mid-Atlantic data since all three areas are noted for smog, and that on the Pacific Coast is mainly photochemical. Concentrations of suspended particulate material are lower for the Pacific Coast than for the other areas, and this conforms with the fact that the number concentrations of large particles, which may make an overwhelming contribution to mass concentrations, are relatively much lower in photochemical smog than in that of the coal-burning communities. Table 26.5 is even more informative.

Table 26.4 Suspended Particulate Matter—Urban Stations—Region and Grand Totals

Station	Years	No. of Samples	Micrograms per Cubic Meter				
			Min.	Max.	Arith. Avg.	Geo. Mean	Std. Geo. Dev.
New England total	57–58	595	20	326	100	86	1.739
Mid-Atlantic total	57–58	714	23	607	146	125	1.772
Mideast total	57–58	516	27	745	123	103	1.698
Southeast total	57–58	578	15	640	125	104	1.689
Midwest total	57–58	967	11	978	158	139	1.629
Great Plains total	57–58	503	22	722	136	120	1.622
Gulf South total	57–58	516	14	630	118	100	1.687
Rocky Mountain total	57–58	247	15	466	99	84	1.809
Pacific Coast total	57–58	704	11	639	136	109	2.026
Grand total	57–58	5340	11	978	131	111	1.772

Source: Cadle, 1965.

Table 26.5 Particulate Concentrations for Selected Cities for 1958 (Values are arithmetic means)

Station Location	A Sus-pended Particle ($\mu g/m^3$)	B Benzene-Soluble Organic Matter ($\mu g/m^3$)	% of A	C Sulfate ($\mu g/m^3$)	% of A	D Nitrate ($\mu g/m^3$)	% of A
Los Angeles	213	30.4	14.2	16.0	7.5	9.4	4.4
San Francisco	80	10.6	13.3	6.2	7.7	2.6	3.3
San Diego	93	12.2	13.1	7.7	8.3	4.2	4.5
Denver	110	11.0	10	6.1	5.5	2.3	2.1
New York	164	14.3	8.7	23.0	14.0	2.2	1.3
Pittsburgh	167	13.0	7.8	15.1	9.0	2.6	1.6
Cincinnati	143	13.7	9.6	12.2	8.5	2.6	1.8
Louisville	228	18.0	7.9	20.6	9.1	4.9	2.1

Source: Cadle, 1965.

Note that the percentages of benzene-soluble organic matter and nitrate are higher, and those of sulfate are lower for the first four cities, which have primarily photochemical smog, than for the second four. Table 26.6 summarizes the nonurban data, and notes the low total and low benzene-soluble concentrations. The low percentage of benzene-soluble material reflects the fact that a very large percentage of the particulate material in nonurban stations consists of mineral matter or very high molecular weight organic material that is primarily of vegetable origin.

Carcinogenic compounds are found in smog almost every-where but especially so in smog of the coal-burning variety. Poly-nuclear hydrocarbons seem to be especially effective carcinogens in polluted atmospheres, and one of these that is used by many lab-oratories as a tracer for such carcinogens is 3,4-benzpyrene. This compound melts at 180°C and boils at about 500°C at atmo-spheric pressure. Thus it exists as particles, almost always associ-ated with particles of other materials. It is formed by the com-bustion of numerous organic substances, and this is the reason it is found in most or all city smogs. Sawicki et al. (1960) made a study of the concentrations of 3,4-benzpyrene in air in many places. For the most part, the concentrations were found to be tens and hundreds of micrograms per thousand cubic meters of air. The concentrations in Los Angeles and San Francisco were much lower

Table 26.6 Particulate Concentrations for Nine Nonurban Stations for 1958 (Values are arithmetic means)

Station Location	A Suspended Particle ($\mu g/m^3$)	B Benzene-Soluble Organic Matter ($\mu g/m^3$)	% of A	C Sulfate ($\mu g/m^3$)	% of A	D Nitrate ($\mu g/m^3$)	% of A
Acadia National Park, Me.	27	2.5	9.3	5.6	21.	0.9	3.3
Baldwin Co., Ala.	27	2.7	10.	3.5	12.9	0.9	3.3
Bryce Canyon Park	83	2.2	2.7	1.9	2.3	0.4	0.48
Butte Co., Idaho	23	1.4	6.1	2.2	9.6	0.6	2.6
Cook Co., Minn.	44	2.4	5.5	4.0	9.1	0.5	1.1
Florida Keys	36	2.0	5.6	4.7	13.1	1.2	3.3
Huron Co., Mich.	44	1.6	3.6	6.5	14.8	1.5	3.4
Shannon Co., Mo.	37	2.0	5.4	6.5	17.5	1.8	4.8
Wark Co., N. D.	28	1.9	6.8	3.2	11.4	0.9	3.2

Source: Cadle, 1965.

than in most cities, and this was true of western U.S. cities in general. As would be expected, the concentrations in nonurban regions were very much lower than those in urban areas.

Much less is known about the particle size distribution of the particles in smog of the coal-burning variety than in photochemical smog. However, what we do know suggests that unless fog is formed the distributions for the two types of smog are very similar except at the very large end of the scale. There are many types of equations that can be used to describe the size distributions of particles. A type that is often very useful for particles in the atmosphere has the form

$$\frac{dN}{dr} = ar^b \tag{26.1}$$

where dN is the number of particles per unit volume in the increment of size range dr; a and b are constants. The function is established by the values of the constants a and b. For example, Steffens and Rubin (1949) obtained a value of -4.5 for b for particles in Los Angeles smog. Junge (1963), plotting $dN/d(\log r)$ against r for particles in the natural atmosphere, obtained a slope of about -3 on a logarithmic scale over much of the size range. Since

$$dN/d(\log r) \equiv r\, dN/dr,$$

this slope corresponds to a value of about -4 for b, in fair agreement with the result that Steffens and Rubin got for photochemical smog.

The type of plot used by Junge is especially useful for particles in the atmosphere (see Figures 26.1, 26.2). A very wide range of sizes and concentrations can be represented, and it is relatively easy to derive the concentrations of particles in any size range from such a plot. Thus, if we define

$$n(r) = dN/d(\log r) \text{ cm}^{-3} \tag{26.2}$$

the number of particles ΔN per cm^3 between the limits of the interval $\Delta(\log r)$ is obtained from the curves by

$$\Delta N = n(r)\, \Delta(\log r) \tag{26.3}$$

The corresponding log radius-surface(s) and log radius-volume (v) distributions are then obtained by

$$s(r) = dS/d(\log r) = 4\pi r^2\, dN/d(\log r) \tag{26.4}$$

and

$$v(r) = dV/d(\log r) = (4/3)\pi r^3\, dN/d(\log r) \tag{26.5}$$

If ρ is the density of the particles,

$$m(r) = \rho v(r) \tag{26.6}$$

This graphical method of representing the size distribution of atmospheric aerosol particles has achieved wide acceptance. The relationship is almost the same regardless of where the results are obtained. The results of Figure 26.3 show this for light scattering calculated from the size distributions while those for Figure 26.4 were based on light-scattering measurements using an integrating nephelometer. The similarity of curves, such as those shown in

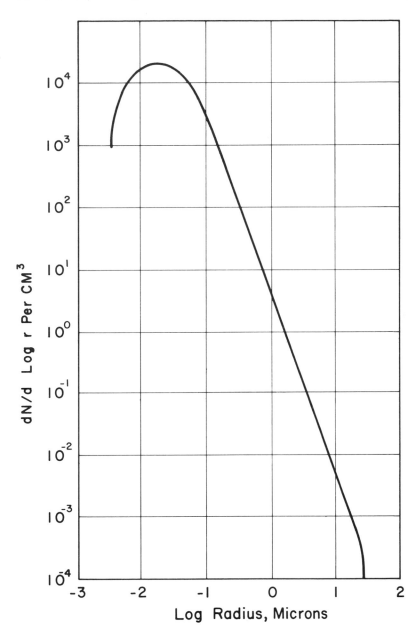

Figure 26.1 Particle size distribution in continental air

Source: U.S. Air Force, 1970.

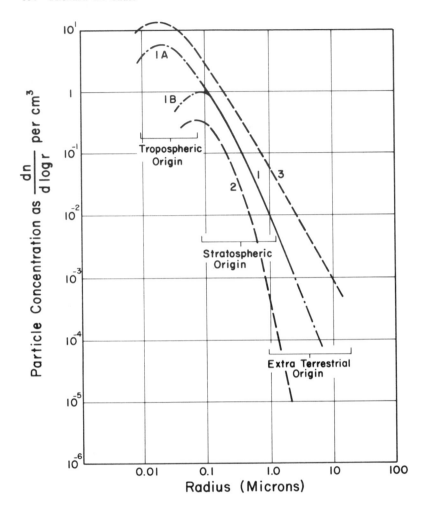

Figure 26.2 Average size distribution for stratospheric aerosols
Note: Curve 1A is for the lower stratosphere and 1B for altitudes above 20 km.
Curves 2 and 3 are estimated confidence limits.

Source: Junge, Chagron, and Manson, 1961.

Figures 26.1, 26.2, and of similar curves for polluted atmospheres
and the results obtained by Charlson, Ahlquist, and Horvath
(1968) demonstrate that a similar size distribution exists for particles in the atmosphere in both smoggy and nonsmoggy environments—at least away from major sources of pollution.

A number of theoretical attempts have been made to quantitatively explain this similarity. They are generally based on the
idea that most of the particles emitted into the atmosphere are at

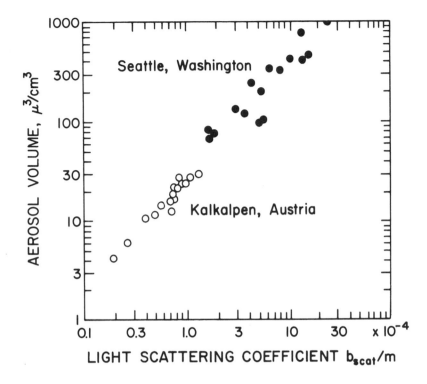

Figure 26.3 The dependence of scattering coefficient (m⁻¹) on volume of aerosol particles (μ^3/cm³) calculated from measured size distributions
Note: The solid circles are based on Seattle, Washington, data and the open circles on Kalkalpen, Austria, data.

Source: National Air Pollution Control Administration (NAPCA) (1969).

the small end of the size distribution scale. These coagulate to form larger particles that eventually settle out. This overall dynamic process produces the observed size distribution. Note, however, as mentioned before, that this type of distribution does not occur when the smog contains fog.

The visibility decrease which accompanies smog is due almost entirely to the particles which it carries, and these in turn decrease the visibility in two ways. Both mechanisms decrease the contrast between any object which is observed and its surroundings. One mechanism involves the attenuation of light from the source, and the other involves the decreasing contrast resulting from the scattered radiation from sunlight in the direction of the observer. Numerous definitions of visibility have been proposed. A particularly useful one is "visual range," which is defined as the

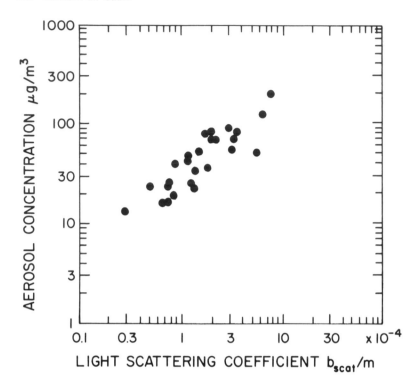

Figure 26.4 Measured dependence of mass of aerosol particles per volume (μg/m³) on the light-scattering coefficient (m⁻¹) in Seattle, November–December 1966.

Source: NAPCA (1969).

distance at which the contrast between a black object and the adjacent sky becomes equal to ε, which is the least fractional difference in intensity the eye can detect. The visual range is often determined by means of instruments employing an assumed or average value for ε (often 0.02). The Weather Bureau defines the "prevailing visibility" as "the greatest visibility which is attained or surpassed around at least half of the horizon circle, but not necessarily in continuous sectors."

The color of smog is largely due to light scattering by the particles which it contains. Thus photochemical smog, in which most of the particles are small relative to the wavelength of light, is blue by scattered light but appears brown when one looks through it.

The composition of the particles in photochemical smog has

been studied extensively. In a light smog, minerals and other in-organic material may make up over 50 percent of the material. A number of studies have been made of the organic fraction of smog, usually after collecting the particles on a filter. For example, the collected material has been extracted with a nonpolar organic solvent, and it has been shown that the dissolved organic material has an infrared spectrum very similar to that which is produced by reacting ozone with certain olefins. The individual particles in photochemical smog have been studied extensively after collecting them by impaction techniques. Many of these have been found to be dark brown, gummy, water-insoluble droplets consisting largely of organic material. Some were droplets that evaporated slowly over a period of several days while a tough film formed over the surface. Some of these were largely organic, while others were largely aqueous. Similar droplets were collected from automobile exhaust gases and from smoke from burning wood. Probably such droplets are generally produced by the combustion of organic ma-terial. When the droplets are largely aqueous, apparently they are prevented from evaporating rapidly by the organic material that they contain. This is probably largely in solution and is adsorbed at the surface by Gibbs adsorption, at least when the droplets are first formed. This organic material at the droplet surface is slowly oxidized to form the tough film which has often been ob-served and further prevents evaporation of the droplets. Inorganic hexagonal particles (plates) are occasionally collected by impaction on slides. The refractive index of such particles is slightly greater than 1.43, so these are possibly very complex compounds of sodium and various halides including fluorine. Needle-shaped particles are often collected from all kinds of smog, photochemical as well as that of the coal-burning variety. These almost always turned out to be gypsum ($CaSO_4 \cdot 2H_2O$).

The exhaust gases from automobiles, and therefore smog, contain particles having high concentrations of lead. The lead of course is the result of the presence of tetraethyl lead in the gas-oline. The gasoline also contains organic bromine compounds that are designed to serve as scavengers for the lead. The theory is that the lead in the gasoline will combine with the bromine of these compounds to form nonvolatile lead bromide which will condense out on the walls of the exhaust system of the automobile and ultimately be swept off into the atmosphere as large particles. Recent studies, however, of both the nature of the lead particles in smog and the nature of the lead particles in the exhaust gases indicate that a large percentage of the particles escape into the atmosphere in very fine form. In a form, in fact, such that they are drawn well down into the respiratory system of human beings where they may cause damage over long periods of time. Thus, while it was felt for many years that lead from automobile exhaust gases was not a problem, recent concern over the possibility of lead poisoning from this source has developed. Studies at the Ethyl Corporation have showed the presence of compounds in exhaust gases such as $PbCl \cdot Br$ and $PbO \cdot PbCl \cdot Br \cdot H_2O$ and smaller amounts of lead ammonium halide complexes. Extremely large particles, that is, those with diameters greater than 0.5 mm, were high in iron oxides and contained lead principally as the sulfate and as $PbO \cdot PbSO_4$.

Bulk sample analyses of aerosol particles collected from smog, undertaken by Robinson and his coworkers (1963, 1964), showed that only about 10 percent of the total lead in the air is soluble in water at 40°C. Thus the more soluble lead compounds such as $PbCl_2$, $PbBr_2$ and $PbSO_4$ probably account for only a small part of the atmospheric lead aerosol. These workers found that a typical size distribution for the atmospheric lead aerosol had a mass median diameter of 0.2μ, with 25 percent of the mass being accounted for by particles less than 0.1μ and another 25 percent for particles greater than 0.5μ.

The concentrations of particles having diameters exceeding about 0.2μ in dense photochemical smog range from about 3,000 to 10,000 particles per cm^3.

Chemical Reactions

As stated earlier, many of the unpleasant properties of photochemical smog result from the products of a series of chemical reactions that are triggered by sunlight. The triggering reaction, or at least the most important triggering reaction, seems to be the photochemical decomposition of nitrogen dioxide into nitric oxide and atomic oxygen. The atomic oxygen can react with molecular oxygen to form ozone, but it also reacts with nitric oxide to form nitrogen dioxide and oxygen. So if these were the only reactions occurring, the amount of ozone that would be produced is very much smaller than the amount of ozone that is often detected in photochemical smog. But other reactions are occuring. The atomic oxygen reacts with various hydrocarbons, such as olefins, and in doing so forms free radicals, and these free radicals undergo a very large sequence of reactions some of which involve the oxidation of nitric oxide to nitrogen dioxide; others may produce ozone, and many of them produce very unpleasant organic compounds such as the aldehydes, formaldehyde, and acrolein, or the nitrogen-containing compound **PAN**. Formaldehyde, acrolein, and **PAN** are probably the compounds that are primarily responsible for the eye irritation that is produced by smog, but no single compound seems to be largely responsible. These reactions are summarized as follows:

$$NO_2 + h\nu \rightarrow NO + O \tag{26.7}$$

$$O + O_2 + M \rightarrow O_3 + M \tag{26.8}$$

$$O_3 + NO \rightarrow NO_2 + O_2 \tag{26.9}$$

$$O + Olefins \rightarrow R + R'O$$

$$\text{or} \quad \overset{\displaystyle O}{\underset{\displaystyle R\!-\!\!-\!\!-\!\!-\!R'}{\bigwedge}} \tag{26.10}$$

$$O_3 + Olefins \rightarrow Products \tag{26.11}$$

$$R + O_2 \rightarrow RO_2$$

$$RO_2 + O_2 \rightarrow RO + O_3 \tag{26.12}$$

R is an organic radical.

Reactions 26.10 and 26.11 are just the first steps in a long series of reactions producing the organic acids, aldehydes, ketones, and organic nitrogen-containing compounds. Reaction 26.12 may be the reaction primarily responsible for the formation of ozone. Although the foregoing sequence of reactions is probably the most important sequence, other very important reactions that are triggered by sunlight are undoubtedly also important. For example, it has been found that aldehydes can take the place of nitrogen dioxide in studies of laboratory-prepared synthetic smogs. It has been suggested that the aldehydes absorb sunlight; they then are converted to an electronically excited singlet state, surface crossing occurs, and they become triplet electronically excited species. These triplet species upon collision with oxygen molecules excite the latter to a singlet state; the singlet oxygen upon collision with molecules of other organic substances undergoes reactions that are in some ways similar to Reactions 26.10 and 26.11. There is considerable question though about the importance of this sequence, since very recent studies have indicated that the excited molecular oxygen may not react sufficiently rapidly with organic species in smog for such reactions to be important in producing unpleasant products.

Numerous investigations have shown that aerosol is formed when many, if not most, six-carbon and larger straight chain, branched chain, and cyclic olefins as well as a number of aromatic hydrocarbons along with NO_2 in air, in the ppm concentration range, are irradiated. If automobile exhaust–air mixtures are irradiated, the irradiation increases the concentration of particles in the mixtures; but it is important to realize that automobiles would cause some smog even in the absence of irradiation. The exhaust gases themselves contain high concentrations of particles, for example, lead compounds, as already mentioned, and also they contain various aldehydes, including formaldehyde, that in their own right are obnoxious for various reasons.

If sulfur dioxide is added to a synthetic smog mixture and the mixture is irradiated, the sulfur dioxide is very rapidly oxidized to sulfur trioxide which in turn reacts with any water vapor that may be present to form sulfuric acid droplets. This oxidation is much more rapid than would just be produced by the irradiation of sulfur dioxide mixed into air and may very well result from the reaction of the peroxy radicals, such as those shown in equation (26.12), with the sulfur dioxide. If ammonia gas is present, as is often the case in polluted atmospheres, the sulfuric acid droplets will react with the ammonia to form ammonium sulfate. Ammonium sulfate is often found to be an important constituent both of smog and of the aerosol in relatively uncontaminated nonurban atmospheres.

Single-Source Pollution

Almost any industrial activity is a potential or actual producer of air pollution. The pollutants may be either solids or gases or even noise. If the pollutants bother only workers in a plant, the problem is classified as industrial hygiene. If the pollutants are outside the plant, then they are termed air pollution. Processes carried out in many industries make them particularly subject to air pollution problems. There is no point in trying to discuss or list all of such industries, but it may be appropriate to mention a few.

A major type of industrial air pollutant consists of fluorine and various gaseous and solid fluorine compounds. These may include hydrogen fluoride, aluminum fluoride, sodium fluoride, silicon tetrafluoride, and hydrofluosilicic acid. The fluorides do considerable damage to plants and grazing animals. Damage to agricultural plants has been a problem in many states including California, Florida, Idaho, Montana, New Jersey, Oregon, Tennessee, Utah, and Washington. Numerous industrial operations produce fluorides. These include the production of ceramics, certain fertilizers, and metals. Aluminum production is a well-known example, the fluoride-containing fume coming from the electrolytic baths. The open-hearth method of steelmaking is another. Fluorides may come from the ore itself or from the fluorspar which may be added to increase the fluidity of the slags.

The public utility industry is a great producer of this type of air pollution, that is, single-source pollution. The problems arise almost entirely from combustion products and from the handling and storage of fuels. The public utility most concerned is the electricity-generating industry, most of which uses steam for power. Steam is produced in boilers by burning fuel, and the steam is used to drive turbines which in turn drive generators. The fuels generally are chosen on the basis of availability and cost, and are practically always coal, natural gas, or fuel oil, although some plants are now being powered by nuclear energy. Both gaseous and particulate pollutants are produced. The former includes sulfur dioxide, nitric oxide, carbon monoxide, and carbon dioxide; the latter, sulfuric acid droplets, fly ash, smoke (including carbon and gums), and at times, small cinders. Since an excess of air is usually used, there is seldom much combustible smoke. Control of the gaseous pollutants is generally achieved by controlling the type of fuel and control of the particles by the use of mechanical collectors such as cyclones, electrostatic precipitators, or both. The disposal of the collected fly ash can be a serious problem. A utility system may produce as much as a million tons per year. Fortunately there are a number of industrial uses including ready-mixed concrete, building blocks, bricks, and filtering media. The most serious problems occur when coal is used as the fuel. When gas is used there is little pollution, but both fuel oils and coal produce ash and gaseous products.

Trends

The results of the Public Health Service Network, mentioned earlier, indicate that there has been no obvious trend in the concentrations of particulate pollutants in the atmosphere over a number of years. (See Chapter 25.) Other studies indicate the same thing. For example, collections of rainfall made at weather stations throughout the United States and over a large number of years have been analyzed at the National Center for Atmospheric Research. The results suggest that although there may be trends in the concentrations of various pollutants that were determined in the rainfall samples from a specific city, on a nationwide basis very little if any trend can be observed. Other aspects of this study, however, were rather interesting in that it demonstrated that lead concentrations were highest where most gasoline was consumed and validated the concept that most of the lead in the atmosphere, at least in a country such as the United States, does result from gasoline consumption. Similarly the sulfate concentrations were highest where the most high-sulfur content fuel was burned.

Dust, which may be defined as the particles larger than 10 μ in diameter, exists in the atmosphere in very low number concentrations but does contribute to a considerable extent to the mass concentrations. Since such large particles have appreciable settling velocities and impact readily at low velocities, they are usually determined gravimetrically following collection by deposition in a so-called dustfall jar. Only relative significance should be attached to the resulting data, and only then if conditions are carefully standardized. Collections in dustfall jars have been made in a number of cities for many years. The results do seem to indicate a trend downward in concentration, perhaps because of less soft coal burning and better control of industrial effluents. Thus this may be an effect of the changes from coal-burning smog to photochemical smog.

Worldwide Pollution

Not only are we polluting the atmospheres of our cities, but we are polluting also the atmosphere on a worldwide basis. There is increasing interest as to the extent of this: for example the change in turbidity of the atmosphere may change the climate and, if some of the particles serve as condensation or freezing nuclei, these also may have an effect on the climate. There are some indications that the turbidity resulting from airborne particles is increasing, but we cannot be sure at this time that this actually results from man's activities. The particles responsible are of course very much smaller than those collected by dustfall measurements.

References

Bartel, A. W., and Temple, J. W., 1952. Ozone in Los Angeles and surrounding areas, *Industrial and Engineering Chemistry, 44:* 857.

Cadle, R. D., and Allen, E. R., 1970. Atmospheric photochemistry, *Science, 167:* 243–249. Table 26.1 Copyright, 1970, by the American Association for the Advancement of Science.

Gas Appliance Manufacturers Association, Inc. (GAMA), 1960. *GAMA Statistical Highlights, Ten Year Summary, 1950–1959* (New York: GAMA). Table 26.2 reprinted with permission of GAMA.

Haagen-Smit, A. J., Bradley, C. E., and Fox, M. M., 1953. Ozone formation in photochemical oxidation of organic substances, *Industrial and Engineering Chemistry, 45:* 2086.

Junge, C. E., 1963. *Air Chemistry and Radioactivity* (New York: Academic Press).

Junge, C. E., Chagnon, C. W., Manson, K. E., 1961. *Journal of Meteorology, 18:* 81.

Magill, A. H., Hutchison, D. H., and Stormes, J. M., 1952. *Proceedings of the Second National Air Pollution Symposium* (Menlo Park, California: Stanford Research Institute).

National Air Pollution Control Administration (NAPCA), 1969. *Air Quality Criteria for Particulate Matter,* publication No. AP-49 (Washington, D.C.: NAPCA).

Nellan, A. H., Dunlap, W. B., Glasser, C. J., and Landes, R. A., 1950. Effect of Atmospheric O_3 on tires during storage, *Rubber Age, 66:* 659.

Robinson, E., Ludwig, F. L., DeVries, J. E., and Hopkins, T. E., 1963. *Variations of Atmospheric Lead Concentrations and Type with Particle Size,* Stanford Research Institute Project No. PA-4211 (Menlo Park, California: Stanford Research Institute).

Robinson, E., and Ludwig, F. L., 1964. *Size Distributions of Atmospheric Lead Aerosols,* Stanford Research Institute Project No. PA-4211 (Menlo Park, California: Stanford Research Institute).

Sawicki, E., Elbert, W. C., Hauser, T. R., Fox, F. T., and Stanley, T. W., 1960. Benzo(a)pyrene content of the air of American communities, *American Industrial Hygiene Association Journal, 21:* 443.

Steffens, C., and Rubin, S., 1949. *Proceedings of the First National Air Pollution Symposium* (Menlo Park, California: Stanford Research Institute).

U.S. Air Force, 1960. *Handbook of Geophysics* (New York: The Macmillan Co.).

27

Turbidity Measurements   G. D. Robinson

Defining and Measuring Turbidity

The turbidity coefficient is a measure of the extinction of solar radiation in the atmosphere. Several definitions have been used. Volz constructed a simple photometer operating at wavelength 500 nm and defined turbidity B, using the following equation:

$$I/I_0 = 10^{-(\alpha_a + \alpha_s + B)m}$$

where I is the observed intensity of solar radiation at $\lambda = 500$ nm, I_0 is the intensity of solar radiation of this wavelength outside the atmosphere, α_a is the known absorption coefficient of atmospheric gases (at this wavelength ozone is the only significant gaseous absorber), α_s is the scattering coefficient of atmospheric gases at $\lambda = 500$ nm (that is, Rayleigh scattering), m is the optical air mass (and approximately equal to cosec h, where h is the solar elevation), and B is the turbidity coefficient. Volz's coefficient B can readily be related to the turbidity indices used by Schüepp, Angström, and Linke. Turbidity measurements are made by direct observation of the sun, or indirectly by observations in cloudless conditions. However, if there is present undetected cloud, such as very tenuous cirrus, it will contribute to the turbidity coefficient. Unrecorded cirrus not detectable as such from the ground is perhaps a common phenomenon.

Measurement by standardized methods of a turbidity index such as with Volz photometers is probably the simplest sound method of maintaining a continuous record in relative terms of the particulate load of the atmosphere. The standardized network of measurements being built up by NAPCA has been in existence long enough to establish geographical patterns of total particulate loading but not long enough to establish secular trends (Flowers, McCormick, and Kurfis, 1969).

The following are several other methods of determining the extinction of radiation by particles.

Solar Radiation Records

Some meteorological services maintain a network of solar radiation measuring stations. The records are of varying quality and must be used with caution. It is, however, general experience that the measured energy reaching the ground is less than that calcu-

Prepared during SCEP.

lated from the transmission properties of a clean atmosphere. If, as is common, there are separate records of the total radiation (sun and sky) and of diffuse (sky) radiation alone, it is possible to estimate separately the scattering and absorption due to atmospheric particles at times when there is no cloud. At a few stations in the world (Mauna Loa is one) a record of total and diffuse radiation in broad spectral bands defined by filters has begun. Diffuse radiation in the solar infrared is a sensitive indicator of particulate loading, because of the very low molecular scattering at these wavelengths (Lettau and Lettau, 1968; Robinson, 1962).

Aircraft Measurements

The *absorption* of solar radiation in the atmosphere between two levels is established by measuring the upward and downward flux of solar radiation at the two levels, which can be done by upward- and downward-facing instruments on aircraft or balloons. Absorption not only gives a relative measure of particulate load but is a direct indication of the heating rate required by dynamical meteorologists. There have been relatively few investigations of this type (Fritz, 1948; Robinson, 1966).

Satellite Methods

An extension of the aircraft method combines satellite measurement of the incident- and earth-reflected solar radiation with surface measurements around the subsatellite point. Current satellite instrumentation is not well adapted to this method (Hanson, Vonder Haar, and Suomi, 1967).

Searchlight and Laser Methods

The backscattering from a searchlight or laser beam is measured. At all heights investigated, it is found to be in excess of the computed Rayleigh scattering. The method provides a height profile of the distribution of particles. If it is known that the nature (size distribution and composition) of the particulate population does not change appreciably, this method could provide a basis for monitoring secular trends which would be especially useful in the lower stratosphere.

Localized Methods

There have been a few investigations of the transmission properties of restricted regions of the atmosphere, using path lengths of order from tens of centimeters to hundreds of meters. Total extinction of a light beam is determined, and the scattering from the beam is integrated, providing a measure of both absorption and scattering coefficients (Waldram, 1945).

Finally it is possible to determine the local radiative characteristics by detailed examination of the particulates—size, distribution, total number, composition, and refractive index. This is hardly practicable on a wide scale, but essential on a limited scale to give insight in interpreting the results of measurements on the bulk loading of the atmosphere.

The Nature of Turbidity Records

The Volz photometer network has not been operating long enough to establish trends. There are, by accident, two stations where reliable turbidity measurements have been made over a span of more than fifty years. (Both stations were involved in the problem of creating and preserving radiation standards.) The stations are Davos, Switzerland, and Washington, D.C. (Smithsonian Institution). Both stations show a secular increase of turbidity. At Washington, D.C., much of the increase is undoubtedly due to local urbanization; this is not so at Davos (McCormick and Ludwig, 1967).

Other solar radiation records are of very variable quality. Many have not been properly standardized. (They include the stations of the U.S. Weather Bureau.) There are several well-conducted stations with twenty to twenty-five years of records, but most of them are in urban locations (Brussels, London, Vienna, Leningrad) and reflect the trends of local urban pollution; for example, the London record clearly documents the implementation of a Clean Air Act. Some of the rural African stations used in an early study (Robinson, 1962) have disappeared or deteriorated in quality. Other South African, Australian, and Japanese stations might repay study, and there is good material within the Soviet Union. Mauna Loa is an ideal base-line station, but the quality of the past record has been questioned.

The few aircraft, ground/satellite, and nephelometer type measurements that have been reported clearly show the potential importance of particulate absorption but are really too few to allow any generalized conclusions. There seems little doubt that at some levels in the atmosphere this absorption is often comparable with that due to water vapor, even in air which would normally be described as unpolluted. It could therefore be of importance in the atmospheric radiative balance, in questions of atmospheric stability, and in atmospheric models.

Particulate loading of the atmosphere has an effect on the earth's albedo—that is, on the available solar energy. If particles scatter radiation without absorption they can only increase the albedo and decrease available energy. However the evidence is that they normally absorb radiation to some extent, and that in some cases they might decrease the albedo. Some computations for a typical 1,000-foot layer of heavy industrial pollution show that for an underlying surface albedo less than 0.2 available solar energy is decreased, for higher surface albedos solar energy is increased. These computations suggest much detailed investigation will be needed to resolve even the broad overall effects of a change of atmospheric particulate loading. In particular it may prove advisable to supplement monitoring of the turbidity coefficient with extraterrestrial observations of the global albedo (Atwater, 1970).

References

Atwater, M. A., 1970. Planetary albedo change due to aerosols, *Science, 170* (3953): 64–66.

Flowers, E. C., McCormick, R. A., and Kurfis, K. R., 1969. Atmospheric turbidity over the U.S.—1961–1966, *Journal of Applied Meteorology, 8:* 955.

Fritz, S., 1948. The albedo of the ground and atmosphere, *Bulletin of the American Meteorological Society, 29:* 303–312.

Hanson, K. J., Vonder Haar, T., and Suomi, V., 1967. Reflection of sunlight to space and absorption by the Earth and atmosphere over the U.S. during Spring 1962, *Monthly Weather Review, 95:* 354.

Lettau, H., and Lettau, K., 1968. Shortwave radiation climatonomy, *Tellus, 21:* 1.

McCormick, R. A., and Ludwig, J. H., 1967. Climate modification by atmospheric aerosols, *Science, 156:* 1358.

Robinson, G. D., 1962. Absorption of solar radiation by atmospheric aerosol, as revealed by measurements at the ground, *Archiv für Meteorologie, Geophysik and Bioclimatologie, B.12:* 19.

Robinson, G. D., 1966. Some determination of atmospheric absorption by measurement of solar radiation from aircraft and at the surface, *Quarterly Journal of the Royal Meteorological Society, 92:* 263.

Waldram, J., 1945. Measurement of the photometric properties of the upper atmosphere, *Quarterly Journal of the Royal Meteorological Society, 71:* 319.

28

The Measurement of Aerosol Particles on a Global Scale

Vincent J. Schaefer

With the increasing industrialization of most parts of the earth the pollution of the air above and downwind of such areas is increasing at an alarming rate (Schaefer, 1969). Since many particulate pollutants are not appreciably affected by gravity (Schaefer, 1970), they move with the air for long distances. Thus the regions where pollution-free air may be found are steadily decreasing. The last remaining extensive areas having clean air are the mid-oceans and the polar and subpolar regions. Some smaller areas are also fairly clean, but these too are becoming contaminated. We have found that the base level of particles in such areas in terms of submicron particles (Aitken nuclei) show concentrations ranging from 200 to 500 particles per cubic centimeter. This we now believe is the global background.

Since the natural air-cleaning mechanisms caused by precipitation in the form of snow, rain, and hail have a finite limit, it is of great importance that we monitor the air quality in some of the remaining clean areas to determine if our attempts at air pollution control can reverse the trend now moving toward air pollution on a global scale.

Our studies have indicated that a measurement of Aitken nuclei provides an excellent index of the degree of air contamination in an area. Based on more than ten years' experience of making such measurements, we have developed the chart shown in Table 28.1.

Hogan has found (1970) that measurements of the concentration of such particles at a specific location permits him to classify that region for its air pollution characteristics. Figure 28.1

Table 28.1 Pollution Range (Indicated by Condensation Nuclei, CN/ccm)

Type of Environment	Concentration	Degree
Oceanic and polar air	$<1,000$/cc	Clean
Country air	$1,000-5,000$/cc	Light
Suburban air	$5,000-50,000$/cc	Medium
Urban and industrial air	$>50,000$/cc	Heavy

Prepared for SCEP.

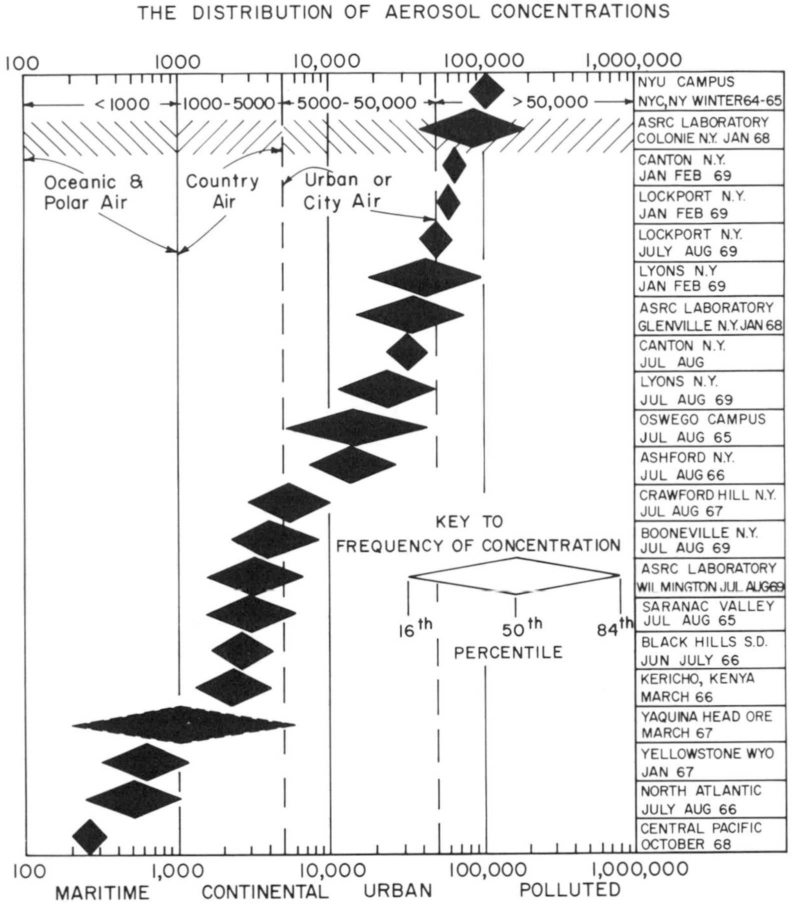

THE DISTRIBUTION OF AEROSOL CONCENTRATIONS

Figure 28.1 Observed concentrations, number of particles/cm³

illustrates his findings. Figure 28.2 includes data obtained in my studies compared to some of Hogan's. We have measured many other features related to airborne particles but find that this simple measurement (plus another which will be described later) is adequate to provide an excellent approximation of the air quality of a specific region.

Therefore we propose that without further delay short-term (two to four weeks) measurements of Aitken nuclei be made at key locations throughout the world. Such measurements are easily made using instruments such as the highly portable Gardner

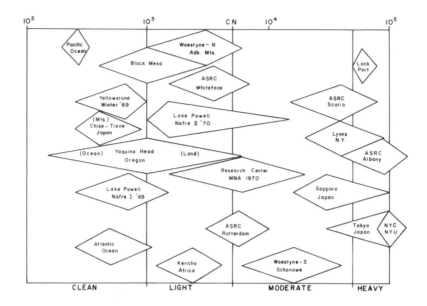

Figure 28.2 Observed concentrations, number of particles/cm³

Small Particle Counter or the new Rich Counter. The simplicity and rugged features of these counters make it possible to train an observer in a few minutes. A useful measurement takes less than a minute.

Fortunately, our records already contain measurements made in such places as Japan, Kenya, the middle of the Pacific and the Atlantic, several regions in Europe, and many places in the United States. Colleagues are measuring Aitken nuclei in other parts of the world using the same type of instrument used in our studies. These counters follow the basic design of the Pollak Counter, which has become the standard reference instrument for small-particle measurement throughout much of the world.

During the past four years we have conducted ten transcontinental low-level flights from New York to California, measuring air pollution on a continental scale. During these flights we measured Aitken, cloud, lead, and natural ice nuclei. A total of more than 2,000 in-flight spot checks were made, and during our last few flights we obtained nearly continuous measurements from coast to coast. The evidence obtained from these measurements, from our other bench-mark stations and other experiences have convinced us it is of critical importance that atmospheric particles be measured on a global scale without further delay.

In 1948 I started measurements of ice nuclei at the Mount Washington Observatory. These measurements made by Mount Washington observers at three-hour intervals for more than twenty years now represent the most important reference points for such studies in existence. Measurements at Mount Washington continue, with an effort now under way to tie the earlier observational procedures to an automatic NCAR counter.

For the past four years we have worked with cadets of our Maritime College and with their cooperation have obtained routine Aitken nuclei measurements during round-trip training cruises from New York to the Mediterranean and the Scandinavian countries (Hogan et al., 1967). After graduation a number of them have continued making measurements for us on a voluntary basis while serving as officers in the Merchant Marine. At the present time we also have one of our SUNYA graduate students in the western Aleutians making nuclei measurements for us with the cooperation of the Air Force.

We propose establishing a global observation program that could be operational in less than a year. It would be relatively inexpensive and would quickly provide information and experience of the kind which would provide a guideline for the establishment of a more comprehensive network if deemed desirable.

The basic pattern we propose would consist of a band of stations extending around the Northern Hemisphere within 42°–45° of latitude. We have identified at least ten stations in this belt whose personnel have indicated a willingness to cooperate with us. In addition, we have seven more stations scattered throughout the United States related directly to our Research Center which would be utilized. Twenty-one Light and Life Saving Stations of the U.S. Coast Guard located along the Great Lakes, the ocean coasts, and at sea are now under active consideration in cooperation with Coast Guard officials. These would tend to form several north-south lines. At least ten more observation spots would be located on moving vessels including the three we have already established.

Supplementing measurements of Aitken nuclei, we propose the immediate deployment of a global network of sedimentation foils to be exposed for month-long periods. By exposing a carefully prepared strip of aluminum foil in a specially designed holder, the sedimentation and diffusion of airborne particles at a specific location can be accurately determined. This is a carefully tested method developed many years ago in Europe. However, with the techniques now available for weighing, microanalysis, electron probe evaluation, and micrography, this method for air pollution evaluation is of great value.

The slides are coated with a very sticky but highly stable silicone adhesive that remains effective for more than sixty days. Using a microbalance, the increase in weight can even be established for a single day in an area having light to moderate air pollution. The holder is designed in such a way that it is never contaminated by birds and under normal circumstances is mounted on a wooden or metal stake. During a test in New York State all activities were carried out by concerned citizens. In addition to measuring the fallout on all of the slides, about 15 percent of them were measured for lead, sulfur, and benzopyrene. At the present time, slides are exposed in the Grand Canyon, on Lake Powell, and at other locations in northern Arizona. A global distribution of a thousand such slides could be made almost immediately.

Acknowledgments

This general plan of action is basically that proposed by Mr. Austin Hogan and Dr. Volker Mohnen of the Atmospheric Sciences Research Center staff.

References

Hogan, A. W., 1970. Report on aeorosol climatology, prepared for New York State Department of Health, Division of Air Resources, May 1970.

Hogan, A. W., Bishop, J. M., Aymer, A. L., Harlow, B. W., Klepper, J. C., and Lupo, G., 1967. Aitken nuclei observations over the North Atlantic Ocean, *Journal of Applied Meteorology, 6:* 726–727.

Schaefer, V. J., 1969. Inadvertent modification of the atmosphere by air pollution, *Bulletin of the American Meteorological Society, 50:* 199.

Schaefer, V. J., 1970. The effect exerted on the atmosphere by very small particulate matter and gases, testimony to U.S. Senate Committee, Washington, D.C., March 16, 1970.

29

Instrumentation for   A. J. Drummond
the Continuous Monitoring
of Solar Irradiance
in Ground-Based Studies

This chapter describes the solar radiation data acquisition system recently acquired by the National Air Pollution Control Administration (NAPCA) for installation at the Mauna Loa Observatory of the National Oceanographic and Atmospheric Administration (NOAA) in Hawaii. The solar radiation spectral sensors and associated readout have been designed to provide continuous monitoring of solar irradiance from the ground. These measurements are very important in meteorological investigations and for monitoring, at the earth's surface, the forward scattering by aerosol particles. Measurements made with this system can be coupled with selected measurements made in free air to provide accurate information on the parallel absorption properties of aerosols.

A block diagram of the data acquisition system is given in Figure 29.1. The instrumentation essentially consists of (1) a sun-tracking multichannel radiometer (pyrheliometer) for measuring

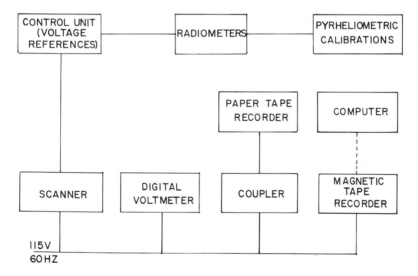

Figure 29.1 Block diagram of the NAPCA solar radiation data acquisition system

Prepared for this volume.

direct solar spectral radiation at normal incidence, (2) a series of precision radiometers or pyranometers for sampling the global (sun + sky on a horizontal surface) radiation in certain similar discrete wavelength intervals, (3) multichannel low-level digital data acquisition with both paper printing and magnetic tape recording for use with the sensors and permitting computer handling of the data, and (4) sensor and readout calibration equipment. The complete system is capable of extension to additional information input without recourse to redesign or new development requirements. All components have been fully developed, are available commercially, and have operated satisfactorily both under test conditions and routinely.

Details of the system components are as follows.

Multichannel Radiometer (Pyrheliometer)

This is a robust radiometer development well suited for field operation in remote locations such as Mauna Loa Observatory where climatic conditions may be adverse at times.

The radiometer has thirteen independent radiation-sensing channels and is constructed of aluminum, with an arrangement of copper coiled tubing around the detector enclosures; these (water) cooling coils are in good thermal contact with the radiometer body (and thus the heat sink of the thermopiles). The purpose of the water cooling is to maintain a uniform stable temperature for thermal zero reference at all operational times, especially for the low signal channels. This zero is easily checked with the aid of the built-in remotely operated (solenoid) shutter that can temporarily shield all channels simultaneously from the solar beam. The radiometer is installed on an automatic equatorial mount (solar tracker).

The series of thirteen thermopile detectors are of the wire-wound-plated type, with about fifty thermojunctions and a receiver of thin Mylar whose surface is suitably blackened to be nonwavelength selective over the solar spectral range, namely, 200 to 4,000 nm; it is apertured so that the (circular) field of view (whole angle) is approximately $10°$. The general thermopile characteristics include: sensitivity 0.1 mV/mW cm^{-2}; response time ($1/e$ signal) 0.5 sec; temperature dependence of sensitivity ± 0.5

percent over a range of $-10°$ to $+30°C$ (but in water-cooled operation this decreases to ±0.2 percent); impedance 800 ohm.

The standard arrangement of the radiometer channels is for the accommodation of eight narrow bandpass filters, three broad bandpass filters, and two quartz windows (for total flux measurements). In each channel between the filter/window and its thermopile there is inserted either a quartz lens or disk, to serve as a blocking filter against infrared radiation emitted within the channel. The narrow-bandpass filters are of the vacuum-deposited interference type and are constructed to withstand mechanical and thermal shock and solarizing; they are weatherproof and represent the present best state of the art in such development. The broad-bandpass filters are of Schott colored filter glass. An important advantage of the combined use of narrow- and broad-bandpass filters is the facility for intercomparison of solar filtered regions and the checking of the narrow-bandpass filters (integrally) with the broad (differentially). Where necessary, narrow-band channels incorporate optical (lens) amplification to increase the signal to the desired level. In the model discussed here, the adopted selection of radiometer channel responses is given in Table 29.1.

With the exception of channels 1 and 2 (0.2 to 0.4 mV output),

Table 29.1 Adopted Selection of Radiometer Channel Responses

Channel	Filter Limits	Wavelength Reference
	nm	nm
1	290–380	335 ultraviolet
2	380–450	415 ⎤
3	450–500	475 ⎥
4	500–550	525 ⎬ visible
5	550–600	575 ⎥
6	600–700	650 ⎦
7	>200	Total
8	700–1000	850 ⎤ infrared
9	1200–1800	1500 ⎦
10	>380 (GG22)	visible + infrared
11	>530 (OG1)	upper visible + infrared
12	>700 (RG8)	infrared
13	>200	Total

all other narrow channels have outputs of about 0.5 to 2 mV; the total and broad channels have outputs of about 5 to 8 mV (at Mauna Loa at noon).

However, other filter bandpasses and arrangements of radiometer channels are possible. All channel optical components are replaceable/interchangeable. Those indicated here were selected mainly on a preliminary evaluation basis.

Pyranometers

These are precision radiometers and, like the pyrheliometer units, incorporate a fast wire-wound-plated thermopile element, with the exception of the ultraviolet radiometer (photometer) that employs a photocell sensor. Each instrument delivers about 10 to 15 mV for full-scale recorder input (that is, Mauna Loa solar noon). The thermopile characteristics are similar to those of the multichannel pyrheliometer thermopiles (see the previous section). Adherence to the Lambert cosine law is about ± 1 percent (and ± 3 percent for the photometer) for a range of solar elevation of $10°$ to $90°$.

The following pyranometers are employed:

1. Total flux (global) radiometer (precision spectral model with quartz hemispheres).
2. The same as (1) with outer filter hemisphere of Schott glass, namely,

λ (nm)

GG22 >380
OG1 >530
RG8 >700.

3. Ultraviolet photometer, with wavelength limiting filter for total UV ($\lambda 295$ to 380 nm).

If the total global flux (1) and the RG8 filtered flux (2) are simultaneously recorded, or nearly so in the case of a single pyranometer by rapid interchange of the outer hemisphere, the natural illumination can be derived without recourse to direct measurement with an illuminometer. Also, consideration of the direct intensity and the global intensity, simultaneously measured, yields the respective diffuse sky component.

Data Acquisition and Recording

The basic objectives are as follows:

1. Recording, by a scanner feeding a guarded data amplifier which, in turn, feeds a six-digit integrating voltmeter (DVM), of up to twenty-five channels of information, with the capability of expansion to at least double this figure (without further system development) should additional measurement requirements arise in the future.

2. Acceptance of radiation signals (without any electrical amplification) ranging from about 200 μV to 15 mV, with thermopile impedance of approximately 800 ohm, and their final display on paper printout and also on seven-track magnetic tape, with a measurement accuracy of 1 percent, in general, as aided by system built-in precision voltage references (see the next paragraph).

3. Inclusion of precision voltage references, namely, 0.25, 0.5, 1, 5, 10, 15, and 20 mV (with impedance matching that of the detectors), and an electrical zero (that is, the sensor short-circuited).

4. Capability of accepting and displaying relatively small signals (for example, of the order of 10 to 20 μV) indicative of radiometer channel thermal zero (as produced by shuttering), which infers a measurement resolution of 1 μV with a $\pm$ sign.

5. Automation of command system on-off, date of record with specific timing indications and selected shutter zero operation, and control of data scans with digital clock technique.

6. Fastest complete scan time (that is, radiation sensors plus the electrical insertions, in this instance) for one cycle of about 5 sec, but with capability for considerably slower scan times (simple switching).

7. Magnetic tape record in standard digital computer format.

The standard-type amplified digital voltmeter has a 0 to 100 mV range, visually displayed in bipolar form. The coupler translates digital voltmeter data into a serial form (bits of information) that is compatible with computer data handling. The paper printer and the magnetic tape recorder are standard commercial units. The auxiliary control unit provides the reference voltages and radiometer shutter signals, and also houses the controls for the digital clock and the manual data option unit. The complete data acquisition and recording system is housed in a standard console rack.

Instrument Calibration

1.   *Radiation sensors.* The basic radiometric working standard is a manually operated Ångström-type electrical compensation pyrheliometer capable of reproducing the International Pyrheliometric Scale (IPS) to within 0.5 percent. All instrumentation was delivered fully calibrated and with (where pertinent) radiometer filter factors to allow for window energy losses incurred through reflection and absorption.

2.   *Readout.* Precision voltage references, as described.

30

Notes on Monitoring
Air Quality

A. P. Altshuller and
Richard D. Cadle

This chapter discusses several aspects of monitoring for each of several pollutants in the atmosphere: hydrocarbons; nitric oxide and nitrogen dioxide; carbon monoxide; heavy metals; organic chlorides; ammonia in the troposphere; sulfates, nitrates, and organic particles in the troposphere; tropospheric ozone; and stratospheric particulate matter. The aspects treated include the reasons for monitoring; the sources, routes, and reservoirs of the pollutant; the monitoring technology; and the costs.

Hydrocarbons

Reasons for Monitoring

Certain types of hydrocarbons are considered significant as reactants capable of producing aerosols in the size range contributing to global effects.

Sources, Routes, and Reservoirs

Hydrocarbons are emitted by both urban and biosphere sources. Methane is not of concern because it is not a source of aerosols. Terpenes from forested areas can be oxidized to form aerosols. The average level of 10 ppb has been estimated for a biosphere terpene concentration. Although conversion of terpenes to aerosols can occur in the presence of molecular oxygen, the added presence of traces of nitrogen oxides may accelerate aerosol formation. Terpenes also react rapidly with ozone. The lifetimes of terpenes can be estimated to be a few hours or less. Therefore, terpene concentrations are likely to be undetectable at monitoring sites distant from their natural sources. Only a very small fraction of the hydrocarbons from urban pollution have structures such that they form organic aerosols in significant amounts. These olefins are probably equivalent to only a few percent of the global terpene emissions.

Aromatic hydrocarbons (C_6 to C_{11}) in the presence of nitrogen oxides also produce organic aerosols. This class of hydrocarbon can contribute as much as half of the urban hydrocarbon emissions on a mass basis. Aromatic hydrocarbons react appre-

Prepared during SCEP and edited after SCEP.

ciably slower than terpenes. Based on available rates of photo-oxidation, aromatics can be anticipated to be present in air masses several hundred miles from urban sources. No experimental results are available indicating aerosol formulation from parafinnic hydrocarbons. Acetylene also should not contribute to aerosol formation.

Little is known about the mechanisms for consumption of hydrocarbons outside of urban areas. Reactions with ozone are fast for terpenes but very slow for aromatic hydrocarbons for the ozone concentration expected at ground level. It is reasonable that the same free radical-type photooxidation reactions important in urban areas also contribute in the biosphere. Oxidative microbiological processes appear to be of considerable significance, and if we also consider transport to the stratosphere, the ozone concentrations may be high enough to consume aromatic hydrocarbons gradually. Above the ozone layer, vacuum ultraviolet photochemical reactions may rapidly destroy these hydrocarbons. In these regions, reactions with O and OH also should contribute the conversions of aromatic hydrocarbons.

Limited monitoring efforts are justified for hydrocarbons. No monitoring for *individual* hydrocarbons is done even in urban areas. However, a continued group of research studies of urban atmospheric reactions does provide an increasing group of gas chromatographic analyses for a wide range of individual hydrocarbons. Terpenes should be monitored in samples collected in and around forested areas. Sampling for terpenes at remote global stations should be attempted, but the presence of these hydrocarbons is more likely to be detected at sites closer to emission sources.

Spatial resolution cannot be decided without a preliminary sampling program, but a set of more than ten or twenty sampling sites cannot be justified. One sample should be collected over two weeks. Accuracies of ± 20 percent may be obtainable if the concentration technique is workable.

Monitoring Technology
1.   *In Situ*
The most practical technique would be sampling of the higher molecular weight hydrocarbons of interest on a solid substrate

such as silica gel, transporting the samples back to a control laboratory, and analyzing on laboratory gas chromatographs equipped with flame ionization detectors.

2. *Remote, Ground-based*

Not likely to be feasible because of numbers of different hydrocarbons with somewhat variable spectral characteristics.

3. Aircraft sampling with solid substrates or large flexible plastic containers should be possible and some tropospheric and stratospheric sampling should be carried out.

4. Satellite measurements not likely to be feasible because of the number of different hydrocarbons with somewhat variable spectral characteristics.

Costs

Gas chromatographic analysis in the laboratory is highly developed, and adequate data processing techniques are available from urban air pollution activities. For analysis in a central laboratory the cost for ten to twenty stations would be of order:

a. Capital Investments—$25,000

b. Annual Operations— 10,000

Nitric Oxide and Nitrogen Dioxide

Reasons for Monitoring

By various kinetic mechanisms, nitric oxide is oxidized to nitrogen dioxide which reacts to form nitric acid which in turn is converted to particulate nitrate. These particulate nitrates are mostly in the particle size range contributing to global turbidity effects.

Sources, Routes, and Reservoirs

Urban air pollution contributes appreciable amounts of nitrogen oxides. The rate of conversion to particulate nitrate appears to be slow so conversions occur at least on a regional scale. It has been estimated that the biosphere over land areas contributes about an order of magnitude more nitrogen oxides to the global atmosphere than does urban pollution. However, the biosphere values are estimates based on only a few experimental measurements. Based on these measurements, biosphere nitrogen oxide levels are in the 2 to 6 ppb range. Nitrogen oxide levels over the oceans have been estimated to be about an order of magnitude lower than concentration levels over land. Substantial amounts of nitrogen dioxide can be lost from the atmosphere in precipitation or dry fallout. Very little is known about the details of formation, transport, and chemical conversion on a global scale.

Air pollution monitoring at urban sources in the United States will be intensified because of air quality standards being established for nitrogen oxides in 1971. A measurement program to follow the transport and chemical transformations on a regional scale around urban centers also is being developed. Monitoring is needed over various representative land-use areas as well as over the oceans. Aircraft monitoring to determine the transport of nitrogen oxides through the troposphere into the stratosphere is desirable.

Spatial resolution should take into consideration the need to estimate biosphere contributions as well as upper atmosphere mechanisms of conversion of nitrogen oxides to nitrates. One sample every one or two weeks should be sufficient at each of twenty ground-level monitoring sites. Aircraft sampling for nitrogen oxides also should be included in the overall aircraft sampling program. Accuracies of about ±20 percent would appear sufficient for initial monitoring efforts.

Monitoring Technology

GROUND-BASED *IN SITU*

CHEMICAL

Nitrogen dioxide can be determined by colorimetric methods. These methods are suitable for sampling periods up to two hours. Nitric oxide is analyzed by the same procedure as nitrogen dioxide after oxidation by a chemical substrate. The same schedule should be used for nitric oxide as for nitrogen dioxide.

PHYSICAL

Prototype instruments are presently being developed capable of measuring nitric oxide and nitrogen oxides ($NO + NO_2$) in the 1 ppb and above range. The technique used involves measurement of the chemiluminescent spectra of nitrogen oxides titrated with atomic oxygen and of nitric oxide titrated with ozone.

GROUND-BASED REMOTE

Long-path instruments based on correlation spectrometry involving the visible spectrum of nitrogen dioxide have been developed and evaluated in urban areas.

SATELLITE, AIRCRAFT, BALLOON

1. *In Situ*

Aircraft samples could be used provided the long sampling period required is compatible with flight objectives.

2. *Remote*

In principle correlation spectrometry for nitrogen dioxide could be used in aircraft or satellite applications. Improved measurements provide the only possible approach to remote sensing of nitric oxide from aircraft or satellites.

SENSOR AND DATA PROCESSING ASPECTS

1. The available colorimetric techniques require one-half-hour to two-hour integration times to obtain adequate sensitivity but cannot be used for longer periods safely because of stability problems. Collection efficiency and stability of the collected sample almost always are problems with colorimetric procedures. In addition, the nitric oxide oxidation to nitrogen dioxide is subject to significant uncertainties because of variability in conversion efficiencies with concentration and relative humidity conditions.

2. The chemiluminscent instruments are in prototype development. The technique is very promising on the basis of excellent specificity, linear response over a wide range and very good signal to noise characteristics. Additional R&D is required to improve sensitivity to the 0.1 ppb to 1 ppb range.

3. The only presently available long-path or remote correlation instrument for nitrogen dioxide requires evaluation concurrently with the colorimetric procedures at nonurban sites. Additional R&D is required to determine whether nitrogen dioxide correlation spectrometer can be used over unpolluted areas to determine nitrogen dioxide. Aerosol effects represent a serious limitation on global applications of this technique. An active R&D program would have to be developed for an improved spectrometer for nitric oxide. A very high resolution system is essential to avoid the serious overlap of the water vapor and nitric oxide vibrational-rotational bands.

Colorimetric procedures could be utilized either in an inexpensive field laboratory or in a central laboratory. Chemiluminescent instruments would be installed at field sites.

It is not possible at present to specify the optimum monitoring system for nitrogen oxides.

Costs—20 Stations

CHEMICAL

a. Capital Investment—$ 50,000
b. Annual Operation —   10,000

PHYSICAL

a. Capital Investment—$150,000
b. Annual Cost   —   10,000

There are no means available at present to determine the *optimum* monitoring system.

Carbon Monoxide

Reasons for Monitoring

It has been suggested that carbon monoxide could be increasing on a global basis because of the increased contribution from vehicular emissions. However, isotopic carbon measurements on carbon monoxide can be interpreted to indicate that biosphere sources of carbon monoxide may produce much higher amounts of carbon monoxide than arises from urban pollution. Carbon monoxide levels over unpolluted land and oceanic sites have been averaging in 1969–1970 in the 0.1 to 0.2 ppm range. Whether or not the steady-state concentrations is increasing, there still is the need to determine the effective sink regions. If, as seems likely, a significant fraction of the carbon monoxide is being transported into the stratosphere ozone layer, it could be influencing the overall ozone reaction mechanism in the stratosphere.

Sources, Routes, and Reservoirs

Vehicular emissions are an important source. Results of measurements of carbon monoxide over the Atlantic Ocean have been interpreted to show that the oceans are a net source of carbon monoxide. The estimated source strength of the oceans is still small compared to the contribution from urban pollution. It appears that decomposition of green vegetation may be a source of carbon monoxide. Certain micro-organisms can consume carbon monoxide.

Urban monitoring systems for carbon monoxide already are in operation and will be expanded. The location of nonurban sources and sinks and estimates of these strengths is very incomplete. Measurements continue to be needed over land masses and the oceans. Seasonal variations should be closely followed to determine the influence of biological processes of formation and decay or organic matter. Carbon monoxide measurement should be done from aircraft both in the troposphere and stratosphere.

Spatial resolution as provided by ten stations is minimal to cover oceans, polar regions, as well as representative land areas in temperate and tropical land masses. Aircraft sampling could assist in supplementing this network by low-level sampling. Time resolution of one sample over two weeks per site should be adequate. An accuracy of at least ±5 to ±10 percent is essential to see trends.

Monitoring Technology
GROUND-BASED *IN SITU*

No suitable wet chemical methods are available. A highly sensitive and specific gas chromatographic technique is available for immediate use. A modification of the mercury vapor-type analyzer has been used for carbon monoxide analysis in unpolluted areas. Carbon monoxide can be analyzed by mass spectrometry after removal of carbon dioxide by oxidation of carbon monoxide to carbon dioxide. Nitrous oxide also must be removed along with hydrocarbons. This method has been used to obtain total carbon monoxide as part of the isotopic carbon monoxide work.

An automated chromatograph capable of monitoring carbon monoxide (and methane) from 0.05 ppm up has been developed, evaluated in the field, and it is commercially available. The sensitivity has been shown to be completely adequate for global monitoring requirements. The mercury vapor analyses and mass spectrometric instruments are prototypes whose use has been limited to research studies. The gas chromatograph has been automated so it can be used either in a field installation or central laboratory. The mercury vapor-type instrument requires a scientific worker present as does the mass spectrometer. The mass spectrometer is a laboratory instrument. The mercury vapor-type instrument could be operated intermittently in the field when manpower is available.

Costs—10 Stations
1. Capital Investment

Gas chromatograph—$10,000 for central laboratory operation, $60,000 for operation at field site.

Mercury Vapor Analyser—not commercially available
Mass Spectrometer—$25,000–$50,000 at central laboratory.
2.   Annual Operation
Gas Chromatograph—$ 5,000 at central laboratory
$10,000–$15,000 at field sites.
Mercury Vapor Analyser—no estimate
Mass Spectrometer—$10,000 at central laboratory

The use of two gas chromatographs at a central laboratory with cylinder collection of samples at sites would be the most economical as well as analytically acceptable for global monitoring. Methane also could be analyzed for no additional cost except for data processing.

Heavy Metals

Reason for Monitoring

Heavy metals are introduced into the air as particles by automobiles (lead compounds in the exhaust gases), by many industries such as smelters of various kinds, and by power plants. Lead and certain other heavy metals can be harmful to mankind and animals if inhaled; they may be harmful to plants and animals in streams, lakes, and oceans, and when they reach the earth's surface, for example by the scrubbing action of rain, they may have harmful effects on plants. The U.S. Public Health Service considers the heavy metals sufficiently important that a large number of metals are monitored routinely by its National Air Sampling Network. This network was established in 1953 and consists of 185 urban and 51 nonurban stations in all fifty states, Puerto Rico, and the District of Columbia.

Sources, Routes, and Reservoirs

Biological considerations seem to indicate that the hazards are sufficiently great that monitoring of heavy metals should be undertaken of both the routes and reservoirs, that is, of air (the route) and soil, lakes, and oceans. Monitoring of all but the air is discussed in other papers from the SCEP Monitoring Committee.

Since the heavy metals are associated with atmospheric particles, their residence time in the atmosphere is relatively short, probably a half-life of a few days. Thus, a large sampling network is advisable, perhaps if possible the 100 station network suggested by the CACR report to the World Meteorological Organization (WMO). It should be emphasized, however, that even one sampling point in a remote location is better than none at all. Ten percent accuracy is satisfactory. (One sample should be taken at each station every two weeks.)

Monitoring Technology

The particles containing the metals can be collected by filtration. Several types of filter material might be used, and some preliminary testing at a nonurban station would be helpful in making a selection. Glass-fiber filters such as those used by the U.S. Public Health Service, cellulose filters (IPC Filters) such as those used on the Air Weather Service flights for the Atomic Energy Commission, or polystyrene-fiber filters of the type made and used by the National Center for Atmospheric Research (Cadle and Thuman, 1960) could all be used, but the polystyrene filters are probably the best because of their high purity compared with the others. The actual analyses can be done directly on the filters by neutron activation, by ashing polystyrene filters and determining the heavy metal concentration by emission spectroscopy, or by extracting the filters with water and determining the concentration of metal ions in solution by atomic absorption techniques or by emission spectroscopy. The combination of polystyrene fiber filters, extraction, and atomic absorption spectroscopy will probably give the most reliable results. The actual filtration can be made using "high volume samplers" which are high-speed vacuum pumps equipped with filter holders. The analyses will have to be conducted at a central laboratory such as the one mentioned in connection with O_3 monitoring. It is very unlikely that remote monitoring will ever be appropriate because of the difficulties inherent in determining detailed composition of particles in this way.

Probably ten to fifteen metals would be monitored. Once the samples have been obtained and prepared for analysis, the determination of the concentrations of a number of metals is relatively easy. A list of appropriate metals is not given here, but lead, mercury, copper, zinc, and vanadium should probably be included. Ecologists and biologists may suggest others.

Costs

Most of the costs have been included elsewhere, but are repeated here.

Costs	Each	Total
1. Cost of central laboratory	$400,000*	$ 400,000*
2. Laboratory for manufacturing polystyrene filters	100,000	100,000
3. Sampling equipment	2,000	200,000
4. Cost (annual) of operating stations	50,000*	5,000,000*

* This assumes that many other substances would be monitored at the stations and the cost is for all monitoring.

Organic Chlorides

Reasons for Monitoring

Organic chlorides (chlorinated hydrocarbons) include a wide variety of products. They are all more or less toxic and even in years long past have caused concern for this reason. During recent years insecticides of this type, especially DDT, and the polychlorinated biphenyls which have a number of industrial uses, for example as plasticizers, have been causing concern. The chlorinated hydrocarbons ultimately accumulate in the oceans, and as much as 25 percent of the DDT produced to date may have been transferred to the oceans. On the other hand, probably only a small fraction of the polychlorinated biphenyls produced are now present in the sea.

The amount of DDT now present in marine biota, although probably only about 0.1 percent of the total production, has had an impact on the marine environment. Examples are the decline in productivity of marine food fish and the accumulation of unacceptable levels of chlorinated hydrocarbons in their tissues. Fish-eating birds have experienced reproduction failures and population declines.

If the use of organic chlorides increases, or even if it continues at its present rate, severe degradation of our environment may result. The monitoring of the concentrations of these toxic substances in our environment would seem to be important.

Sources, Routes, and Reservoirs

The routes of transport of organic chlorides from the place where they are used may be via the atmosphere directly to the oceans (for example, scrubbing the aerosol particles from the air by rainfall over the oceans), from the air to the land to the oceans via rivers, or directly to the oceans. But the air and the land should be considered to be reservoirs as well as routes (American Chemical Society, 1969).

This problem (of organic chlorides) seems to be so important that monitoring should probably be carried out of the atmosphere, soil, oceans, and biological material. However, this discussion is related to atmospheric monitoring.

DDT enters the atmosphere in several ways, including aerial drift during application, evaporation from water surfaces, and evaporation from plants and soils. Many other chlorinated hydrocarbons may enter the atmosphere in similar ways. For example, polychlorinated biphenyls may enter the air when plastics containing them are burned. They are probably also partially converted to hydrogen chloride, carbon dioxide, and water.

Most of the organic chlorides of concern are of low volatility, and these tend to be associated with airborne particles. Since the latter have a short half-life in the air (probably a few days), the same is probably true for the organic chlorides. Thus, following the CACR recommendation to the WMO, attempts should be made to have about 100 ground stations (assuming satellites will not be available to do the job for several years, if ever). One sample at each station every two weeks is sufficient, and 20 percent accuracy will suffice.

Monitoring Technology

Many of the organic chlorides being used have an appreciable vapor pressure. The analytical methods for the most part assume that the chlorinated hydrocarbons are associated with airborne particles and start by the collection of such particles on a filter. However, when the concentrations are very low, as in the ambient atmosphere, this assumption may lead to gross errors since the chlorinated hydrocarbons may be largely in the gas phase. Thus, it is recommended that at least at first both the gas and particulate phases of the atmosphere be analyzed for these substances. The

Costs

Costs	Each	Total
1. Cost of central laboratory	$400,000*	$ 400,000*
2. Laboratory for manufacturing polystyrene filters	100,000	100,000

Costs (contd.)	Each	Total
3. Sampling equipment	$2,000†	$200,000†
4. Cost (annual) of operating	50,000*	5,000,000*
5. Research into the development of more sensitive analytical techniques‡	500,000	500,000

* This assumes that many other substances would be monitored at the stations, and that the cost is for all monitoring.
† This may have to be doubled if monitoring for gaseous as well as particulate chlorinated hydrocarbons is necessary.
‡ This may not be necessary. See preceding discussion.

usual analytical technique involves the collection of particles on a filter, extraction of the filter, and a chromatographic analysis of the extract. Special techniques may have to be developed for the gas-phase analysis.

Ammonia in the Troposphere

Reasons for Monitoring

The main reason for monitoring is that ammonia combines chemically with the sulfuric acid droplets resulting from sulfur dioxide oxidation and hydration to form ammonium sulfate particles (Cadle and Robbins, 1966). The nature of the particles in the atmosphere has a strong effect on the extent to which they scatter and absorb solar radiation, and also their behavior as nuclei.

Sources, Routes, and Reservoirs

Both natural and man-operated sources of ammonia exist, but there are many uncertainties (Junge, 1963). For example, it is not absolutely established whether the oceans are sources or sinks for ammonia; the soil can be both, depending upon the pH. Alkaline soils tend to release ammonia and acid soils to absorb (react with) it. In general, natural ammonia results from the decomposition of amino acids by bacteria. Man-produced ammonia results from some combustion processes, from the raising of cattle, and from agriculture. The reservoirs are the earth and the oceans reached via the air. Monitoring should probably only be done in the ambient air, for the reasons already mentioned, at locations remote from sources.

Ammonia is probably short-lived in the atmosphere (a matter of days) (Junge, 1963) so if possible it should be monitored at about 100 stations every two weeks. Ten percent accuracy should be sufficient.

Monitoring Technology

Very simple, effective sampling and analysis techniques involving bubblers, reagents, and colorimeters (photometers) are available that can achieve the desired accuracy. Both sampling and analysis can be performed in the field at little additional cost over that for operating field stations for other trace constituents (see, for example, the discussion of ozone monitoring). As for most other trace atmospheric substances, satellite monitoring technology for ammonia has not been developed but might be preferable to ground stations, if available.

Sulfates, Nitrates, and Organic Particles in the Troposphere

Reasons for Monitoring

These types of substances (in addition to the aerosols themselves) are major constituents of atmospheric aerosols and some of them are man-made. They may affect the climate through their contribution to turbidity, their absorption of radiation, and also by serving as condensation nuclei.

Sources, Routes, and Reservoirs

The human sources of sulfate and nitrate are the combustion of sulfur-containing fuels and nitrogen fixation during combustion. The former produces largely sulfur dioxide which is oxidized and hydrated in air to form sulfuric acid droplets. The latter produces largely nitric oxide (NO) which is oxidized and hydrated to form HNO_3 and nitrates (Cadle and Allen, 1970). There are numerous natural sources of sulfur dioxide and sulfates such as volcanoes, the oceans, and forest fires. The latter and lightning are natural sources of oxides of nitrogen (Junge, 1963). Organic particles are formed by chemical reactions in smog (Cadle and Allen, 1970) and emitted directly by various combustion processes (natural and man-made). The routes and reservoirs are thus rather complicated, but we recommend here only monitoring in the ambient atmosphere.

Monitoring Technology

Probably the best way of doing this for sulfate and nitrate at present is at surface monitoring stations collecting particles on polystyrene fiber filters (Cadle and Thuman, 1960), as discussed in the section on heavy metal monitoring. The filters are extracted and the extracts analyzed using colorimetric techniques (Lazrus, Lorange, and Lodge, 1968; Brewer and Riley, 1965). The monitoring for organic particles could use glass fiber filtration followed by extraction with benzene, evaporating the benzene and weighing the residue. Total particle concentrations are obtained by weighing the filters before and after sampling. Since these substances are atmospheric particles, their residence time in the atmosphere is relatively short and a large sampling network is advisable, perhaps 100 stations. It should be emphasized, however, that even one sampling point in a remote location is better than none at all. Ten percent accuracy is satisfactory. One sample should be taken at each station every two weeks.

Costs

The costs are the same as for monitoring metals and are itemized below:

Costs	Each	Total
1. Cost of central laboratory	$400,000*	$ 400,000*
2. Laboratory for manufacturing polystyrene filters	100,000	100,000
3. Sampling equipment	2,000	200,000
4. Cost (annual) of operating stations	50,000*	5,000,000*

* This assumes that many other substances would be monitored at the stations and the cost is for all monitoring.

Tropospheric Ozone

Reasons for Monitoring

Tropospheric ozone has several possible effects, as follows:
1. Reactions with terpenes to form aerosol.
2. Reactions with nitric oxide and nitrogen dioxide to form nitrate aerosol.
3. Plant damage effects on truck crops, forest species, and some ornamentals.

Sources, Routes and Reservoirs

The sources and routes of ozone in nonurban areas can be associated with the following processes:
1. Transport from urban areas.
2. Downward transport from stratosphere.

3. Formation from reactions of biosphere-produced terpenes and nitrogen oxides.

4. Combinations of 1, 2, or 3.

As ozone reacts on the land biosphere with terpenes, nitrogen oxides, and plant surfaces or decomposes on the surfaces of particles, it slowly disappears. Despite these multiple sources for consumption, ozone has been detected by damage to crop or forest species from 50 to 100 miles from urban sources. Normally, ozone is destroyed if near the surface during nighttime hours. Some available results suggest that ozone can persist aloft overnight.

Ozone concentration levels over nonurban land areas have been reported in the 5 to 40 ppb range. Measurements of ozone over the Atlantic Ocean between 33°N latitude and the equator and from 76° W and 47° W longitude ranged from below 1 ppb up to 15 ppb. The average diurnal ozone concentration irrespective of position averaged between 5.0 and 5.7 ppb. These ozone levels over the ocean were attributed to vertical transport from the stratosphere. Monitoring of ozone in the Antarctic region has given averages of 20 to 30 ppb.

Monitoring of ozone in urban centers is already conducted in some areas in the United States, and monitoring will increase. Monitoring of ozone is very limited elsewhere. Monitoring would seem the most useful in the biosphere with only limited measurement sites over oceans or in polar regions. The study of damage to pine, tobacco wrapper, and other species in forested or agricultural areas usually has not involved concurrent meteorological work to determine whether the ozone was transported from distant cities, from the stratosphere or was generated by photochemical reactions of terpenes and nitrogen oxides of local origin.

Spatial resolution of ten stations is minimal to cover various forests, agricultural areas as well as oceans and polar sites.

While time resolution is not important in reservoirs, it is important on routes. Since integrated samples have not proved satisfactory for ozone determination, instrumental analysis should be utilized. The available instrumentation can give continuous monitoring for ozone.

An accuracy of ±10 percent is adequate since the emphasis should be on frequency of active levels of ozone. Calibration accuracy is also a limiting factor for an unstable substance such as ozone.

Monitoring Technology

GROUND-BASED *IN SITU*

No generally acceptable procedure exists for integrated ozone measurements by collection using solution or other methods of holding samples because of the low stability of the system. Instrumental methods can be used, but the colorimetric KI and the carbon iodine (Komhyr) detectors are nonspecific for ozone. These instruments respond to ozone and to varying amounts of nitrogen dioxide, chlorine, and other strong oxidizing agents. Sulfur dioxide, unless removed, suppresses the ozone reaction with these types of instrumentation. The chemiluminescent instrumentation now available for ozone is specific and highly sensitive to ozone.

BIOLOGICAL

A number of plant species are highly sensitive to ozone. Tobacco wrapper Bel W3 is sensitive down to 30 ppb of ozone, and it has been frequently utilized at urban and nonurban sites as a biological indicator for ozone.

GROUND-BASED REMOTE

Ozone has been measured over long path lengths by use of either its strong ultraviolet absorption bands or its infrared absorption in the 9.6-μ region.

SATELLITE, AIRCRAFT, BALLOON

1. *In Situ*

Aircraft measurements have been made by use of the available instruments for ozone measurement (KI, Komhyr, chemiluminescence).

2. Remote

If measurements are made using a path through the entire atmosphere, there are great difficulties in separating out the absorption of the low ozone concentrations in the troposphere. The required technology is available for the *in situ* instrumentation. Additional R&D may be required on the open-path infrared instrumentation.

The available instrumentation, particularly the chemiluminescent instrumentation, is adequate, and data processing can be handled satisfactorily. A remote ground-level infrared ozone instrument is presently being developed with a computer readout system. The ozone instrumentation would be operated at the field sites.

The optimum monitoring system for all requirements except remote measurement is the chemiluminescent instrumental technique. If the absence of other interfering substances can be assured, the Mast (KI) and Komhyr instruments are convenient and compact.

Costs—10 Stations

a. Capital Investment—$20,000–$30,000

b. Annual Operation—   5,000

Stratospheric Particulate Matter

Reasons for Monitoring

There are two reasons. One has to do with the emissions of supersonic transports (SSTs) and the other with the other emissions resulting from man's activities. The present estimates of emissions of SSTs may not be accurate, and monitoring will be needed to determine what the actual concentration increase following the introduction of SSTs will be.

The SSTs are not the only source of man-made particles in the stratosphere. The tropopause is not an impermeable barrier between the stratosphere and troposphere, and an important fraction of the particles emitted by man into the troposphere reaches the stratosphere. Also, sulfur dioxide entering the stratosphere will be oxidized by atomic oxygen to SO_3 which is immediately hydrated to form droplets of sulfuric acid (Cadle and Allen, 1970).

Sources, Routes, and Reservoirs

To a large extent, these will be the same as for water vapor in the lower stratosphere. However, the sampling, if from aircraft using filters, will be batch rather than continuous. Each filter collection will probably require about one-half hour of flying time. Thus, considerable averaging will occur. Measurement accuracy will probably have to be ±1 ppb by weight.

Monitoring Technology

Cadle and his associates at NCAR are at present determining the concentrations of particles at about 18 km using filters carried by RB 57-F aircraft, though for research purposes. Thus, the required technology is well in hand. It may be appropriate to use two types of filters on each flight. Glass-fiber filters can be extracted with organic solvents following a flight to furnish concentrations of organic particles (largely from the SSTs), while the extremely high purity polystyrene-fiber filters (Cadle and Thuman, 1960) of the type now being prepared for such purposes at NCAR may be used for sampling for $SO_4^=$, SI, and NO_3^- (Cadle et al., 1970). The Si measurements are included to serve to indicate the amounts of soil mineral matter present. By weighing filters before and after flights, total particle loadings can be obtained.

Remote sensing by laser radar (lidar) can also be done, and probably ground-based lidar stations (three) should be used to supplement the aircraft flights (Cadle, 1966).

Costs

The main capital investment will be for (a) the clean rooms and other equipment for manufacturing the polystyrene-fiber filters ($100,000); (b) a laboratory for conducting the analyses ($400,-000, assuming that the laboratory will also be used to analyze samples collected at 10 to 100 ground-based stations, for servicing instruments, and for data processing); and (c) the three lidars ($450,000). The annual operation, excluding aircraft flights, would cost on the order of $50,000.

Water Vapor in the Lower Stratosphere

Reasons for Monitoring

Water vapor is a major exhaust product of jet engines such as those carried by SSTs. If, as has been estimated, in the 1985–1990 time period the SSTs in the world's airlines fly 7 hr/day in the stratosphere, and each engine produces about 41,400 lb/hr of water vapor, the SSTs will increase the 3 ppm H_2O in the stratosphere to 3.2 ppm H_2O. This may have an effect on world temperature and on the ozone concentration.

Sources, Routes, and Reservoirs

It does not seem appropriate to monitor at the source (the jet aircraft). However, it may be appropriate to monitor in the vicinity of major air routes, as well as in regions remote from such routes.

Monitoring Technology

Several techniques based on direct sensing are at present available (Mastenbrook, 1968). The instrumentation is borne by balloons, rockets, or airplanes; the latter are probably preferable, especially in view of the RB57F aircraft of the U.S. Air Weather Service. Several types of instrumentation might be used. Mastenbrook (1968) has used balloonborne frost-point recorders, while an instrument adopted at NCAR for use on RB57F aircraft is based on changes in the electrical properties of a surface as a result of the adsorption of water vapor. The earliest reliable observations, over Western Europe and Africa, were made from aircraft of the (British) Meteorological Research Flight, using observer-operated frost-point hygrometers (the Dobson-Brewer hygrometer).

Eventually, satellites may be able to do this monitoring, but the needed technology has not been developed. There has been some successful indirect sensing from aircraft, but not on a routine basis.

The costs for monitoring by aircraft are difficult to estimate, but if the flights are available anyway, as at present, the capital investment could be as little as $10,000 for instrumenting two aircraft. The instrumentation needs to be carefully calibrated and the calibrations repeatedly checked. Thus, instrument maintenance and analysis might cost $20,000 per year. Obviously, if special flights had to be financed, the costs would increase astronomically.

Probably four flights per year in the Northern and Southern Hemispheres would suffice. Since the instruments are continuous-reading devices, a given flight could cross both well-traveled and little-traveled regions. The accuracy should probably be about 0.2 ppm.

RB 57-F aircraft instrumented either with NCAR or Mastenbrook devices plus calibration, maintenance, and data analysis support perhaps constitute the optimum monitoring system. The type of analytical device needs further study to determine which would be the preferable monitoring instrument since all are now research tools, and the Mastenbrook device has been developed for balloons rather than for high-altitude aircraft.

References

American Chemical Society (ACS), 1969. *Cleaning Our Environment: The Basis for Action* (Washington, D.C.: ACS).

Brewer, P. G., and Riley, J. P. P., 1965. The automatic determination of nitrate in sea water, *Deep Sea Research, 12:* 765–772.

Cadle, R. D., 1966. *Particles in the Atmosphere and Space* (New York: Reinhold).

Cadle, R. D., and Allen, E. R., 1970. Atmospheric photochemistry, *Science, 167:* 243–249.

Cadle, R. D., Lazrus, A. L., Pollack, W. H., and Shedlovsky, J. P., 1970. The chemical composition of aerosol particles in the tropical stratosphere, *Proceedings of the American Meteorological Society Symposium on Tropical Meteorology,* Honolulu, June 3–11, 1970.

Cadle, R. D., and Robbins, R. C., 1961. Kinetics of atmospheric chemical reactions involving aerosols, *Faraday Society Discussions, 30:* 155–161.

Cadle, R. D., and Thuman, W. C., 1960. Filters from submicron-diameter organic fibers, *Industrial and Engineering Chemistry, 52:* 315–316.

Junge, C. E., 1963. *Air Chemistry and Radioactivity* (New York: Academic Press).

Lazrus, A., Lorange, E., and Lodge, J. P., Jr., 1968. New automated microanalyses for total inorganic fixed nitrogen and for sulfate ion in water, *Advances in Chemistry* series, No. 73 (Washington, D.C.: American Chemical Society), pp. 164–171.

Mastenbrook, H. J., 1968. Water vapor distribution in the stratosphere and high troposphere, *Journal of Applied Meteorology, 25:* 299–311.

Several papers in the previous section discuss the various ways in which particles and turbidity might affect the heat balance of the atmosphere. There is another subtle effect which particles might have on weather and perhaps even climate by acting as nuclei to initiate the formation of cloud droplets or the freezing of super-cooled water drops. There has been considerable speculation about how air pollution may inadvertently "seed" clouds with such nuclei, but it is not at all clear whether the net effect of such seeding will be an increase or decrease in rainfall.

Two SCEP participants, Drs. Richard D. Cadle and Christian E. Junge, prepared a short paper during the Study for the Work Group on Climatic Effects. This paper, which is reproduced here, provides a summary of the present state of knowledge of this field. The following paper by Dr. Smith is a more detailed review of the process of nucleation and the role of pollutants in the formation of condensation and ice nuclei. By examining evidence of observed precipitation changes that have been associated with pollution, he is able to draw some tentative conclusions about potential local, regional, and global effects of such nucleation.

One phenomenon of water vapor condensation formation which almost everyone has observed is the contrail (condensation trail) of jet aircraft. A question that has puzzled meteorologists is whether these are persistent only when natural cirrus clouds would have subsequently occurred anyway. In a previously unpublished paper, Drs. Machta and Carpenter suggest that certain parts of the atmosphere have indeed exhibited real increases in cirrus cloudiness since the onset of commercial jet aircraft.

Another working paper of SCEP concludes this section on particles and clouds. Written by Dr. Julius London, it outlines some of the methods that have been used to observe clouds from satellites and suggests ways in which such observations could be improved. It should also be noted that the paper by Dr. Seymour Edelberg in Part X contains a section on the monitoring of clouds.

31

Atmospheric Nuclei   Richard D. Cadle and
                                Christian E. Junge

Condensation Nuclei

Atmospheric condensation nuclei can be defined as those air-borne particles on which water condenses to form the droplets of clouds and fogs. For the most part they are the "large" particles of atmospheric particulate matter (that is, those of 0.1 to 1.0 μ radius). Water-soluble, hygroscopic particles are especially active, but any particle in this size range may serve as a condensation or cloud nucleus (Fletcher, 1962; Mason, 1957).

Unless the atmospheric aerosol is dominated by single sources (as for instance the sea spray at sea surface, the mineral dust in dust storms, or large industrial sources), the individual aerosol particles represent a complex and variable mixture of water-insoluble and water-soluble inorganic and organic matter. This is the result partly of complex conditions during particle formation processes, and partly of a variety of aerosol modification processes within the atmosphere, all of which enhance the "internal" mixture of the aerosol particles. As a result, almost all particles after being in the atmosphere for some time have accumulated water-soluble material which enables them to behave approximately as soluble nuclei. For this reason the dominant factor in determining the supersaturation (in the range of less than a few percent relative humidity) required to initiate true condensation (and not just stable growth by rising humidity) is the size of the particles, unless the fraction of soluble material of the particles becomes less than 10 percent total. It is therefore to be expected that cloud nuclei spectra, measured recently at various places by diffusion cloud chambers, and aerosol size spectra should be closely related to each other.

Under a wide variety of geographical conditions, the size spectra of natural aerosols have an upper limit of about 100 μ radius with rapidly increasing concentrations down to 0.1 μ radius. The size distribution, if plotted on a log basis, has a maximum between 0.1 and 0.01 μ radius with decreasing concentrations below 0.01 μ.

Theoretical and experimental data indicate that in a form-

Prepared during SCEP.

ing water cloud the largest particles are activated first until a cloud droplet concentration of about 50 to several hundred per cm^3 is reached, depending on the cooling rate of the air and other cloud parameters. Under continental conditions, a large fraction of the particle population is left inactivated, since the total number of aerosol particles is much larger than the number of cloud droplets.

In very clean tropospheric air masses (upper troposphere, ocean areas), the total number of aerosol particles per cm^3 varies between about 100 and several hundred, which is about the number of cloud droplets formed under normal conditions, leaving no or almost no particles inactivated. Any increase in aerosol concentration and thus in cloud nuclei concentration under clean air conditions can result in an increase in concentration of initially formed cloud droplets, making the cloud, if it is a warm cloud, colloidally more stable and less inclined to produce rain. It is this process by which modifications of the global aerosol population may result in changes of cloud cover and rainfall. Although this process is well understood for individual clouds, it is not possible to predict the consequences in terms of the effect on circulation and climate, since on a regional and global basis requirements of the water and heat budgets will be the controlling factors. Therefore, even estimates as to the direction and degree of such changes are not possible at the present time. Because of our lack of knowledge of this possible influence of atmospheric aerosols, any effect of air pollution on the particle concentration in relatively clean air should be considered very seriously with respect to effects of global circulation and climate.

Over the oceans sea spray aerosols are constantly formed; but since their total concentration is less than about 10/cm^3, they do not play an important role as cloud nuclei except for the production of a few large cloud droplets which may be important in triggering the warm rain formation process. Most of the aerosol particles, even over the ocean, have another origin; since $(NH_4)_2SO_4$ is a common constituent of these particles, formation from SO_2 and NH_3 is likely. This is apparently true for most of the troposphere except for certain continental areas where the formation of other particles, industrial or natural, may dominate. This aerosol composition suggests that even in very clean tropospheric air new aerosol particles may be continuously formed and that the observed concentrations of a few hundred/cm^3 may represent a dynamical equilibrium between production and decrease in number by coagulation. This concept for clean air aerosols, for which recently some observational support over the Atlantic Ocean was obtained, suggests that the natural aerosols cannot be treated as conservative populations because gas reactions may be involved.

If this concept is correct, the question of atmospheric aerosols and its relation to global air pollution will be even more complex than hitherto expected.

Summarizing, we may say that aerosol size distributions and cloud nuclei spectra should be related to each other for a wide range of chemical compositions; that increase of particle concentration in clean air areas (which is the larger portion of the troposphere) may result in changes of cloudiness and rainfall which, however, cannot be estimated at the present time; that the atmospheric aerosol most likely represents a complex dynamic system, whose relation to global air pollution cannot be properly assessed because of lack of understanding of this system.

Freezing Nuclei

Freezing nuclei may be defined as those airborne particles that lead to the change of phase of water substance from the vapor or liquid phase to the solid (ice) phase. Thus, they may, in the former case, be acting by the inverse of sublimation, or they may involve freezing by contact with a supercooled water drop. Most freezing nuclei are in the size range of large particles as just defined, but are quite specific with regard to chemical (crystalline) makeup of the particles, and there are large variations in effectiveness between various substances. In the natural atmosphere, clay particles seem to constitute most of such nuclei, at least over continents, but other substances, such as extraterrestrial particles, may also play a role.

The number concentrations of freezing nuclei in the troposphere vary greatly with time and location. Studies at the National Center for Atmospheric Research showed variations of from 0 to 20,000 nuclei per meter^3 within a radius of a few hundred miles on one day over Texas. On the same day, the concentrations of stratospheric freezing nuclei varied from about 15 to 45 nuclei per meter^3 (Cadle et al., 1969).

Measurements of concentrations of freezing nuclei downwind of highly industrialized cities indicate that such cities are important sources and may, therefore, modify the downwind climate. However, there is some uncertainty with regard to the sources of these particles. Schaefer (1966) has suggested that they consist of lead iodide, formed by the reaction of lead compounds emitted by automobiles with iodine vapor in the air. However, measurements of freezing nuclei made in and downwind of Los Angeles, where there is no lack of automobile exhaust, indicated low or zero concentrations of freezing nuclei. Possibly industrial plants such as steel mills are more important than automobiles as sources of freezing nuclei.

Other Effects of Contaminants on Cloud Droplets

There are a number of possible effects of pollutants on cloud droplets about which little or nothing is known. For instance, traces of organic material adsorbed on the surface of droplets may markedly affect their electrical properties, whether or not they coalesce on contact with one another, and the rate of evaporation. The albedo of clouds may also be reduced in polluted atmospheres, as suggested by measurements over England. Research to determine the importance of such effects is badly needed.

References

Cadle, R. D., Bleck, R., Shedlovsky, J. P., Blifford, I. H., Jr., Rosinski, J., and Lazarus, A. L., 1969. Trace constituents in the vicinity of jet streams, *Journal of Applied Meteorology, 8:* 348–356.

Fletcher, N. H., 1962. *The Physics of Rainclouds* (Cambridge: University Press).

Mason, B. J., 1957. *The Physics of Rainclouds* (Oxford: Clarendon Press).

Schaefer, V. J., 1966. Ice nuclei from automobile exhaust and iodine vapor, *Science, 154:* 1555–1557.

32

Pollutants and Cloud Theodore B. Smith
Condensation Processes

An understanding of the physical processes involved in nucleation and an appreciation of the properties of natural nucleus
populations is essential to a judgment of the likelihood of modification of precipitation processes by pollutant aerosol. On the
other hand, there are broad thermodynamic arguments which
suggest that conceivable levels of pollution are not likely to affect
climate on a global scale by way of nucleation effects alone. To
avoid an unbalanced accumulation of detail, only a brief overall
description of the nucleation process is included in the body of
this report: the necessarily much fuller treatment by me and my
colleagues is in Appendix 1* (8).

There is much more likelihood of detectable, even serious,
modification of precipitation processes on a local and perhaps
on a regional scale, and some of my conclusions on these matters
are mentioned here and repeated in Appendix 1.

General Nature of the Nucleation Process
Certain atmospheric particulates act as cloud condensation nuclei (CCN) or ice nuclei (IN). CCN are those particles in the
atmosphere on which water vapor condenses to form cloud droplets. IN are those particles which have the special property of
nucleating the ice phase in clouds either by nucleating super-
cooled droplets or by serving as centers upon which ice is deposited directly from the vapor phase.

The concentration in the air of those CCN which are active
at the maximum supersaturation existing in a cloud determines
the concentration of cloud droplets. This is one factor controlling
the efficiency with which cloud droplets can grow by coalescence
to form raindrops in a warm cloud. Thus, if the concentration
of effective CCN (and therefore cloud droplets) is high the average size of the droplets will be small and their growth to raindrop
size will be difficult. This is thought to be the situation for clouds
forming in continental interiors. On the other hand, in maritime

Reprinted from G. D. Robinson, *Long-Term Effects of Air Pollution—A
Survey,* prepared for the National Air Pollution Control Administration (Hart-
ford: Center for the Environment and Man, Inc.), pp. 25–31.
* All section or appendix references in this paper refer to the original report.

air masses the concentration of CCN is quite small, the average size of the cloud droplets is therefore larger than in continental clouds, and raindrops are produced more readily. The addition of CCN to a cloud mass therefore increases the time required for the development of precipitation. On the other hand, if relatively small numbers of highly efficient CCN (so-called giant nuclei) are introduced into a cloud they may serve as preferential centers for condensation and these droplets may increase rapidly in size to form raindrops. (This, of course, is the principle behind the seeding of warm clouds with giant salt particles in order to enhance precipitation.)

In the case of clouds which extend above the 0°C level, the growth of ice particles by the Bergeron-Findeisen process (distillation from supercooled droplets to ice particles) provides another mechanism by which precipitation particles may be formed. The concentration of ice particles is related to the concentration of active ice nuclei in the cloud. Below a certain critical concentration of ice nuclei, the ice particles can grow to precipitation size fairly readily. However, if the ice nuclei exceed this critical concentration the formation of precipitation may be hindered. The nucleation of the ice phase in clouds has an additional and important effect, namely, that it releases a significant quantity of latent heat which increases the buoyancy of the cloud. Under certain environmental conditions, this can result in the "explosive" growth of the cloud. These ideas have received general confirmation in experiments in which supercooled clouds are seeded with artificial ice nuclei.

Nature and Origins of Cloud Nuclei

Since the maximum supersaturations which exist in clouds rarely exceed 1 percent, only the larger (say > 0.05 μm) and generally hygroscopic particles in the air act as CCN. Typical concentrations of CCN are 10^2 cm^{-3}, whereas the total concentration of particles may be of the order 10^5 cm^{-3}. The surface of the earth and the ocean are thought to be sources of CCN. Recently, certain industries (e.g., paper mills) and artificial and natural fires have been identified as sources of very effective CCN.

Natural ice nuclei are very rare. Typically, only 1 particle in 10^{11} in the atmosphere is effective as an ice nucleus at a temperature of $-10°C$. The exact mode of action of an ice nucleus is still a matter of dispute. The relatively few studies which have been made of the composition of natural ice nuclei indicate that silicate minerals from the earth's surface are dominant. Certain industries (e.g., steel mills) emit large quantities of ice nuclei into the atmosphere. Measurements made at three widely separated sites (Hawaii, Alaska, Washington) indicate that under certain conditions ice nuclei may be advected over distances of thousands of miles.

Pollution and Nucleation—Condensation Nuclei

It is estimated that globally only a few percent of the CCN are man-made, but in localized urban areas the number of artificial nuclei may exceed the natural population. Pollution aerosol may also play a part in the activation or de-activation of nuclei by coagulation with natural nuclei. Pollutants may produce nuclei by secondary reactions in the atmosphere.

The effectiveness of additional CCN in modifying precipitation processes depends in practice more on their size range and that of the natural population to which they are added than on overall numbers. They change the rate at which the coagulation process becomes effective in producing rain, not the nature of the process. Some computations in realistic cases are set out in Appendix 1 where it is concluded that, "the results can be viewed from the perspective of the overall mechanism involved in the modification of the coalescence process. It is generally assumed, to a first approximation, that the dynamics and lifetime of the warm cloud are not affected by changes in the coalescence growth rate. This means that precipitation changes can result only from changes in the storage of cloud water, either increased or decreased. Since the water supply of most of the large clouds exceeds the cloud storage capacity under natural conditions, it is only the smaller and marginal clouds which are likely to be affected by the changes in particulate concentrations."

Pollution and Nucleation—Ice Nuclei

We have seen that natural ice nuclei are not abundant; it also appears that very few pollutant particulates possess the ice-nucleating property. Details of some measurements are given in

Appendix 1 which suggest that urban (specifically Los Angeles) air may sometimes contain fewer ice nuclei than the surrounding rural air—presumably because of deactivation by coagulation. On the other hand, large numbers of IN have been detected in the effluent from steel mills, and some urban atmospheres (specifically Seattle) have been found to contain an excess of IN. Two further possible sources of pollutant IN are examined in Appendix 1, where it is suggested that they are probably not significant. The first, suggested by Schaefer, results from reactions between lead compounds emitted in automobile exhaust gases and iodine vapor, to produce PbI, which is a highly efficient IN. The second is the exhaust particulates of jet aircraft.

Direct measurement within the exhaust of a jet engine detected changes in ice nucleus content which were not large enough to be considered significant, and the importance of Schaefer's mechanism has not been confirmed by independent observation in the atmosphere, perhaps because iodine vapor, even in the minimal concentrations which would be significant, exists only as a rare local pollutant. To quote from Appendix 1, "The foregoing studies indicate that the effect of air pollution on ice nuclei concentrations may be quite variable, depending on the type of pollution, concentrations, etc. Values may range from the 1000/liter found in the French industrial regions to changes of less than an order of magnitude to, finally, the Los Angeles area where there is a tendency for lowest ice nuclei values to be associated with heavy pollution. Schaefer's comments on iodine and lead reactions in automobile exhausts can be viewed in the perspective of the measurements made in the two metropolitan areas of Seattle and Los Angeles. In the one case, area concentrations may have been increased by a factor of six as a result of automobile and/or industrial sources. In the other case, with industrial sources more restricted than in most areas, the ice nuclei counts in pollution tend to be lower than in unpolluted regions. These comments support the general conclusion that local effects of the iodine and lead reaction may occur but there appears to be no substantial evidence of a widespread contribution from this source.

Long-term records of ice nuclei variations essentially do not exist. Partly this results from the modifications and improvements in ice nuclei measurements which have occurred in the past 15 years. Grant at Colorado State University, however has measured ice nuclei routinely at Climax, Colorado, for the past 10 years. Fortunately, these measurements have been consistent in terms of observational technique. Grant (private communication) reports no definable change in ice nuclei counts over the 10-year period which might be considered as a background trend. It remains possible, however, that areas such as the eastern sections of the United States or portions of industrial Europe might be experiencing gradual increases in ice nuclei content but that an efficient natural removal process might serve to make the trend indiscernible at Climax. The Climax data also fail to show any pronounced influence from polluted areas to the west such as Los Angeles. This is in keeping with measurements made in the Los Angeles area itself which fail to show widespread ice nuclei effects, even in the source region itself."

Observational Evidence of Precipitation Changes

Local Effects

Substantial evidence of precipitation modification is meager but significant. The two outstanding and oft-quoted examples are an increase in reported rainfall at La Porte, Indiana (1), and local decreases in Australia (2). Changes in precipitation of 25 to 30 percent have been found in these areas. Hobbs, Radke, and Shumway (3) provide suggestive data relating precipitation increases in Washington to the location of major industrial complexes. Increases of over 30 percent were found for the period 1947–1966 compared to 1929–1946. Miller (4) found apparent increases in precipitation of the order of 15 percent over Long Island and downwind of New York City. The remaining examples of precipitation effects have been summarized by Changnon (5) and Peterson (6) and generally show changes in precipitation of the order of 10 percent or less. It is useful to examine these cases in the perspective of the preceding sections.

Changnon (5) has found that much of the increase in reported precipitation at La Porte occurs during the warm season and that the number of thunderstorm days is increased significantly and concludes that midwest urban areas produce significant increases in convective activity. Principal precipitation effects come from an increase in the number of days with 0.25 inch of rain or more. The conclusion was reached that thermal and frictional effects were primarily responsible for the urban effects in the midwest and that La Porte represented a unique situation due to an unusual combination of urban, industrial, and lake contributions.

These conclusions are in agreement with the available results from advertent modification experiments. Increases in precipitation of as much as 30 percent in the annual rainfall in an area such as La Porte seems to require thermodynamic seeding effects, i.e., through the release of additional convective activity. Microphysical effects such as have been observed at Climax should appear primarily in terms of increased number of light precipitation days and should, as well, be more related to stratiform cloud types. Ice nuclei, according to Langer (7) were measured in concentrations of about 30/liter at $-20°C$. Results of seeding programs at Flagstaff and elsewhere suggest that this concentration may not be sufficient to produce the dynamic seeding effects found in the advertent cumulus seeding programs. Condensation nuclei effects, although not discussed at length in the La Porte example, have not been shown to result in such large precipitation increases. It is concluded that the thermal and frictional effects, resulting in frequent updraft regions in a localized area and the stimulation of convective motions in this area, are the most likely causes of the pronounced increases indicated by the La Porte observations. A corollary of this conclusion is that it should be possible to identify those days on which convective activity is so poised that additional stimulation is sufficient to release the latent convective instability in the preferred area. This identification problem has been considered in the case of cumulus seeding with considerable success.

The Australian study by Warner has shown apparent decreases in precipitation of the order of 25 percent. These effects have been attributed to the production of large numbers of small cloud droplets and a consequent decrease in coalescence growth rate. Cloud droplet concentrations averaged about 900 per cm^3

in the inland areas where cane smoke might have affected the cloud microstructure. Peak droplet concentrations were over 2500 per cm³.

It has been indicated earlier that the principal effect of a slowdown in the coalescence growth rate would be to prevent precipitation from developing in the smaller clouds whose lifetimes are not particularly long. It has been estimated that a decrease of 25 percent in total precipitation would result if precipitation were eliminated from clouds with diameters less than 3 to 4 km. Larger clouds (with longer lifetimes) might produce rainfall in much the same manner, regardless of cloud droplet concentration. There is evidence in Warner's data that the number of showery days has been decreased substantially but that the amount of rain per shower day has not changed in a similar manner.

The concentration of cloud droplets reported by Warner is not extreme in comparison to numbers found in other areas. In Flagstaff, concentrations of 700–1000/cm³ are frequent and occasional values of 2000/cm³ are observed. In Flagstaff, however, substantial coalescence rain occurs in clouds of 3 to 4 km diameter in spite of the large droplet concentrations. Number of droplets/cm³ should not be the ultimate predictor of the occurrence of coalescence precipitation, and it is entirely possible that a sufficient number of large particles may have been present at Flagstaff to initiate the precipitation. It is concluded, therefore, that the Australian data, if statistically sound, may show the result of added condensation nuclei but that a droplet distribution devoid of large particles would be required to produce such a marked change in precipitation as has been reported.

Hobbs, Radke, and Shumway (3) indicate that centers of increased precipitation appear downwind of local industrial sources such as smelters or paper mills. Precipitation increases are quoted at 30 percent for several areas when comparing the industrially active, recent years (1947–1966) with the earlier years of 1929–1949. The authors point out, however, that the entire northwest apparently experienced heavier rain in the 1947–1966 period than in the earlier era. The value of 30 percent includes both the effects of the wetter areal trend and the possible effects of industrial sources. No attempt has been made to separate these two effects and a cursory glance at the data suggests that no more than a 15 to 20 percent increase should be attributed to the industrial effects. The authors attribute the apparent increase to the effects of condensation nuclei released from the industrial plants.

Observations are decribed in the article by Hobbs et al., which indicate numerous occasions of cumulus cloud development downwind of the particular plants in question. On one occasion, a cloud street of 30 km in length was reported downwind of a paper mill in a manner which can hardly be attributed to condensation nuclei. Evidence is also given to indicate that the precipitation in the 1947–1966 period was more convectively generated than in the earlier period. The size of the apparent increase (second in magnitude only to that reported from La Porte), the cloud observations, and the convective nature of the precipitation all suggest that dynamic effects on the precipitation mechanisms should not be ruled out of consideration. The similarities to the La Porte situation in terms of cloud developments and precipitation increases are striking. In addition, calculations such as have been possible to date do not support such large precipitation increases as 15 to 20 percent (in annual rainfall) as a result of stimulation of the condensation–coalescence process. It is concluded, therefore, that the Washington data may afford another example of a thermodynamic or frictional effect on precipitation in an environment somewhat different from that existing at La Porte.

Data on Long Island (4) suggest precipitation increases of around 15 percent downwind of New York City compared to surrounding areas. No detailed examination of the concurrent environment conditions has been given. The pattern of the increase, with isohyets symmetrical with respect to the island's longest axis, is suggestive of dynamic effects rather than microphysical causes. It could be hypothesized that convection over the island with reference to the cooler surrounding water surfaces might contribute to such an isohyetal pattern. As in the La Porte case, the effect of New York City could be to initiate the convection in a consistent location.

The remaining examples of precipitation changes are of the order of 10 percent or less. Little physical documentation of the environment conditions is usually given and the increases themselves are small enough to be more subject to doubt than the preceding cases. Under these conditions, it is possible to comment on plausible reasons for the precipitation effects if, in fact, they can be shown to be significant.

Regional Effects

In view of the possible influence of nuclei on warm cloud precipitation, as shown by the Australian data, it was decided that long-term trends in summer rainfall in several areas of the United States should be examined. In the light of probable increases in pollution levels over a period of years in the eastern sections of the country, widespread effects on precipitation might be expected to appear first in summer precipitation amounts in these

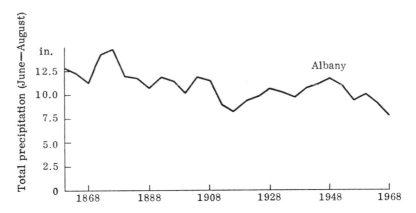

Figure 32.1 June–August precipitation trends

areas. June, July, August rainfall was plotted for a period of years for several stations in the eastern part of the United States. Data for Albany and St. Louis are shown in Figure 32.1. Data for Nashville, Cincinnati, and Philadelphia were also examined but show no particular trends over the past 100 years or so.

The most pronounced summer rainfall trend found at any of the five stations is shown in the curve for Albany in Figure 32.1. Summer precipitation in this area appears to have declined 20 to 25 percent over the period of the last 100 years. It can be argued that this trend was particularly pronounced during the past 20 years, but such short-term trends may occur as a result of a number of causes. At St. Louis, the general trend was downward until the mid-1930's, after which evidence of rising summer precipitation amounts is apparent.

Long-term trends in areal precipitation amounts could result from a variety of causes. Along with the possibilities of nuclei and moisture effects, it is quite probable that true climatic trends, identified with long-term circulation changes, may also play a role in determining precipitation trends. As we have seen such long-term changes may be associated with environmental changes, or may be a result of "almost-intransitivity." It is important, therefore, that identification of such areas as Albany be followed by more detailed studies to identify possible causes.

References

1. S. A. Changnon, "The La Porte Weather Anomaly, fact or fiction?" *Bull. Amer. Meteor. Soc., 49,* pp. 4–11, 1968.

2. J. Warner, "A reduction in rainfall with smoke from sugar-cane fires—an inadvertent weather modification," *J. Appl. Meteor., 7,* pp. 247–251, 1968.

3. P. V. Hobbs, L. F. Radke and S. E. Shumway, "Cloud condensation nuclei from industrial sources and their apparent influence on precipitation in Washington State," *J. Atmos. Sci., 27,* pp. 81–89, 1970.

4. J. F. Miller, "Precipitation regimes over Long Island, N. Y.," ESSA Office of Hydrology, 1967.

5. S. A. Changnon, "Recent studies of urban effects on precipitation in the U.S.," *Bull. Amer. Meteor. Soc., 50,* pp. 411–421, 1969.

6. J. T. Peterson, "The climate of cities. A survey of recent literature," NAPCA Pub. No. AP-59, 1969.

7. G. Langer, "Ice nuclei generated by steel-mill activity," Amer. Met. Soc. Proc. 1st Nat. Conf. on Weather Modification, pp. 220–227, 1968.

8. T. B. Smith, A. I. Weinstein, and A. J. Alkezweeny, "Nucleation aspects of inadvertent weather modification," MRI 170 FR-912 Meteorology Research, Inc., Altadena, California.

33

**Trends in High Cloudiness
at Denver and
Salt Lake City**

Lester Machta and
T. Carpenter

Introduction

It is well known that contrails form behind jet aircraft. In temperate latitudes the present commercial jets frequently fly just below the tropopause or at altitudes of about 33,000 to 38,000 feet. Here the air is at or near the coldest in the column in which aircraft might fly and can hold the smallest amount of moisture. The temperature, which controls the amount of moisture a parcel of air can hold, decreases up to the tropopause. The moisture property preserved in any nonsaturated mixing process is the humidity mixing ratio or the mass of water vapor per unit mass of air in which it is embedded. The saturated humidity mixing ratio, a measure of the capacity of the air for moisture, increases with height above the tropopause even though the temperature remains constant as is frequently the case. Further, with few exceptions, the relative humidity decreases from the troposphere to the stratosphere. Thus, the present jet flight altitude may be the most vulnerable part of the atmosphere in which man can propel himself insofar as cloud formation from added moisture is concerned.

Very large amounts of moisture are emitted from the exhaust of a modern commercial airliner; about 1.25 pounds for each pound of fuel consumed. A large four-engine aircraft like a DC-8 or Boeing 707 uses about 46,000 pounds of fuel per hour of cruise flight. While this amount of water from an aircraft may appear to be high, it will form extensive clouds rather than transient contrails only when the air is very close to saturation, a relative humidity of over 99 percent. Under such almost saturated conditions, it is frequently argued, nature is on the verge of forming clouds anyhow, and the artificial cirrus clouds develop only momentarily sooner than would naturally occur. Thus, while we may a priori suspect an increase in cirrus cloudiness when commercial jet aircraft arrived on the scene in about 1958, there are many who would justifiably discount their importance in enhancing the incidence of cirrus clouds. This study attempts to see if

Prepared for SCEP and expanded after SCEP.

there have been changes in the pattern of cirrus clouds after commercial jets began operation. It must be viewed only as a progress report.

Observed Changes in High Tropospheric Cloudiness

The records of cloud observations at Denver and Salt Lake City have been examined to detect changes associated with the onset of commercial subsonic jet aircraft operations. The work, as of this writing, is still in progress and, while suggestive of a correlation between jets and high clouds, is not conclusive. Two stations, Denver and Salt Lake City, were chosen for several reasons: (1) few low or middle clouds permit frequent observations of high clouds; (2) the two stations are on or near major air lanes; if any changes in high cloudiness were likely, they would show up at these stations; (3) records of airway weather observations were available on magnetic tapes. The second reason, being near air lanes, suggests that even if changes are proved for Denver and Salt Lake City, they may not occur more generally. But, it may be noted that there are places in the United States where flights are even more numerous than over Denver and Salt Lake City.

The data for the following analysis, cloud measurements at the two airport stations, are notoriously difficult to observe quantitatively. This defect, it is hoped, may be overcome by averaging many observations and by subjecting the results to independent checks. Second, the relationship between commercial jet aircraft and increased high cloudiness, if true, depends exclusively on coincidence in time. There may be trends in high clouds due to other causes such as changes in circulation patterns caused by other than aviation activities. Finally, even if a convincing correlation between clouds and aircraft by this analysis is demonstrated, it may not be due to added moisture. The atmospheric disturbance by the aircraft, that is the convection by the aircraft motion or engine heat, might also be the cause of the enhanced cloudiness.

Figure 33.1 shows the average annual high cloudiness at Denver for each year beginning in 1940. From 1940 to 1948 observations were made at a military base, Lowry Field, and thereafter at Stapleton Field, the commercial airfield serving Denver. Figure 33.2 shows the same high cloud history at Salt Lake City

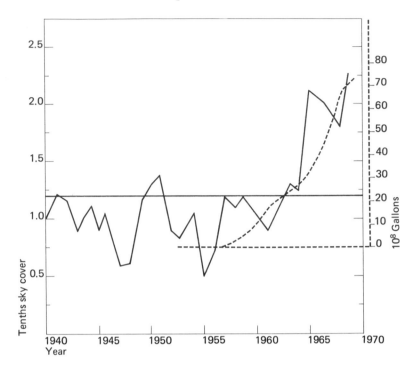

Figure 33.1 The history of the annual high cloudiness at Denver, Colorado
The dashed line shows the growth in jet fuel consumption by domestic commercial jet aircraft

but beginning in 1949. In both cases only the three-hourly observations with no low or middle clouds entered the annual averages. On Figure 33.1 the gross consumption of jet fuel has also been plotted; the gross fuel consumption probably reflects the relative frequency of jet flights over the two stations.

The two graphs, Figures 33.1 and 33.2, shows an increase in high cloudiness after the onset of commercial jet operations in about 1958; the growth at Denver is more marked than at Salt Lake City. Thus, this correlation suggests, but does not prove, that commercial jet activity may indeed have increased the amount of high clouds. However, the conclusion should be tempered by other analyses of the cloud data. These are:

1.   At both Denver and Salt Lake City, the relationship for those observations with one- or two-tenths of low or middle clouds with commercial jet aircraft activity is as good as, or better than, that shown in Figures 33.1 and 33.2.

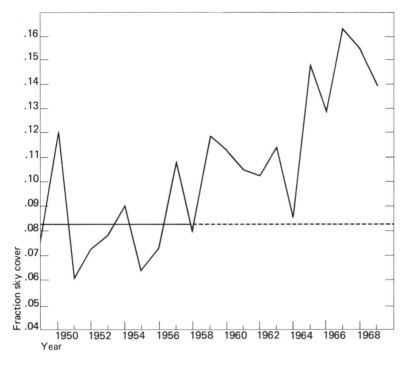

Figure 33.2 The history of annual high cloudiness at Salt Lake City, Utah
Source: Carpenter.

2.   The observations of high clouds (with no low or middle cloudiness) were analyzed for Denver by the time of the day: daylight hours, dark hours, and sunrise and sunset hours. It is likely that the sunrise and sunset hours provide the most reliable high cloud observations. All three sets of data parallel one another although the post-1958 growth is not quite as distinct for the more reliable sunset and sunrise hours as for Figure 33.1.

3.   The trends in the annual mean cloudiness of all three layers (low, middle, and high) are individually similar in each of the four seasons.

4.   The history of middle clouds, in the absence of low cloudiness, has been analyzed in the same fashion as shown for Figures 33.1 and 33.2. Middle cloudiness at Salt Lake City shows no trend of significance; if anything there is a very slight downward trend until about 1960. But the middle cloudiness at Denver exhibits a disturbing, marked decrease of about 0.1 sky cover from 1944

to 1958 and remains about constant thereafter. This decrease in middle cloudiness shows that one can observe marked long-term changes without any apparent explanation. The immediate implication is that the upward trend in high clouds between 1958 and 1969 might be due to some similar cause such as circulation or observational procedural changes. However, the most probable change in observation practice as the cause of the increase of high clouds after 1958 does not appear to be borne out by the changes in middle and high clouds. Aircraft pilots can advise weather observers about the cloud heights. For this reason, one might presume that the upward trend in high clouds might coincide with a decrease in middle clouds as the pilots tell the observer the true cloud level. At both stations, there was no apparent positive or negative correlation between the trends of middle and high clouds after 1958; the mean middle cloudiness remained about constant with time. The smaller amounts of middle clouds after 1958 would have offered more opportunity for observing high clouds. But, one might also have more frequent clear skies, which would decrease the mean amount of high clouds. The greater opportunity to see high cloudiness need not increase its mean annual value.
5.   There have been no significant trends in low cloudiness.

With one significant exception the auxiliary analyses do not contradict the growth of cirrus cloudiness after 1958 at either Denver or Salt Lake City. The analyses cannot more specifically assign the cause to increased jet aircraft activity. The one disturbing finding is the sharp decrease of middle cloudiness between 1944 and 1958. As noted, this unexplained decrease raises serious doubts as to the justification for attributing the post-1958 rise in cirrus cloudiness to aviation activities. Finally, one must repeat that the increase after 1958 is ascribed to aviation activity only by its coincidence in time and the visual evidence of contrails behind aircraft.

Assuming that the cirrus increase is due to subsonic jet aircraft, one might expect a continued increase in such cloudiness with time. Future projections call for a sixfold increase in passenger-miles by the period between 1985 and 1990, but improved technology may result in an aircraft efficiency such that the amount of moisture will increase by only threefold.

It is believed that the case for aviation causing an increase in cirrus cloudiness at Denver and Salt Lake City is sufficient to justify further studies. These should be aimed at proving or disproving the hypothesis and, if true, to determine the geographical extent of the increased cirrus clouds. Lastly, if the growth is due to aircraft, one must describe the optical and other properties of the artificial clouds to permit an assessment of the influence of the clouds on the weather and climate.

Reference

Carpenter, T. Meteorological statistics of the Environmental Sciences Services Administration.

34

Observations　　　　　　　　Julius London
of the Distribution and
Variation of Cloudiness

The distribution of cloudiness is an important atmospheric parameter because of its central role in affecting the radiation budget of the atmosphere—and therefore the evolution of climate.

Clouds reflect solar radiation (the reflectivity differs according to cloud type) and, in general, act as very efficient radiators for terrestrial radiation. While the geographic and temporal distribution of clouds, total amount as well as type and height, is an important climatic indicator per se, this information is particularly necessary for computations of the radiation budget and concomitant dynamics of the atmosphere. Also, because cloud transmissivity of solar radiation is higher than that for terrestrial (infrared) radiation, they are important contributors to the so-called "greenhouse effect."

Before 1960 our knowledge of the cloudiness distribution was derived almost entirely from ground-based land and ship observations. Most of the observations were made over populated land areas and narrow, but frequently used, shipping lanes. In general, relatively few observations were available in the tropics and in the Southern Hemisphere.

The ground-based observations show that the earth is covered about 50 percent by clouds with possible global variations from about 48 to 52 percent. There are also suggested large seasonal variations in the tropics and subpolar latitudes and marked longitudinal variations at all latitudes. There are, however, no reliable data on the year-to-year variation of total global cloudiness, although the results of some studies have suggested a recent increase in cirrus clouds. (See Chapter 33.)

Since 1960, satellite observations have been used to give some indications of the global cloudiness distribution. Because of the resolution and video systems used (where there is a need to set a proper threshhold brightness for cloud detection) individua small cloud groups are generally smeared out, and often thin cirrus clouds are not detected (although cirrus clouds could be inferred from infrared sensors). If the brightness contrast is low, the threshhold will be set at a low value and satellite-observed clouds

Prepared during SCEP.

may be reported in higher amounts than actually present. Some simultaneous observations by different satellite systems indicate that this problem cannot be neglected if correct cloud populations are to be reported. Also, it is sometimes difficult to distinguish clouds from highly reflecting surface systems such as ice fields or desert areas. Despite these limitations, satellite observations have already shown the existence of unique cloud cluster formations and large cloudiness variability in the tropics and have provided considerable information on relative cloudiness in the Southern Hemisphere where there was almost no prior data.

Since 1966, daily (and with the launch of ITOS-1 in 1970, twice daily) global cloud-cover maps are available *on a routine basis.* Detailed ground-based cloud-cover data should be used to calibrate the satellite observations so that inferences made from satellite information in sparse data areas be correctly interpreted. This coordination will help in standardization of the satellite observations so that absolute value of the total cloud cover and their variations can be determined.

Comparison of satellite cloud photographs with ground or aircraft observations shows that there is a marked tendency for low-resolution satellite observations to give relatively poor cloud definition in distinguishing individual clouds or small cloud groups. That is, clouds tend to be "smeared" together in some of the satellite pictures. Spatial (horizontal) resolution of the order of 500 meters is therefore needed in order to get "correct" cloud population counts. We understand that this is within the capabilities of satellite systems presently in operation and those planned for the near future.

Clouds appear in the atmosphere in different forms and at different levels. This information is, of course, exceedingly important in heat budget and climatic change studies. However, there is no practical method known, at present, to determine precisely the distribution of cloud amounts by type and height. Models have, therefore, been developed of these distributions based on the regular meteorological synoptic and aircraft reports. It would be highly desirable if surface and satellite systems (making use of visible, infrared and microwave sensors) can be coordinated to derive the necessary picture of the three-dimensional cloud patterns.

A method for obtaining standardized global cloud observations will provide the information for studies of the month-to-
month, year-to-year, and trend variations, as well as latitudinal
and hemispheric differences in cloudiness and cloudiness patterns
(whether natural or man-made). It would be helpful if all this
information were processed in such a form that it can be readily
available for improved radiation studies and the analysis of climatic trends, both "natural" and inadvertent.

With the advent of large rockets, high-flying jet aircraft, and
nuclear explosions we have the capability of changing the upper
atmosphere's composition. This is possible because there is relatively little substance there to begin with, and, as shown in the
paper prepared for SCEP by Dr. Martell, contaminants can remain there for years. However, it is important to consider each
contaminant carefully before drawing conclusions about the significance of "pollution" of the upper atmosphere. This summary
reviews some of the factors that must be considered, and the other
papers in this section amplify various aspects.

Increases in stratospheric water vapor may cause potential
changes in weather and climate through a slight greenhouse effect
which would warm the lower atmosphere and through the increased formation of clouds in the colder parts of the stratosphere.
Three aspects that relate directly to a fuller understanding of the
implications of such climate modifications are discussed in the
paper by Dr. Machta: observed increases in water vapor reported
by H. J. Mastenbrook; speculations on the causes of the observed
increases; and possible increases due to SST operations in the
stratosphere.

Dr. Cadle's paper summarizes our present knowledge of particles in the stratosphere and our theories concerning why they
are there. Much of our information about these particles has
been gained quite recently, and, as Dr. Cadle points out, there
are still some important unanswered questions. It seems likely
that, as a result of a growing interest in the influence of SSTs on
stratospheric aerosol content (generated in large part by the SCEP
Report), there will be an intensification of efforts to study this
intriguing subject.

The fact that large volcanic eruptions in the tropics inject
"dust" into the stratosphere has been recognized since the Royal
Society published its celebrated study of the Krakatoa eruption
(the report is dated 1888), but of course there was no way to
measure stratospheric temperature in those days. When Mount
Agung erupted in 1963, however, there were regular radiosonde
observations of stratospheric conditions being made at many stations throughout the world. Australian meteorologists first called
attention to the stratospheric temperature rise following Agung,

and in the final paper of this series Professor Newell summarizes some of the data obtained. Significantly, although there was a noticeable change in the direct sunlight reaching the earth's surface following the Mount Agung eruption, it appears that no change in the *surface* temperature or "climate" has been linked to this event.

35

Residence Times and Other   Edward A. Martell
Factors Influencing Pollution
of the Upper Atmosphere

Introduction

The possible consequences of the pollution and modification of
our thin upper atmosphere involve many complex questions that,
at present, can be answered only qualitatively and require much
further attention and research.

The important sources of high-atmosphere pollution include
(1) relatively stable gaseous pollutants from industrial and urban
areas (carbon monoxide, methane, hydrogen, and so on) which
mix up into the stratosphere where they are oxidized and contribute to the accumulation of carbon dioxide and water vapor,
(2) water vapor, surface materials, radioactive debris, and trace
elements that are injected into the upper atmosphere by high-
yield thermonuclear explosions in the troposphere or at the
earth's surface, (3) exhaust products of large rockets and supersonic aircraft, (4) chemical releases, and (5) reentry burnup products from rockets, satellites, and space vehicles.

Possible long-term climate change due to increases in water
vapor and carbon dioxide is one of the more pressing questions
relating to upper atmosphere contamination. Other undesirable
effects are those that can result from the introduction of relative
small amounts of foreign materials which may cause some modification of the ionosphere and adversely affect radio communications, as well as atmospheric changes which may interfere with
scientific observations—effects that already have been experienced
in the past. The evaluation of these and other possible pollutant
effects requires an improved knowledge of the background distribution of trace constituents and their role in the upper atmosphere plus better information on residence times, mixing
patterns, and transport mechanisms at high altitudes. Some of
the more relevant published literature is now briefly reviewed.

Many aspects of the problems of upper atmosphere pollution and modification have been considered (Kellogg, 1964; National Academy of Sciences-National Research Council, 1966;
and Pressman, Reidy, and Tank, 1963). With regard to the ques-

Prepared for SCEP.

tion of rocket pollution of the high atmosphere, Kellogg (1964) concludes:

Doubling of the CO_2, H_2O or NO content would require per year on the order of 10^3 to 10^5 Saturn-type rockets, each injecting 100 tons of exhaust above 100 km. On the other hand, a few hundred small rockets per year, each containing 10 kg of the chemical, would probably double the Na content; similarly, less than two such rockets per year would be expected to double the Li content.

Implications of the contamination of the stratosphere by supersonic transport aircraft are discussed elsewhere (National Academy of Science—National Research Council, 1966; Newell, 1970). The influence of increased atmospheric carbon dioxide and stratospheric water vapor on temperatures at the earth's surface have been considered by Manabe and Wetherald (1967) using a model atmosphere with radiative convective equilibrium and with a given distribution of relative humidity.

In the following sections, present knowledge of meteorological factors and residence times which control the distribution of contaminants in the upper atmosphere, the background distribution of water vapor in the stratosphere, and related topics are briefly reviewed.

Meteorological Factors and Residence Times

Circulation and mixing processes that control the redistribution of contaminants in the upper atmosphere vary widely with altitude, latitude, and season. (The atmospheric nomenclature used here is the IUGG nomenclature adopted in 1960 and shown in Figure 35.1.) Tropospheric air rises slowly across the very cold (about −80°C) tropical tropopause layer where aerosols and water vapor are largely removed by ice particle growth and sedimentation. Transport in this region has been discussed by Brewer (1949) and by Reed and Vlcek (1969). Within the tropical stratosphere atmospheric constituents are spread poleward by the competing processes of eddy mixing and slow mean motions (Martell, 1968; Newell, Wallace, and Mahoney, 1966) with very little exchange between hemispheres at stratospheric levels. Rates of meridional and vertical eddy mixing in the stratosphere vary widely with time and space and are highest at the higher latitudes in winter (Newell, Wallace, and Mahoney, 1966). The redistribution

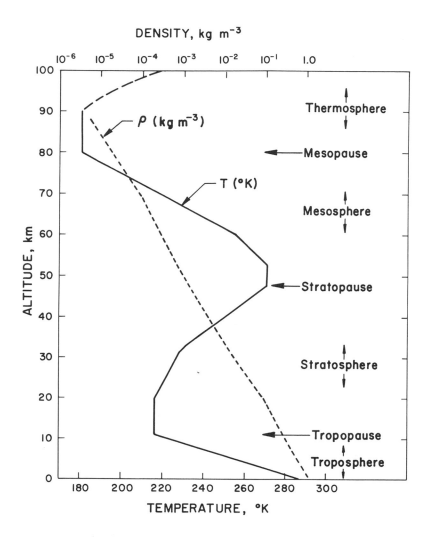

Figure 35.1 Temperature and density profile according to the U.S. Standard Atmosphere (1962) and the IUGG nomenclature (1960)

of stratospheric contaminants will, of course, follow these complex patterns of circulation and mixing.

Rates of eddy diffusion, which control the spread of cloud sources or line sources of pollutants in the atmosphere also vary widely in the upper atmosphere. Eddy diffusion rates in the stratosphere, mesosphere, and lower thermosphere are discussed in a number of publications (Brewer, 1949; Colegrove, Hanson, and

Johnson, 1965; Johnson and Wilkins, 1965; Kellogg, 1956; Kellogg, 1964; Martell, 1970a; and Williamson and Houghton, 1965, and briefly reviewed elsewhere (National Academy of Sciences-National Research Council, 1966).

The mean residence time of contaminants (the average time the contaminant remains in a given region or layer of the atmosphere) increases appreciably with altitude between the tropopause and stratopause (Figure 35.1), with estimates of about one month at the tropopause, about one to two years at 20-km altitude and about four to twenty years at 50-km altitude. Because of the relatively very rapid rate of vertical eddy mixing throughout the mesosphere (50 to 80 km) and lower thermosphere (80 to 100 km), contaminants injected at these levels mix rapidly throughout these layers down to the stratopause (50 km), a volume which contains only 0.1 percent of the total mass of the atmosphere.

Estimates of upper-atmosphere residence times are based on observations of nuclear bomb debris as well as natural atmospheric radioactivity, and on the results of two high-altitude radioactive tracer experiments, rhodium-102 tracer produced in a high altitude thermonuclear explosion on August 11, 1958, and cadmium-109 tracer produced in another high-altitude thermonuclear explosion in July 1962. Estimation of atmospheric residence times on the basis of the concentration of one or more radioactive cosmic ray spallation products are given by Lal and Peters (1962) and Bhandari, Lal and Rama (1966). The use of the ratios of polonium-210 and lead-210, radioactive decay products of atmospheric radon, to estimate the residence times for particulate aerosols in the lower stratosphere and troposphere is discussed by Burton and Stewart (1960) and Peirson, Cambray, and Spicer (1966) and has been questioned by Martell and Poet (1969). Results for the high-altitude tracers, rhodium-102 and cadmium-109, are discussed in numerous publications (Kalkstein, 1962; Leipunskii et al., 1970; Martell, 1970a; Salter, 1964; and Telegadas and List, 1964).

Water Vapor in the Stratosphere and Mesosphere

The normal distribution of water vapor in the lower stratosphere has been determined in a number of investigations employing good techniques of measurement on balloons up to about 30 km altitude (Mastenbrook, 1968; Neporent et al., 1967; and Williamson and Houghton, 1965). (See also Chapter 36.) The results show that the water vapor mixing ratio in the lower stratosphere varies between 2.1×10^{-6} and 3.5×10^{-6} g H_2O per g air, with the lowest mixing ratio occurring during late winter and spring. The observations by Neporent et al. (1967) and Mastenbrook (1968) indicate that the mixing ratio of water vapor increases with altitude in the stratosphere. This trend is confirmed and explained by recent direct measurements of water vapor and methane in an air sample collected near the stratopause by means of a rocketborne cryogenic air sampler (Martell, 1970a and Scholz et al., 1970). These measurements show that the atmospheric methane concentration, averaging about 1.4 ppmv (parts per million by volume) in the troposphere, is ≤ 0.05 ppmv near the stratopause. Thus it is shown that methane is destroyed chemically as it mixes upward in the stratosphere, perhaps by reaction with atomic oxygen in accordance with the suggestion of Cadle and Powers (1966). The expected products are water and carbon dioxide (other hydrocarbons released in the stratosphere also would be rapidly oxidized to water and carbon dioxide). Based on these considerations, the concentration of stratospheric water vapor must increase with altitude from about 1 to 2 mg H_2O per kg air in the lower stratosphere to 3 to 4 mg H_2O per kg air in the upper stratosphere (Martell, 1970c).

The very low water-vapor mixing ratio of the lower stratosphere is accounted for by ice crystal formation in air moving up across the "cold trap" at the tropical tropopause. This removal process is bypassed by two types of events which have taken place repeatedly in the past: (1) major volcanic explosions such as Krakatoa (1883), Katmai (1912), and Agung (1963); and (2) major thermonuclear weapons tests, particularly the 1954 *U.S. Castle* tests at 11°N and the massive 1961 and 1962 Soviet tests at 75°N. In such events massive quantities of water, aerosols, and trace gases are injected directly into the lower and middle stratosphere, well above the cold trap. It is quite possible that some of these events have increased the water vapor mixing ratio of the stratosphere to a significant extent for long periods over large areas near the latitude of injection. It is unfortunate that stratospheric measurements of water vapor and other trace constituents were not made systematically following recent events of this nature. Plans should be made to do so when such events occur in the future.

Concluding Remarks

Background information on the upper-atmosphere distribution of water vapor, hydrogen compounds, carbon dioxide, and other minor constituents should be improved by periodic observations of their distribution, to observe their natural variations in time and space and/or to monitor their accumulation from pollutant sources. Because carbon dioxide exhibits large seasonal and regional variations in the troposphere it may be easier to follow its increase with time by measurements in the upper stratosphere and the mesosphere.

Now that techniques are available for direct measurements of trace gases and aerosols at rocket altitudes (Martell, 1970b) it should be possible to obtain better values for the residence time of trace constituents in the upper stratosphere and mesosphere. This can be done by monitoring the rate of decrease in the present high excess of bomb-produced tritium near the stratopause (Scholz et al., 1970) or by measurements following future high-altitude tracer experiments.

Estimates of upper atmosphere mean residence times, one to two years at 20-km altitude and four to twenty years at 50 km, are based on radioactive particle tracers. Because gravitational sedimentation influences the atmospheric redistribution of particles, the estimated residence times may be significantly shorter than those applicable to water vapor and trace gases at these levels. Such differences should be evaluated in future atmospheric tracer experiments.

It is desirable to investigate the distribution of the important minor constituents in the upper atmosphere and to gain a fuller understanding of their influence on chemical and physical processes to the extent practical before their natural abundances are significantly increased by contributions from pollutant sources. Some examples of natural minor constituents of the upper atmosphere which already are masked by the products of thermonuclear tests include cosmic-ray-produced carbon-14 and tritium, as well as atomic lithium, which is observed in the twilight airglow near the mesopause.

Large-scale thermonuclear tests have given rise to a wide range of temporary environmental changes and effects. One possible effect that has not been adequately studied is the extent to which large thermonuclear explosions at the earth's surface and in the troposphere may have increased the water content of the stratosphere and its possible temporary influence on climate in periods following such tests.

References

Bhandari, N., Lal, D., and Rama, 1966. Stratospheric circulation studies based on natural and artificial radioactive tracer elements, *Tellus 18* (2): 391–406.

Brewer, A. W., 1949. Evidence for a world circulation provided by the measurements of helium and water vapour distribution in the stratosphere, *Journal of the Royal Meteorological Society, 75:* 351–363.

Burton, W. M., and Stewart, N. G., 1960. Use of long-lived natural radioactivity as an atmospheric tracer, *Nature, 186* (4725): 584–589.

Cadle, R. D., and Powers, J. W., 1966. Some aspects of atmospheric chemical reactions of atomic oxygen, *Tellus, 18:* 176–186.

Colegrove, F. D., Hanson, W. D., and Johnson, F. S., 1965. Eddy diffusion and oxygen transport in the lower thermosphere, *Journal of Geophysical Research, 70:* 4931–4941.

Johnson, F. S., and Wilkins, E. M., 1965. Thermal upper limit on eddy diffusion in the mesosphere and lower themosphere, *Journal of Geophysical Research*, 70: 1281–1284; Correction to Thermal upper limit on eddy diffusion in the mesosphere and lower thermosphere, *Journal of Geophysical Research, 70:* 4063.

Kalkstein, M. J., 1962. Rhodium-102 high altitude tracer experiment, *Science, 137* (3531): 645–652.

Kellogg, W. W., 1956. Diffusion of smoke in the stratosphere, *The Journal of Meteorology, 13:* 241–250.

Kellogg, W. W. 1964. Pollution of the upper atmosphere by rockets, *Space Science Reviews, 3* (Dordrecht: D. Reidel Publishing Co.), pp. 275–316.

Lal, D., and Peters, B., 1962. Cosmic ray produced isotopes and their application to problems in geophysics, *Progress in Elementary Particle and Cosmic Ray Physics, 6,* edited by J. G. Wilson and S. A. Wouthuysen.

Leipunskii, O. I., Konstantinov, J. E., Fedorov, G. A., Scotnikova, O. G., 1970. Estimation of the mean residence time of radioactive aerosols in the upper layers of the atmosphere based on fallout of high altitude tracers, *Journal of Geophysical Research, 75:* 3569–3574.

Manabe, S., and Wetherald, R. T., 1967. Thermal equilibrium of the atmosphere with a given distribution of relative humidity, *Journal of Atmospheric Sciences, 24* (3): 241–259.

Martell, E. A., 1968. Tungsten radioisotope distribution and stratospheric transport processes, *Journal of Atmospheric Sciences, 25:* 113–125.

Martell, E. A., 1970a. Transport patterns and residence times for atmospheric trace constituents vs. altitude, *Advances in Chemistry* series, No. 93: 138–157 (Washington, D.C.: American Chemical Society).

Martell, E. A., 1970b. High altitude air sampling with a rocketborne cryo-condenser, *Journal of Applied Meteorology, 9:* 170–177.

Martell, E. A., 1970c. Hydrogen compounds in the stratosphere and meso-sphere, IACC Symposium, Tokyo, September 1970.

Martell, E. A., and Poet, S. E., 1969. Excess atmospheric ^{210}Po—Pollutant or natural constituent?, *Transactions of the American Nuclear Society, 12:* 484–485.

Mastenbrook, H. J., 1968. Water vapor distribution in the stratosphere and high troposphere, *Journal of Atmospheric Sciences, 25:* 299–311.

National Academy of Sciences-National Research Council, 1966. *Weather and Climate Modification,* Vol. II, Publication No. 1350, pp. 97–105.

Neporent, B. S., Kiseleva, M. S., Makogonenko, A. G., and Shlyakhov, V. I., 1967. Determination of moisture in the atmosphere from absorption of solar radiation, *Applied Optics, 6* (11): 1845–1850.

Newell, R. E., 1970. Water vapour pollution in the stratosphere by the super-sonic transporter?, *Nature, 226:* 70–71.

Newell, R. E., Wallace, J. M., and Mahoney, J. R., 1966. The general circula-tion of the atmosphere and its effects on the movement of trace substances, Part 2, *Tellus, 18:* 363–380.

Peirson, D. H., Cambray, R. S., and Spicer, G. S., 1966. Lead-210 and polonium-210 in the atmosphere, *Tellus, 18* (2): 427–433.

Pressman, J., Reidy W., and Tank, W., 1963. Rocket pollution of the upper atmosphere, IAS Paper No. 63–83, Institute of the Aerospace Sciences, 2 East 64th Street, New York 10021.

Reed, R. J., and Vlcek, C. L., 1969. The annual temperature variation in the lower tropical stratosphere, *Journal of the Atmospheric Sciences, 26:* 163–167.

Salter, L. P., 1964. Note on the detectability of cadmium isotopes from starfish in 1964 ground level samples, in Fallout Program Quarterly Summary Report, HASL-142, January 1, 1964, United States Atomic Energy Commission Health and Safety Laboratory, pp. 303–306.

Scholz, T. G., Ehhalt, D. H., Heidt, L. E., and Martell, E. A., 1970. Water vapor, molecular hydrogen, methane, and tritium concentration near the stratopause, *Journal of Geophysical Research, 75:* 3049–3054.

Telegadas, K., and List, R. J., 1964. Global history of the 1958 nuclear debris and its meteorological implications, *Journal of Geophysical Research, 69* (22): 4741–4753.

Williamson, E. J., and Houghton, J. T., 1965. Radiometric measurements of emission from stratospheric water vapour, *Quarterly Journal of the Royal Meteorological Society, 91:* 330–338.

Zimmerman, S. P., and Champion K. S. W., 1963. Transport processes in the upper atmosphere, *Journal of Geophysical Research, 68:* 3049–3063.

36

Stratospheric Water Vapor   Lester Machta

Introduction

Recently H. J. Mastenbrook (private communication) has reported on a significant increase in the water vapor content of the stratosphere. In this brief note, the data are presented together with some speculation on the likely causes of the water vapor changes.

Increases in stratospheric water vapor may provide two potential changes in weather or climate. The first is a greenhouse effect. The second is possible cloud formation in almost saturated (or very cold) regions.

Instrumentation

The history of stratospheric water vapor measurements is one of controversy and uncertainty; the present view of the writer is not necessarily universally shared by others. The cause of the difficulty lies in contamination by the auxiliary equipment accompanying the sensors. The stratospheric water vapor is measured in parts per million, a degree of dryness achieved with some effort in laboratories. Thus, even minimal desorption of water from nearby equipment produces a large change in apparent concentration. This condition has been alleviated by sampling during descent, keeping the sensor intake at the bottom of the train.

Most sensors are frost-point indicators although a phosphorus pentoxide instrument developed in the United Kingdom has been flown with success. The frost-point instrument used by the Naval Research Laboratory successively warms and cools a mirror surface to straddle the formation of frost which is detected by vacuum phototubes sensing reflected and scattered light. The determinable accuracy is about $1°C$ for the frost point, equivalent to 0.6 ppm (by mass) at 50 mb or 68,000 feet and a mean frost point of $-85°C$ where the humidity mixing ratio is 3 ppm.

The instrument is absolute; at this time Mr. Mastenbrook's principal reservation pertains to a small radial gradient of temperature from the center of the mirror to the edge, and the relative positions along the gradient of the controlling edge of the frost deposit and the thermistor. The instrument is not normally

Prepared for SCEP.

reused; only a few are recovered from the launch point near the Chesapeake Bay Station (about thirty-five miles east of Washington, D.C.). All are carefully calibrated in advance; except for one change to be noted later, all procedures have been followed without modification since 1964, the beginning of the record. Thus, there is no basis to suspect a bias in the observations as a function of time.

There may be a bias in the observations as a function of altitude. Therefore, at the turnaround point near the top of the balloon flight some air may be carried aloft from the wetter troposphere, and indeed the fluctuations on the high side at the peak flight altitudes are likely to be a consequence of such contamination rather than due to real increases in moisture with altitude. After a short period of flushing during descent this effect disappears. A second defect during part of the sampling period may also be noted. During 1964 and 1965 Mr. Mastenbrook failed to telemeter the operation of a ventilating fan. It is possible that on two flights, one of which is conspicuous in Figure 36.1, the fan failed to operate, and the very high concentrations then reflected moisture contaminant in the stagnated sample flow.

Observed Temporal Changes in Stratospheric Water Vapor
Figure 36.1 displays the history of water vapor in the stratosphere derived by Mastenbrook at the Chesapeake Bay Station of the Naval Research Laboratory. The individual observations made at approximately monthly intervals are entered on the graphs. Normally the humidity mixing ratio remains approximately constant with altitude some distance above the tropopause. In the late winter and spring a slight minimum appears between the tropopause and 60,000 feet, above which the uniformity with altitude reappears.

All levels in the stratosphere well above the tropopause reveal the same long-term trend resulting in an increase in water vapor concentration from about 2 to 3 ppm. With the two (or one at the higher altitude shown on the graph) very high readings a statistical analysis through 1969 indicates a lack of statistical confidence in the slope. From the independent evidence (probable fan operation failure), we may discard the very high moisture values and the increase of humidity with time is then statistically

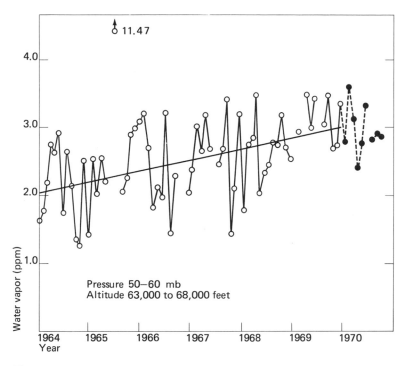

Figure 36.1 Monthly concentrations of water vapor at the Chesapeake Bay Station near Washington D.C., in the lower stratosphere

Source: Mastenbrook, 1970

significant through 1969. The 1970 data shown as crosses were unavailable for the statistical analysis; the 1970 points suggest possible leveling off of the growth rate.

Explanation for the Dry Stratosphere and Its Changes

It is believed that the dry stratospheric air (frost point of about −85°C) results from passage of air from the troposphere to the stratosphere primarily through the equatorial tropopause where temperatures are also about −85°C at certain times. Little or no moisture enters via the warmer (−65°C) temperate or polar tropopause, or there would be much more moisture. This hypothesis was propounded in the early 1950s by two British scientists, Dobson and Brewer, to explain the low-frost-point measurements they observed in the lower stratosphere aboard their reconnaissance aircraft. The data from those flights in the late 1940s and early

1950s were lower than the 2 ppm found by Mastenbrook, although the difference in altitude (the U.K. data were collected much closer to the tropopause) may not make the comparison with higher altitude data valid. The relative humidity in the normal (−65°C) stratosphere is 5 to 10 percent.

Thus, a logical explanation for the observed increase of moisture between 1964 and 1970 would be in warmer temperatures at the lower equatorial tropopause. The 100-mb (55,000 feet) temperatures at several equatorial weather stations show no such warming; if anything there is a slight cooling between 1963 and 1968 at the Canal Zone, 9°N; Singapore, 0°; and Canton Island, 3°S. It is possible that a larger fraction of the moisture in temperate or high latitudes may be entering the stratosphere from below. This could be either the result of higher tropospheric moisture or more intense vertical mixing through the tropopause.

Another suggestion for increased moisture in the stratosphere is direct insertion by vehicles propelled through the lower stratosphere. The input from rockets (burning of fuels or from a deliberate insertion of moisture by a Saturn rocket) seems very unlikely to be significant. Aircraft now fly frequently into the stratosphere at altitudes well above 50,000 feet. The total U.S. commercial jet aircraft production inserted in a narrow band near the latitude of Washington, D.C., could increase the water vapor concentration three times the annual growth of the measurements in Figure 36.1 if one asumes that the peak concentration in the stratosphere near Washington, D.C. is tenfold greater than the global stratospheric average concentration. But the stratospheric water vapor injections are likely to be a very small fraction of the total from routine U.S. commercial flights. It is felt, therefore, that the observed growth of stratospheric moisture is not attributable to aircraft operations in the stratosphere.

Other forms of human intervention can produce an increase in atmospheric moisture. Specifically, the burning of hydrocarbons by all human activities releases moisture into the air. Cooling towers similarly add vast amounts of moisture; one large tower complex, for example evaporates 10^4 to 10^5 gallons per minute in western Pennsylvania. Irrigation and, to a lesser extent, agricultural practices may also represent large man-made contributions of water. These releases take place mainly at or near the

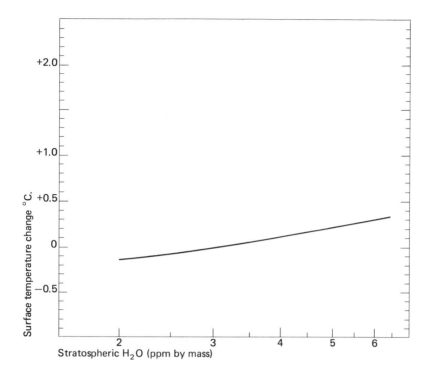

Figure 36.2 Surface air temperature change due to increased stratospheric water vapor with average cloudiness and convective — radiation equilibrium

ground where the residence time of the moisture will be measured in days rather than in weeks or months. Most calculations of the moisture increase from man-made sources diluted into the atmosphere (even a narrow belt of the Northern Hemisphere) do not produce significant concentration increases.

As of this writing, the author offers no plausible explanation for the increase in stratospheric water vapor near Washington, D.C., reported by Mastenbrook.

Reference
Mastenbrook, J., 1970. Private communication.

37

Stratospheric Particles   Richard D. Cadle

Nature of Particles

Above the tropopause the variation of number concentration with height for aerosol particles is entirely different for Aitken nuclei (particles smaller than 0.1 μ radius) and "large" particles (0.1–1.0 μ radius). The concentrations of the former decrease with increasing altitude, strongly suggesting that they are tropospheric particles that have been carried into the stratosphere (Junge, 1961). On the other hand, a highly structured worldwide layer of the large particles has a maximum concentration at an altitude of about 18 km in mid-latitudes and extends for several kilometers above and below the concentration peak.

The earliest studies were optical. For example, Bigg (1956), from measurements of twilight intensity as a function of time, concluded that a dust layer existed between 15 and 20 km. The first study of the composition of the large particles was made by Junge and his coworkers (Chagnon and Junge, 1961; and Junge and Manson, 1961). The particles were collected by aircraft and high-altitude balloons using impaction techniques. The collected particles were analyzed by electron microprobe and x-ray fluorescence methods. Sulfur was the predominant element, but traces of Al, Cl, Ca, and perhaps Fe were also detected. Junge hypothesized that the major anion was sulfate and the major cation was either NH_4^+ or H^+. The weight concentrations of particles were of the order of 2×10^{-15} gram per cm^3. Later studies by Friend (1966) suggested that persulfate as well as sulfate was present, and an investigation by Shedlovsky and Paisley (1966) indicated that no more than 10 percent of the particles could be of extraterrestrial origin.

Collections of stratospheric aerosol particles were made by Cadle, Lazrus, Pollock, and Shedlovsky in 1969 and 1970 (Cadle et al., 1970) using polystyrene and cellulose fiber filters flown on RB 57-F U. S. Air Weather Service aircraft at 18-km altitude. Sampling was undertaken both in the tropics in the vicinity of Central America and in mid-latitudes. Analyses were made by neutron activation techniques and by colorimetric analyses of aqueous extracts of the filters. The latter indicated that no NH_4^+

Prepared during SCEP.

was present in the tropical aerosol and very little in the mid-latitude aerosol. In fact, there was not nearly enough cation present to account for all of the sulfate, suggesting the presence of large amounts of sulfuric acid (Tables 37.1 through 37.3). A surprising finding was that the concentrations of NO_3^- were often as high as the $SO_4^=$ concentrations. Possibly nitric acid is dissolved in droplets of sulfuric acid in the stratospheric aerosols. However, nitrate was found in appreciable amounts only on the cellulose filters. Possibly the nitrate was present in the stratosphere as nitric acid vapor and was absorbed by the very absorbant cellulose filters. Nitric acid vapor has been identified in the lower stratosphere (Murcray et al., 1968). The particulate Cl/Br ratios were surprisingly low, in the vicinity of 15, compared with about 300 for seawater. This raises the interesting possibility that the excess bromine is derived from the organic bromine compounds used as scavengers for lead in gasoline. If so, the stratosphere may be an interesting place to look for worldwide air pollution.

During one of these RB 57-F collections with filters, collections were also made by impaction on electron microscope grids. Most of the particles were droplets that produced a pattern on partial evaporation characteristic of sulfuric acid droplets, namely, a central nucleus surrounded by satellite droplets. An electron microprobe examination showed only the presence of sulfur (nitrogen, if present, would not have been observed).

Particles collected on another flight at 18 km over the United States a few weeks after extensive brush fires in California contained chainlike agglomerates of particles resembling those produced by burning wood (Figure 37.1).

The sulfate particles were probably at least in part produced by the oxidation in the stratosphere of SO_2 by O (O_3 does not react with SO_2) followed by hydration of the resulting SO_3. Some of the particulate sulfate may have also been injected directly by volcanoes.

The stratosphere contains large amounts of radioactivity, both natural and man-made. Much of it seems to be associated with large particles which may have served as nuclei on which the radioactive elements or compounds condensed.

Recently, determinations of vertical profiles of sulfate and chloride concentrations for the upper troposphere and the strato-

Table 37.1 Analyses of Particles Collected on IPC Filters in the Tropical Stratosphere above Central America

| At start of sampling | | Alti-tude (ft × 10⁻³) | Concentrations (µg/m³ ambient) | | | | | | | Radio-activity (Nb⁹⁶) (Dpm/m³ ambient) |
Lati-tude (N)	Longi-tude (W)		$SO_4^=$	Si	Na	Cl	NO_3^-	Mn	Br	
31°40'	99°30'	55	0.11	0.052	0.020	0	0.058	0	0.0025	3.09
30°00'	96°05'	55	0.12	0	0	0	0.042	0	0.0014	2.39
28°00'	93°20'	55	0.13	0.022	0.025	0	0	0	0.0026	6.09
25°30'	90°30'	55	0.075	0	0.037	0.038	0	0	0.0020	0.69
23°00'	88°00'	56	0.10	0.058	0.004	0	0.16	0	0.0019	1.49
20°30'	85°20'	56	0.26	0	0.015	0.034	0	0	0.0030	1.73
19°20'	83°30'	57	0.092	0	0	0	0	0	0.0020	1.19
18°30'	80°00'	58	0.30	0.008	0.015	0	3.4	0	0.0068	8.77
11°00'	79°00'	58	0.11	0.015	0.049	0.017	0.20	0.00047	0.0030	4.84
8°00'	80°00'	58	0.20	0.0057	0	0	0.17	0	0.0018	3.18
4°00'	81°20'	58	0.17	0.040	0.053	0.053	0.058	0.00047	0.0015	1.57
2°00'	82°30'	58	0.14	0.025	0	0	0.049	0	0.0009	1.54
1°30'	83°30'	58	0.32	0.010	0	0.002	0.058	0.00002	0.0021	2.19
3°00' (S)	85°00'	62	0.13	0	0	0	0.021	0	0.0006	1.05
1°00' (S)	82°00'	62	0.29	0.0062	0.002	0.018	0.015	0.00004	0.0015	1.65

Source: Cadle et al., 1970.
Note: All samplings were thirty minutes long. Longitudes and latitudes are for start of sampling. No nitrite, potassium, or ammonium ion was detected in any of the samples, and the NO_3^- concentration was calculated by the method specific for NO_3^-. The upper set of samples was obtained on July 10, 1969 and the lower set on July 23, 1969.

Table 37.2 Analyses of Particles Collected on IPC and Polystyrene Filters in the Mid-latitude Stratosphere

Filter Type	Latitude (N)	Longitude (W)	Altitude (ft × 10⁻³)	Concentrations ($\mu g/m^3$ ambient)								Radioactivity (Nb[95]) (Dpm/m³ ambient)
				SO_4^-	Si	Na	Cl	(NO_3^-)*	NH_4	Mn	Br	
PS	34°30'	102°55'	55	0.32	0.18	0.004	0.042	0(0.0012)	0.0034	0.0036	0.0021	28.4
PS	37°20'	102°30'	58	—	0.19	0.003	0.023	—	—	0.0021	0.0019	37.5
IPC	40°20'	102°15'	58	0.21	0.037	0.054	0.071	0.31	0.026	0.0010	0.0028	40.9
IPC	43°25'	101°50'	59	0.37	0.035	0.030	0.052	0.41	0.0089	0.0009	0.0021	29.2
PS	46°30'	101°31'	59	0.24	0.19	0.003	0.041	0(0)	0.017	0.0025	0.0026	42.6
PS	47°48'	101°45'	60	0.22	0.17	0.002	0.030	0(0)	0.012	0.0012	0.0020	43.2
IPC	44°25'	101°45'	61	0.35	0.084	0.050	0.088	0.36	0.0067	0.0009	0.0030	49.3
IPC	41°00'	102°10'	60	0.32	0.031	0.001	0.046	0.35	0.0040	0.0004	0.0020	32.3
PS	37°30'	102°40'	60	0.36	0.17	0.003	0.051	0(0.0036)	0.043	0.0049	0.0024	59.7

Source: Cadle et al., 1970.
Note: All samplings were thirty minutes long, made on December 4, 1969. Longitudes and latitudes are for start of sampling. No potassium ion or nitrite ion was detected in any of the samples.
* The NO_3^- concentration was calculated from the total combined nitrogen less that combined as NH_4^+ (the value in parentheses), and also by the method specific for NO_3^-.

Table 37.3 Analyses of Particles Collected on IPC Filters in the Upper Troposphere near the Phillippine Islands*

Lati-tude (N)	Longi-tude (W)	Alti-tude (ft × 10⁻³)	Concentrations (μg/m³ ambient)								Sampling Time (min)
			SO_4^-	Si	Na	Cl	(NO_3^-)*	NH_4^+	Mn	Br	
35°04'	139°35'	26	0.041	—	—	—	0.025(0.020)	0.0064	—	—	120
26°21'	120°40'	26	0.051	0.011	0.055	0.085	0.026(0.019)	0	0.00034	0.0014	84
19°32'	120°40'	25	0.10	0.016	0.016	0.052	0.054(0.042)	0.0013	0.00033	0.0018	36
16°30'	117°00'	25	0.15	0.022	0.00072	0.029	0.050(0.022)	0.013	0.00048	0.00096	72
13°00'	124°00'	25	0.073	0.009	0.010	0.013	0.046(0.039)	0.0033	0.00025	0.00063	68
21°00'	127°54'	25 → 39	—	0.014	0.019	0.018	— —	—	0.00032	0.0006	105
11°00'	122°00'	39	0.23	0.0067	0	0.057	0(0)	0	0.00041	0.0006	20
11°00'	119°00'	39	0.16	0	0.019	0.060	0.0074(0.011)	0	0.00026	0.00073	44
16°52'	117°15'	39	0.068	0.0033	0.016	0.052	0.025 (0.026)	0	0.00008	0.00056	47

Source: Cadle et al., 1970.
Note: All samplings were made on November 19, 1969. Longitudes and latitudes are for start of sampling. No potassium ion or nitrite ion was detected in any of the samples.
* The NO_3^- concentration was calculated from the total combined nitrogen less that combined as NH_4^+, and also (the value in parentheses) by the method specific for NO_3^-.

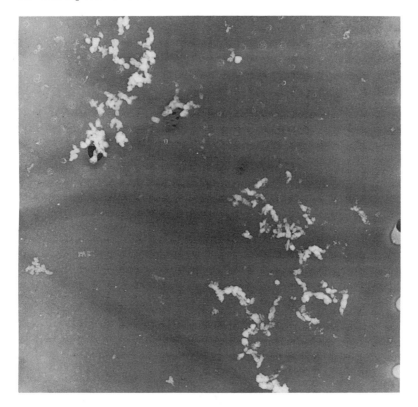

Figure 37.1 Particles collected by impaction from the stratosphere at 18-km altitude on October 21, 1970. The distance across the micrograph is 6 μ.

sphere have been made using RB 57-F aircraft and the polystyrene filters (Shedlovsky and Lazrus, 1970). The sulfate concentrations decreased with increasing altitude up to the tropopause but started to increase rapidly with increasing altitude immediately above the tropopause. Chloride concentrations, however, continued to decrease above the tropopause. These results seem to rule out an oceanic origin for most of the stratospheric particles.

Several studies have been made of the size distribution of stratospheric aerosol particles, the latest by Blifford and Ringer (1969). The shape of the distribution curve is somewhat different from that for most tropospheric particles, but even so, over much of the distribution, the curves for both atmospheric regions can be represented by equations of the form $dN/d(\log r) = ar^{-3}$, where dN is the number of particles per unit volume in the increment of radius $d(\log r)$ and a is a constant.

Injections of Volcanic Eruption Clouds

Following the eruption of Agung volcano in Bali in March 1963 direct measurements of the nature and concentration of particles in the stratospheric aerosol layer showed that marked changes in the concentrations and minor changes in the nature of the particles had occurred. The number concentrations had increased thirtyfold over those obtained by Junge (1961) and Cadle et al. (1970). Mossop (1964, 1965) found that high concentrations of "ash" particles (finely divided lava) were present, but the concentrations of $SO_4^=$ had also increased remarkably.

Newell (1970) has shown that the temperature of stratospheric air increased by as much as 6°C following the eruption and maintained this increased temperature for several years. Presumably this resulted from the absorption of solar radiation by the particles. Very brilliant sunsets were soon observed worldwide and were produced by a layer of particles at an altitude of about 20 km and presumably caused by the Agung eruption. These sunsets persisted for several years, and at the time of writing (1970) concentrations in the stratosphere remain higher than they were prior to the Agung eruption. This may reflect the fact that volcanic activity all over the world has been relatively intense during the last decade, and while most of the eruptions have not injected eruption clouds directly into the stratosphere, they must at least indirectly contribute to the stratospheric aerosol particles as a result of transport of air upward across the tropopause.

This raises the question of the lifetimes of stratospheric particles. The mean residence time of trace constituents in a given region of the stratosphere increases appreciably with altitude: about one month at the tropopause, about one to two years at 20-km altitude, and about four to twenty years at 50-km altitude. (See Chapter 35.)

Unanswered Questions

Important questions that remain to be answered include:

1.   How much of the stratospheric sulfur particles result from man's activities?
2.   Does the excess bromine result from man's activities?
3.   Is there detectable lead in the stratospheric aerosol?

4. Are there high concentrations of NO_3^- in the stratospheric aerosol, as recent results suggest? Or is it mostly in the form of nitric acid vapor?

References

Bigg, E. G., 1956. The detection of atmospheric dust and temperature inversions by twilight scattering, *Journal of Meteorology, 13:* 262–268.

Blifford, I. H., Jr. and Ringer, L. D., 1969. The size and number distribution of aerosols in the continental troposphere, *Journal of Atmospheric Sciences, 26:* 716–726.

Cadle, R. D., Lazrus, A. L., Pollock, W. H., and Shedlovsky, J. P., 1970. Chemical composition of aerosol particles in the tropical stratosphere, *Proceedings of the Symposium on Tropical Meteorology, Paper K-IV (Boston. American Meteorological Society).* Tables 37.1, 37.2, and 37.3 were reprinted with permission from the American Meteorological Society.

Chagnon, C. W., and Junge, C. E., 1961. The vertical distribution of submicron particles in the stratosphere, *Journal of Meteorology, 18:* 746–752.

Friend, J. P., 1966. Properties of the stratospheric aerosol, *Tellus, 18:* 465–473.

Junge, C. E., 1961. Vertical profiles of condensation nuclei in the stratosphere, *Journal of Meteorology, 18:* 501–509.

Junge, C. E., and Manson, J. E., 1961. Stratospheric aerosol studies, *Journal of Geophysical Research, 66:* 2163–2182.

Mossop, S. C., 1964. Volcanic dust collected at an altitude of 20 km, *Nature, 203:* 824–827.

Mossop, S. C., 1965. Stratospheric particles at 20 km altitude, *Geochimica et Cosmochimica Acta, 29:* 201–207.

Murcray, D. G., Kyle, T. G., Murcray, F. H., and Williams, W. J., 1968. Nitric acid and nitric oxide in the lower stratosphere, *Nature, 218:* 78–79.

Newell, R. E., 1970. Modification of stratospheric properties by trace constituent changes, *Nature, 227:* 697–699 (reprinted in this volume, Chapter 38).

Shedlovsky, J. P., and Paisley, S., 1966. On the meteoritic component of stratospheric aerosols, *Tellus, 18:* 499–503.

Shedlovsky, J. P., and Lazrus, A., 1970. Recent data on the stratospheric sulfate layer, Abstracts, 159th American Chemical Society National Meeting, Houston, February 1970.

38

Modification of Stratospheric Properties by Trace Constituent Changes

Reginald E. Newell

It has been suggested that large quantities of water vapour and other pollutants introduced into the stratosphere by supersonic transporters may change the basic properties of the stratosphere (1). But what happens when large changes in trace constituent concentrations occur naturally, for example, after a volcanic eruption? The stratosphere temperature seems to increase by about 5°C and it remains higher than normal for several years. Some of the evidence for this is given in Figures 38.1 and 38.2, which are based on daily upper air data from Australia and New Zealand. Monthly temperature means were formed for each month, then mean monthly values computed from data for the period 1958–62 inclusive. Deviations from these long term values were noted, and a three-month running average obtained. Figure 38.1 gives the time variations of these deviations at selected stations. A major eruption of Mt Agung, Bali (8°S, 115°E), occurred on March 17, 1963, and there have been reports of other eruptions that year. Temperature at the low latitude stations increased almost immediately and remained higher than normal for over two years. Figure 38.2 contains meridional cross-sections of the temperature deviations for particular months based on station data for the longitude region 110°E to 180°E. At the beginning of 1963, temperature was below the five year normal and the time variations had been following the characteristic southern hemisphere biennial pattern (2). The eruption interrupted this pattern. The cross-sections bear some resemblance to those for the products of nuclear weapons tests (for example, see the isolines of tungsten 185 radioactivity in ref. 3), as might be expected. The deviation pattern exhibits a maximum at low latitudes in the lower stratosphere and slopes downwards towards the pole, in some cases at a slope greater than that of the potential temperature surfaces. It is difficult to detect the changes, if any, below 300 mbar. While negative values predominate in the examples shown, it was not possible to draw in a −2°C contour. Similarly, at higher latitudes in the stratosphere the effect is uncertain; large deviations from

Reprinted from *Nature*, 227 (1970), pp. 697–699.

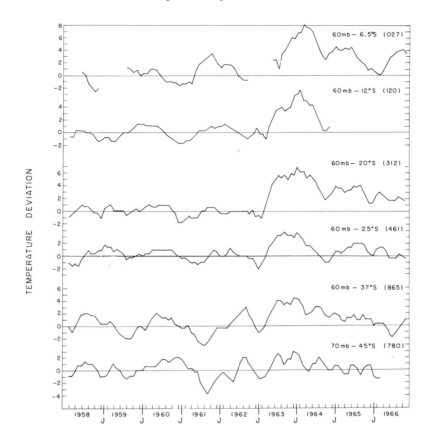

Figure 38.1 Temperature deviations from monthly averages based on the period 1958–1962
Units: degree centigrade. Pressure level shown in millibars.

average are more common there. It has been reported (3) that the Australian Weather Bureau changed its thermistor mounting at the end of 1962. Maps of temperature for January 1963 and 1964 show that the temperature change is present at other longitudes.

If temperature changes of up to 5°–6°C above average are produced within about 2 months, as seen from Figure 38.1, the equivalent heating rate is at least 0.1°C/day, which is a relatively large contribution in the lower stratosphere. The heating is probably caused by absorption of solar radiation by aerosols such as that demonstrated by Robinson (4) for the troposphere. The exact nature of the particles is unknown, although Mossop (5), who

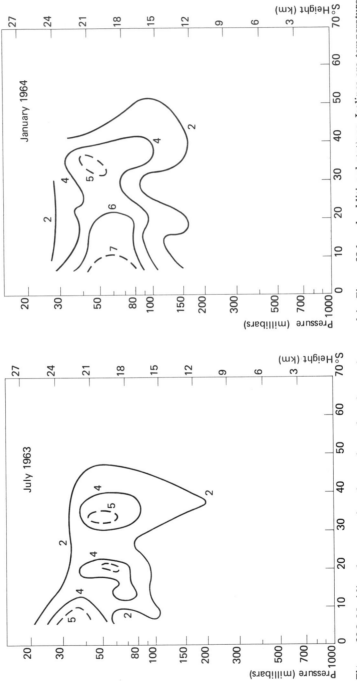

Figure 38.2 Meridional cross-section drawn from data for stations used in Figure 38.1 and additional stations. Isolines are temperature deviations, during month shown from monthly average based on 1958–1962 Units: degree centigrade.

sampled them directly from an aircraft, comments that the dust seems to be coated with sulphuric acid. Volcanic gases contain large quantities of water vapour, carbon dioxide and sulphur dioxide, and fairly rapid formation of sulphuric acid has been suggested for the Krakatoa eruption (6). A strong smell of sulphur was reported at Bali (7). Several workers have suggested that the particles grow *in situ* from the gases introduced into the stratosphere (5, 6, 7, 8, 9, 10, 11).

Station data from the northern hemisphere tropics also show evidence of a similar temperature increase. An overall increase of 5°C in the layer 17–25 km over more than one quarter of the globe will evidently produce other geophysical changes. There are obvious items, such as changes in the mean wind field, the evidence for which is not discussed here. The height of those constant pressure surfaces above the dust layer was raised by up to 200 m in January 1964 at 50 mbar. With no obvious compensatory cooling at higher levels, although doubtless the circulation patterns there will adjust somewhat, one might expect to see larger densities measured by the satellite drag technique. I leave a search for this effect to others more conversant with the problems of satellite data.

Measurements of solar radiation at the surface (12) and scattered light from twilight (13) have been used to trace the cloud from Bali. Temperature deviations may also be used where the data are adequate and we are working on this problem now, using also stratospheric mean wind charts for the period. To assist in this work we would greatly appreciate copies of any reports or reprints dealing with twilight glows, temperature anomalies or surface radiation anomalies. Many people noted the Krakatoa events (see Symons) (6), and the interest in theories of climatic change involving volcanoes was thereby heightened (14, 15). Bali produced temperature changes in the stratosphere much larger than those normally considered in climatic change discussions.

Our work on planetary transport processes is supported by the U.S. Atomic Energy Commission. I thank Robert Crosby for his help and suggestions with the data processing. The Meteorological Services of Australia and New Zealand kindly provided the data.

References

1. Newell, R. E., *Nature, 226,* 70 (1970).

2. Sparrow, J. G., and Unthank, E. L., *J. Atmos. Sci., 21* (1964).

3. Newell, R. E., *Geofis. Pura Appl., 49,* 137 (1961).

4. Robinson, G. D., *Quart. J. Roy. Meteor. Soc., 92,* 263 (1966).

5. Mossop, S. C., *Nature, 203,* 824 (1964).

6. Symons, G. J., ed., *The Eruption of Krakatoa and Subsequent Phenomena* (Rep. of Krakatoa Comm. of Royal Soc., Trübner and Co., London, 1888).

7. Booth, W. P., *National Geographic, 124,* 436 (1963).

8. Junge, C. E., in *Tellus, 18,* 685 (1966).

9. Cadle, R. D., and Powers, J. W., *Tellus, 18,* 176 (1966).

10. Friend, J. P., *Tellus, 18,* 465 (1966).

11. Meinel, A. B., and Meinel, M. P., *Science, 155* (1967).

12. Dyer, A. J., and Hicks, B. B., *Quart. J. Roy. Meteor. Soc., 94,* 545 (1968).

13. Volz, F. E., *App. Opt., 8,* 2505 (1969).

14. Humphreys, W. J., *Physics of the Air* (third ed.) (McGraw-Hill, New York, 1940).

15. Wexler, H., *Bull. Amer. Meteor. Soc., 32,* 10 (1951).

Part IX

A Nonproblem and a
Potential Problem

SCEP carefully considered many environmental problems and attempted to establish some scientific criteria for determining the priorities for each of them. One of the areas considered was the question that has been raised from time to time of whether man might deplete his oxygen supply. There are two main facets to this question: the thought that we might use up the oxygen by burning fossil fuels; and the thought that we might interfere with photosynthesis on a worldwide basis. The following paper by Drs. Machta and Hughes, first published in *Science* (June 26, 1970), reports the results of an extensive program conducted over several years to determine if the oxygen supply were dwindling. The conclusion is that we should be much more worried about other problems. This paper does, however, demonstrate that answers to these critical questions can often be determined if society is willing to support the monitoring and analytical studies.

In the midst of SCEP, when all of the participants had become sensitized to the interplay between seemingly unrelated events occurring on the earth, the following article appeared in the *New York Times* for July 7, 1970 (quoted here only in part):

Brazil Is Challenging a Last Frontier, by Joseph Novitski. (Rio de Janeiro, July 6) . . . The Brazilian nation is battering, clearing and burning its way into the heart of South America to open and settle what may be the last unknown frontier in the world—the immense Amazon Basin.

It is a tenuous process thus far, with about 3.5 million people scattered in the jungle, over vast highlands and in lonely river valleys in an area the size of Alaska, Washington, Oregon, California, Idaho, Nevada, Arizona, Utah, Montana and Wyoming put together. . . .

. . . People and, perhaps more important, money from all over the country are beginning to move into the basin, chasing a Brazilian dream that is at least three decades old. . . .

. . . There are also scattered signs that man's advance into the area, slashing and burning rain forests and hunting freely, might become dangerous to Amazonia's ecology. Much of the soil is not thickly fertile and erodes almost immediately under the heavy rains when it is cleared of forest. . . .

. . . The United Nations Food and Agriculture Organization, in a 1963 report, rates Brazil as "a leading country in reforestation" because of the legislation spurring the rebuilding of forests south of Amazonia that were stripped by colonizers. But there are few officials to enforce reforestation legislation in the vast area of the Amazon Basin. . . .*

Meteorologists have wondered about the effects of some of man's changes to the surface of our planet, and whether they could have an effect on the global atmospheric circulation, and thus the climate. Here was a possible example, timely and real, and Professor Newell wrote the following brief comment that suggests that deforestation of the tropical jungles might indeed have effects that reach far beyond the tropics themselves.

39

Atmospheric Oxygen in 1967 to 1970

Lester Machta and E. Hughes

Abstract

Observations of atmospheric oxygen in clean air between 50°N and 60°S, mainly over the oceans, yield an almost constant value of 20.946 percent by volume in dry air. Since 1910 changes with time over the globe appear to be either zero or smaller than the uncertainty in the measurements.

In May 1966, the late Lloyd Berkner urged the Office of the President's Science Advisor to measure oxygen in the clean atmosphere. He justified his request in a Memorandum for the File, which was prepared jointly with L. C. Marshall and dated 29 April 1966, entitled "Potential Degradation of Oxygen in the Earth's Atmosphere." Here it was noted "that fish in the Newfoundland Banks contain significant . . . quantities of herbicides and insecticides." These, it was argued, derived from unicellular organisms, "the grain of the sea. . . . Thus, in the absence of more precise information, it must be assumed that the concentration of insecticides and herbicides in the fish of the sea arises from initial concentration by the photosynthetic organisms which are also the primary source of atmospheric oxygen. The problem is whether the herbicides and pesticides concentrated by the basic photosynthetic organisms can affect their population, thereby modifying the equilibrium concentration of oxygen in the earth's atmosphere." Each year about 0.05 percent of the atmospheric oxygen is renewed by photosynthesis of which over 60 percent derives from the oceans. It must be noted that the authors found "themselves in the dilemma that the extent of knowledge so far accessible to them is insufficient to demonstrate whether or not the problem is serious now or in the identifiable future."

In response to Berkner's request, the Environmental Science Services Administration and the National Science Foundation collected seventy-eight samples during 1967 and 1968 by the oceanographic vessels *Oceanographer* and *Eltanin,* both over the continental shelf and open ocean. The Analytical Chemistry Division of the National Bureau of Standards developed the method of analysis and determined the oxygen content.

Reprinted from *Science, 168* (June 26, 1970), pp. 1582–1584.

Samples were collected, after drying, in one-liter evacuated stainless steel flasks. Previous experience had shown that samples stored in similar containers suffered no detectable change of oxygen after storage for long periods of time, even at slightly elevated temperatures. The oxygen content of the samples was determined by repeated comparison with a gas of known oxygen content with the use of a modified Beckman oxygen analyzer (1). The oxygen contents of the comparison standards were derived from a primary standard which was determined by a gravimetric technique (2) as 20.959 percent by volume with an accuracy of 1 standard deviation of ±0.006 percent by volume. The mean value of the primary standard and its uncertainty were obtained from 33 results selected from 36 determinations; all 36 results gave the same mean value but a standard deviation of ±0.008 percent by volume.

Analytical results, sampling sites, and dates are given in Table 39.1. Three portions were taken from each sample and each measured at least ten times; the standard deviations for the samples in Table 39.1 derive from the set of three mean values. The averages for all 78 samples is 20.9458 percent by volume of dry air with a standard deviation of ±0.0017 percent by volume. With one exception, 20.9527 ± 0.0012 percent by volume, the larger departures from the mean are generally associated with large standard deviations suggesting the analytical uncertainty as the source of variation. Values obtained during the second three days, 15 to 17 April, also average 20.9458 percent by volume but exhibit a smaller variability between samples, ±0.0005 percent by volume, than those during the first 3 days. Increased familiarity with the instrument and its peculiarities after a thousand or so separate measurements probably explains this reduction in variability, although other less readily explained phenomena are undoubtedly present.

The recommended value for oxygen abundance of dry air from the more reliable oceanographic analyses is 20.946 percent by volume relative to a standard with an accuracy of ±0.006 percent by volume. The precision or geographical variability, or both, is ±0.0005.

In February 1970, Hughes collected ten nearly identical samples from a relatively isolated rural site in western Maryland. The

Table 39.1 Measured Oxygen Abundances, Percent by Volume (dry air)

| | | | Analysis performed during | | | |
| | | | 15 to 17 April 1969 | | 9 to 11 April 1969 | |
Latitude	Longitude	Date collected	Value	S.D.	Value	S.D.
In 1967 by the R.V. Oceanographer						
49°00′N	007° 5′W	4/18			20.9457	±.0006
45°40′N	014°30′W	4/17	20.9463	±.0006		
42°09′N	019°19′W	4/16	20.9450	±.0017		
40°02′N	024°07′W	4/15	20.9450	±.0010		
36°08′N	031°08′W	4/14	20.9460	<±.0001		
35°54′N	018°12′E	5/6	20.9460	<±.0001		
35°54′N	018°12′E	5/6	20.9453	±.0012		
21°00′N	038°30′E	5/22	20.9460	<±.0001		
21°00′N	038°30′E	5/22	20.9457	±.0006		
19°20′N	070°53′E	6/14			20.9427	±.0023
19°10′N	068°20′E	6/18			20.9487	±.0006
19°10′N	068°20′E	6/18			20.9440	±.0020
18°54′N	065°00′E	6/19			20.9450	±.0026
14°27′N	070°12′E	6/5			20.9487	±.0042
11°06′N	060°04′E	6/3			20.9493	±.0059
10°00′N	092°04′W	11/17	20.9460	±.0010		
10°00′N	092°04′W	11/17	20.9457	±.0006		
08°09′N	073°06′E	7/2			20.9457	±.0058
05°08′N	090°01′E	7/6	20.9463	±.0006		
00°49′N	092°00′W	11/22	20.9457	±.0006		
00°00′	106°10′E	8/10	20.9460	<±.0001		
05°12′S	113°16′E	8/12	20.9463	±.0006		
09°52′S	115°06′E	8/14	20.9463	±.0006		
10°01′S	084°59′W	11/16	20.9463	±.0006		
15°02′S	113°01′E	8/17	20.9463	±.0006		
15°02′S	113°01′E	8/17	20.9437	±.0032		
20°00′S	115°01′E	8/18	20.9457	±.0006		
20°00′S	076°54′W	11/9	20.9463	±.0006		
25°20′S	112°01′E	8/20	20.9457	±.0006		
30°00′S	075°00′W	11/4	20.9457	±.0006		
30°25′S	114°50′E	8/23	20.9453	±.0012		
32°18′S	130°42′E	9/4			20.9450	±.0017
32°18′S	130°42′E	9/4	20.9453	±.0006		
34°16′S	149°58′W	10/18			20.9480	±.0020
35°00′S	165°00′W	10/15			20.9450	±.0061
35°00′S	135°00′W	10/20			20.9473	±.0012
35°00′S	135°00′W	10/20			20.9467	±.0015
37°42′S	180°00′W	10/12			20.9470	±.0010
Unknown	Unknown	Unknown			20.9457	±.0006
Unknown	Unknown	Unknown	20.9460	<±.0001		
Unknown	Unknown	Unknown	20.9463	±.0006		

Table 39.1 (continued)

Latitude	Longitude	Date collected	Analysis performed during				
			15 to 17 April 1969		9 to 11 April 1969		
			Value	S.D.	Value	S.D.	
Unknown	Unknown	Unknown	20.9457	±.0006			
Unknown	Unknown	Unknown	20.9457	±.0006			
Unknown	Unknown	Unknown			20.9487	±.0006	
Unknown	Unknown	Unknown			20.9457	±.0012	
Unknown	Unknown	Unknown			20.9453	±.0012	

In 1968 by the R.V. Eltanin

Latitude	Longitude	Date collected	Value	S.D.	Value	S.D.	
33°30′S	126°10′E	9/8			20.9527	±.0012	
34°40′S	124°00′E	9/9			20.9470	±.0030	
35°11′S	121°25′E	9/10			20.9467	±.0006	
35°15′S	137°50′E	7/31	20.9460	±.0010			
35°15′S	137°50′E	7/31	20.9457	±.0006			
35°17′S	138°15′E	8/12			20.9443	±.0021	
35°18′S	137°43′E	10/7			20.9460	±.0010	
36°02′S	116°57′E	9/12			20.9453	±.0023	
36°02′S	116°57′E	9/12			20.9400	±.0052	
38°15′S	160°10′E	6/25	20.9457	±.0006			
38°15′S	160°10′E	6/25	20.9463	±.0006			
40°00′S	134°03′E	8/14			20.9480	±.0010	
41°52′S	117°04′E	9/14			20.9413	±.0042	
42°40′S	150°12′E	6/29	20.9457	±.0006			
43°03′S	147°22′E	6/30	20.9453	±.0012			
43°58′S	160°02′E	6/23	20.9457	±.0006			
44°42′S	145°31′E	7/1	20.9450	±.0017			
47°30′S	128°00′E	8/31			20.9460	±.0017	
47°30′S	128°00′E	8/31			20.9427	±.0021	
47°30′S	128°00′E	9/1			20.9470	±.0020	
47°30′S	128°00′E	9/1			20.9453	±.0012	
52°56′S	135°00′E	7/23	20.9460	±.0010			
53°36′S	122°30′E	9/28			20.9467	±.0021	
55°17′S	160°00′E	6/17	20.9463	±.0015			
57°00′S	160°04′E	6/16	20.9457	±.0006			
58°45′S	117°00′E	9/25			20.9427	±.0081	
59°57′S	167°53′E	6/10	20.9457	±.0006			
60°00′S	135°00′E	7/19	20.9463	±.0006			
60°00′S	128°00′E	8/24			20.9417	±.0035	
60°02′S	140°18′E	7/18	20.9457	±.0006			
60°10′S	145°00′E	7/17	20.9453	±.0006			
60°10′S	145°00′E	7/17	20.9457	±.0006			

In 1970 16 kilometers west of Frederick, Maryland

Latitude	Longitude	Date collected					
39°35′N	77°32′W	2/21	20.944	20.949	20.946	20.946	20.947
			20.944	20.949	20.945	20.946	20.945

average value by the same method of analysis was found to be 20.946 ± 0.0018 percent by volume. In 1967 Hughes (2) reported an abundance of 20.945 percent by volume for a sample collected in the same region.

While new measurements may shed light on any future destruction of oxygen sources by pollution, comparison of present and past observations would indicate whether or not there has been any recent degration of atmospheric oxygen. Many of the past measurements are reported to three decimal places but Paneth (3) contends that even the second decimal place is imperfectly given. Glueckauf (4) took a less pessimistic view and quoted 20.946 percent by volume as the probable value in his 1951 survey of previous work. We do not know the absolute error in past determinations of oxygen by the volumetric-chemical method but feel that there is some confidence in the third decimal.

There are only five measurements or series of measurements of the absolute oxygen abundance in the atmosphere between 1910 and 1967–70. The first of these was an extensive investigation during 1910–12 by Benedict (5) of the air near his laboratory in Boston as well as air obtained in flasks from isolated geographical areas. There are doubts concerning the validity of the flask samples. We, following Krogh (6), have used the mean of 212 direct laboratory analyses; 20.952 percent by volume. The variability among these samples was ±0.007 percent by volume with no correlation to observed variations in weather conditions. Krogh suggested a correction of +0.002 percent by volume for the formation of carbon monoxide in Benedict's apparatus; we believe this correction may be too high. On the other hand, Benedict appears to have reported the abundance of oxygen relative to the total of oxygen, nitrogen, and argon after removal of carbon dioxide. Since this is so, the oxygen abundance in the dry air with normal carbon dioxide would have been 20.946 rather than 20.952 percent by volume.

In 1919, Krogh (6) reported an abundance of 20.948 percent by volume for two analyses of atmospheric air in Denmark. In the early 1930s Carpenter (7) using equipment similar to that of Benedict analyzed over 1000 samples in eastern United States. The results are almost identical with those of Benedict, but it is not clear whether Carpenter used Benedict's uncorrected value, 20.939 percent by volume, as his standard since he was mainly concerned with variability. Carpenter concluded the oxygen abundance is constant, his standard deviation being about 0.003 percent by volume.

Six analyses of a single large sample collected west of Washington, D.C., by Shepherd (8) of the National Bureau of Standards in 1935 ranged from 20.935 to 20.950 percent by volume, the average being 20.946 percent by volume. In 1942, Lockhart and Court (9) reported oxygen abundances in Antarctica averaging 20.92 percent by volume and suggested that the low values might be unique to the location. Glueckauf pointed out that they performed no analysis of normal non-Antarctic air to confirm their procedures. To further cast suspicion, their carbon dioxide abundances were many times higher than that found in recent times. Table 39.1 does not suggest lower values approaching the Antarctica.

All reliable oxygen data since 1910 fall in the range 20.945 to 20.952 percent by volume. The change in atmospheric oxygen since 1910 has been either very small or zero. It is possible that there has been no change even in the third decimal but a more realistic assessment recommends no change in the second decimal place, there being little confidence in differences of the third decimal place. Failure to detect changes in atmospheric oxygen may be consistent with Marshall's view that there is now less cause for alarm about the reduction in photosynthetic production of oxygen (10).

Several "doomsday" predictions for the eventual loss of the oxygen from the earth's gas mantle have been proposed and dismissed. Parson (11), for example, examined and rejected the views that oxidation of ferrous metals in lower oxidation states to their highest oxidation states and escape of oxygen from the top of the atmosphere would deplete the earth of its oxygen.

It is likely that the burning of fossil fuels would slightly diminish the oxygen content of the atmosphere. Benedict and Krogh showed that there were slight decreases in the oxygen abundance within cities coinciding with high carbon dioxide values. In fact, there is an approximate one-to-one correspondence in the opposite changes of carbon dioxide and oxygen gases when expressed as percent by volume.

Between 1910 and 1967 atmospheric oxygen should have decreased by 0.005 percent by volume as a result of the combustion of fossil fuels (12). But the uncertainty in both the 1910 and 1967–70 measurements prevents us from attributing significance to any coincidence between expected and observed changes in atmospheric oxygen. Extrapolating the depletion of oxygen due to fossil fuel burning to the day when all known recoverable reserves are consumed (13) leads to an abundance of about 20.8 percent by volume (14). The direct effects of this lower value on human respiration would be insignificant since it corresponds to an oxygen partial pressure change equivalent to a rise of about 75 m in altitude.

In summary, the 1967–70 abundance of oxygen in clean air, 20.946 percent by volume of dry air, is statistically the same as all the reliable measurements since 1910; the extreme range among reported values is 0.007 percent of volume. The accuracy of the present oxygen standards is ±0.006 percent by volume. This latter uncertainty rather than the geographic or instrumental variability may limit the detection of small changes of atmospheric oxygen in the future. Further periodic examination of the oxygen content is planned after more accurate standards are developed.

Acknowledgments

We thank M. L. Fields for sampling aboard the *Eltanin* and J. Wells aboard the *Oceanographer,* and T. O. Jones of NSF and H. Weickmann of ESSA for arranging for the collections aboard the two vessels. H. Weickmann, W. Komhyr, and T. Ashenfelter also assisted in organizing the collection and analysis program.

References and Notes

1. J. K. Taylor, Ed., *Nat. Bur. Stand. Tech. Note 435,* 15 (1968); *ibid. 505,* 7 (1969).

2. E. Hughes, *Environ. Sci. Technol. 2,* 201 (1968).

3. F. A. Paneth, *Quart. J. Roy. Meteorol. Soc. 63,* 433 (1937).

4. E. Glueckauf, *Compendium Meteorology,* T. F. Malone, Ed. (American Meteorological Society, Boston, 1951), p. 3.

5. F. G. Benedict, *Carnegie Inst. Wash. Publ. No.* 166 (1912). p. 1155.

6. A. Krogh, D. *Kgl. Danske Videnskab. Selskab, Mat.-Fys. Medd. 1,* 1 (1919).

7. T. M. Carpenter, *J. Amer. Chem. Soc. 59,* 358 (1937).

8. M. Shepherd, "U.S. Army Stratosphere Flight of 1935 in the Balloon Explorer II" (National Geographic Soc., Washington, D.C., 1935).

9. E. E. Lockhart and A. Court, *Mon. Weather Rev. 70,* 93 (1942).

10. L. C. Marshall, personal communication, 1970.

11. A. L. Parson, *Nature 156,* 504 (1945).

12. United Nations, *Proc. Int. Conf. Peaceful Uses of Atomic Energy 1,* 3 (1956); United Nations, *Statistical Papers, Series J* (New York), for appropriate years. We use the same carbon-hydrogen composition of fuels given in (*13,* p. 114). The calculation included oxygen loss due to the formation of both carbon dioxide and water.

13. U.S. Government Report, "Restoring the Quality of our Environment" (Government Printing Office, Washington, D.C., 1965), p. 120.

14. E. K. Peterson, *Environ. Sci. Technol. 11,* 1162 (1969). A smaller decrease in oxygen abundance is predicted by Peterson.

40
The Amazon Forest and Atmospheric General Circulation

Reginald E. Newell

The atmospheric heat engine is powered by processes which heat the air where it is warm and cool it where it is cold. There are three main elements that account for most of the heating and cooling: *radiational processes,* which in the troposphere give rise to cooling of the air at essentially all latitudes (in some seasons a differential cooling occurs with larger values at higher latitudes); *latent heat liberation,* which heats the air most in the tropical rain belts and the middle latitude cyclone belts; and *boundary layer heating,* which also varies with latitude and season. The largest contributor toward heating the air is the latent heat liberation, while the largest contribution to cooling is radiation.

At low latitudes most of the latent heat is provided to the air in the regions of mean rising motion centered over South America, Africa, and the "Maritime Continent"—Indonesia, and so forth. Over the oceans in between, at about the same latitudes, there occur mean sinking motions. The three regions of mean rising motion drift north and south with the sun, although there is some phase lag. (By "mean" I am referring specifically to three-month seasonal means based on seven years of data, but the rising motion shows up as enhanced cloud masses in those three regions on individual satellite pictures.)

A legitimate question for present consideration is: "What will happen to the large-scale atmospheric general circulation if the tropical forests are removed over Brazil and Indonesia?—and perhaps over Central Africa?" I think that the answer, as for increasing CO_2, is that we do not know, but we can speculate on the *kind* of influence it will have.

Figure 40.1 shows the three contributions to the total heating (cooling) rate. The global energy generation rates for December to February are (in units of 10^{20} ergs sec^{-1}):

Radiation	23
Latent Heat	87
Boundary Layer	36
Total	146 ($\sim$3 watts m^{-2})

Notice the dominance of the latent heat term.

Prepared for SCEP.

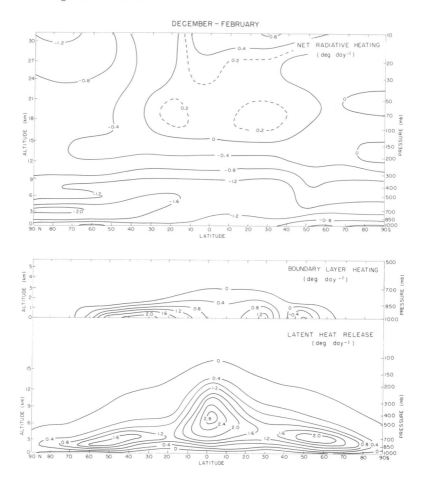

Figure 40.1 Composition of diabatic heating for the atmosphere for December–February
Units: degrees/day.

Source: Newell et al., 1970.

Reduction of the water cycling between the forest and the air, which provides the latent heat high in the column, would clearly alter the pattern of the latent heat forcing function. If one examines the heat balance at say 500 mb and 5°S, one finds that the latent heat term is almost exactly offset by radiation and adiabatic cooling by rising motion (the magnitude of this is not shown in Figure 40.1, but must be about the same as the radiative cooling term). If a change occurs in the latent heat term one would expect changes in other terms also to maintain a balanced budget. It is difficult to even guess what they might be without a proper dynamical model. Boundary layer heating would obviously change due to the change in the surface albedo and conductivity as the forests are destroyed, although it should be borne in mind that the total longitude span over which the change of land surface occurs is small. Most of the contribution to the latent heat term comes from South America, Africa, and the Maritime Continent at this season, so that changing the tropical forests would interfere in a major way with this term. The radiational term would also change on account of less cloudiness.

Water vapor transport patterns, horizontal heat and momentum transport, and the convergence-divergence patterns of all these transports would also be expected to change. The amount of regular upper air information over the tropical continents is so small that we cannot yet establish the normal values of these transports properly. In a very general sense, one would expect a decrease in the generation of zonal available potential energy, as there would be less heating of the air column in middle levels at low latitudes. (This is in the opposite direction we predict for the effect of increasing CO_2.) Here recourse to a numerical model is indicated.

I should emphasize here that the linkages between the various parts of the general circulation are not well understood, even diagnostically. The extent to which latitude displacements of heat sources force changes in the tropics, and vice versa, is particularly uncertain. I would place the study by atmospheric modeling of the Brazilian and other proposed tropical changes a little above CO_2 in my list of priorities, and a little below the stratospheric aerosol problem.

Reference

Newell, R. E., Vincent, D. G., Dopplick, T. G, Feruzza, D., and Kidson, J. W., 1970. Energy balance of the global atmosphere, *Global Circulation of the Atmosphere* (London: Royal Meteorological Society).

Throughout this volume, discussions of monitoring principles and techniques have been presented with respect to specific problems being addressed in the various sections. There were three papers prepared in conjunction with SCEP which could not be comfortably placed within those sections because they present techniques that can be used for many of the critical problems considered by SCEP and which are also of intrinsic scientific and technical interest.

The first of these papers was prepared by Drs. Pate and Cadle after SCEP, but since Dr. Cadle was a participant in the Study this information was available during July. The paper presents numerous methods used to study the atmospheric chemistry of remote and unpolluted areas. These comprise a specialized methodology—quite distinct from air pollution monitoring—which has been developed over the past decade at the National Center for Atmospheric Research (NCAR). These methods, proved by use in the field, are suitable for implementing several of the recommendations of the SCEP Work Group on Monitoring.

The advantages and some special problems which occur when remote sensing is used in gathering data about global and regional environmental problems are discussed in the following paper by Dr. Seymour Edelberg. This paper, which is an elaboration of information obtained before and during SCEP, treats a wide variety of techniques including ground-based, airborne, and satellite platforms with active or passive instruments operating over various portions of the electromagnetic spectrum.

The final paper in this group is an authoritative outline of the possibilities of monitoring air and water pollution from satellites and/or airplanes. It was prepared for SCEP by Drs. Hanst, Lehmann, Reichle, and Tepper of NASA and discusses both the advantages and disadvantages of such systems. Several experiments are also indicated which promise to provide useful data in the detection and monitoring of pollutants.

41
Chemical Analytical Methods　　John B. Pate and
Used by the National Center　　Richard D. Cadle
for Atmospheric Research

Introduction

In this paper we describe the techniques currently used by the
National Center for Atmospheric Research (NCAR) in its investigation of atmospheric chemistry. Contrary to the general belief,
the techniques used in air pollution are not usually applicable to
sampling and analysis of remote or unpolluted areas. Thus, the
body of techniques described herein represents a specialized methodology.

Interestingly, atmospheric chemistry (and contamination control) are specialized areas that are repeating the history of the
development of the methodologies used in industrial hygiene
and air pollution. Thus, during the Second World War, the expansion of the industrial plant and the introduction of numerous
new processes created a major crisis in the sampling and analysis
of in-plant atmospheres. The traditional methods were found to
be neither sufficiently sensitive nor selective enough to be useful
as tools for measuring the impact on the industrial worker. A
broad-scale attack was begun and persisted for some years after
the war, in which the whole of the air sampling methodology used
in industrial hygiene (the part per million range) was reinvesti-
gated. Today, analytical development in industrial hygiene tends
to be quite specific and confines itself to the application of new
analytical techniques to air sampling and analysis or the development of techniques for newly introduced contaminants of interest.

A directly parallel development in the air pollution range
(part per billion range) developed in the postwar period as the
Los Angeles pollution problem was recognized as being an emergency situation. Although drawing heavily on the basic industrial
hygiene methodology, it was soon found that direct extrapolation
of the methodology was not possible. New approaches, new techniques, and a new methodology were found to be necessary. Research in air pollution methodology is today in the same secondary mature stage of methodology development which characterizes
that of industrial hygiene.

Prepared for this volume.

The space program created another major crisis in the methodology of air sampling and analysis. Contamination control or "clean room technology" was developed as a direct result of the extraordinary demands of production conditions. Most of the methodology developed in this area was for the sampling and analysis of particles. A small but important portion of the space program involved further refinement of methodology to handle the problem of accumulation of trace gases and particles in the recycled atmospheres of the space vehicles. Presumably, this latter area will assume increasing importance as the program develops.

With the establishment of NCAR ten years ago, a group was created to study atmospheric chemistry. Although individual techniques and studies were being conducted, no broad base of sampling and analytical methodology was available for the subpart per billion range of trace constituents encountered in remote or unpolluted areas. Almost five years devoted to research in the methodology elapsed before the first valid field samples were collected and analyzed by this atmospheric chemistry group. Today, with the increasing attention being paid to the environment, a number of other groups throughout the country are concentrating on the methodology required. Accordingly, analytical development at NCAR is now concentrating on specific targeted methods required for measuring specific constituents.

This paper details and discusses a battery of methods used by NCAR in studying the atmospheric chemistry of remote and unpolluted areas. Some of the methods described here in detail have been briefly summarized in previous reports (Lodge and Pate, 1966; Pate and Lodge, 1970). Other methods, which have been fully reported (Cadle et al., 1969; Blifford and Ringer, 1969; and Langer, 1969) with particular reference to stratospheric and upper tropospheric sampling, are referred to only to the extent to which they are generally applicable for use in surface and lower tropospheric measurements. Certain techniques are not discussed since they involve a specialized methodology and application, for example, radioactivity and isotope studies (Bainbridge and Heidt, 1966; Martell, 1968, 1970), chemistry of precipitation (Lodge et al., 1968), and ozone soundings (Dütsch, 1966).

Trace Gases and Vapors

The first set of tests described here are those in which a specific and sensitive reagent is reacted with the species to be determined. As the specificity of such a method approaches the absolute, the method is used as a reference standard against which other methods of measuring species are compared. Methods of a second type involve a less specific property (usually physicochemical) which can be used as the basis for a more convenient or continuous measurement. These latter are useful only insofar as it is known that the atmosphere being sampled is free from interferences. Since this can be demonstrated only by the use of a specific reference method, most of the research in method development at NCAR is directed toward such reference methods. In most cases, these are used in the field with sacrifice of quantity of data obtained, for the sake of quality of data.

Bubbler Apparatus

This apparatus is designed to permit an atmospheric sample to be bubbled through an aqueous collecting solution, which scrubs and traps the species to be determined.

The solution may contain a reagent that reacts with the constituent of interest, or a reagent may be added at a later time in the analysis. The apparatus used for sampling is shown schematically in Figure 41.1; it consists of a prefilter, a glass and Teflon bubbler, a flow control device, and a vacuum pump. Prefilters are used only when it has been shown that they do not remove the species being determined, since some materials sorb specific gases (Byers and Davis, 1970).

The bubbler (Wartburg, Pate and Lodge, 1969) is the most important element of the apparatus (see Figure 41.2). It contains a reagent solution that extracts a specific gas as the air sample is pulled through a long glass gas dispersion tube. The tube, seen at the center of the bubbler, has a fritted glass disperser at its lower end; this causes the sample to disperse into small bubbles, and thus helps the liquid reagent to contact the gas efficiently.

The flow rate is controlled by the use of a hypodermic syringe needle plugged into a vacuum manifold that is connected to the vacuum pump (Lodge et al., 1966). The needles act as critical orifices in controlling the air flow. If the pressure drop across the orifice is more than 0.5 atmosphere, the flow rate through the

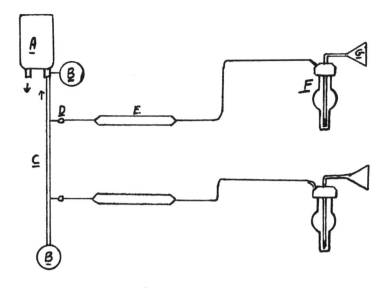

A	Vacuum Pump	E	Orifice Protector
B	Vacuum Gauge	F	Bubbler
C	Vacuum Manifold	G	Prefilter
D	Critical Orifice		

Figure 41.1 Gas sampling apparatus, schematic

orifice will be relatively constant and thus unaffected by changes in air resistance in the sampling apparatus. The orifices are protected by a plastic tube containing silica gel to remove spray from the liquid reagents, which might otherwise clog the needles.

The flow rate of the needles begins to decrease after a few days; thus, as a matter of good practice, new needles are used each day. The needles currently in use give a flow rate of 2 liters/min, thus yielding a 120-liter sample each hour.

If line power is available, carbon vane pumps are used in order to avoid hydrocarbon vapor contamination of the sampling site by oil-lubricated pumps. If a generator must be used, 100-foot extension cords are used to separate the generator and sampling apparatus. Under these conditions, it is preferable to use 12 V pumps and to carry an automobile storage battery into the field.

Sulfur Dioxide

This is determined by a modified West-Gaeke method (Pate et

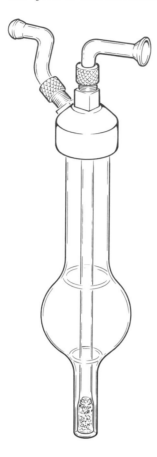

Figure 41.2 Bubbler with Teflon top

al., 1965; West and Gaeke, 1956) using the bubbler apparatus. The gas is removed from the air sample by a solution of sodium tetrachloromercurate; the resulting sulfitodichloromercurate complex is stable to oxidation. In the analysis, the complex is reacted with formaldehyde and pararosaniline to form a colored product. The intensity of the product is measured on a colorimeter, the color formed being proportional to the amount of sulfur dioxide in the sample.

Ammonia

Ammonia is scrubbed from the air sample by a 2 percent sulfuric acid solution. The resulting ammonium ion is reacted in the laboratory with sodium phenate and hypochlorous acid to yield indophenol, a colored product (Tetlow and Wilson, 1964). Since the color formation varies continuously after addition of the reagents, an automated analyzer is used for the analysis, if available. If not, careful timing of reagent addition and color intensity measurement is required for precise results.

Nitrogen Dioxide

Nitrogen dioxide is removed from the air sample by an aqueous reagent solution, containing N-napthylethylenediamine hydrochloride, acetic acid, and sulfanilic acid (Saltzman, 1954).

The nitrogen dioxide forms nitrous acid that diazotizes sulfanilic acid. The diazonium compound then couples with naphthylethylenediamine to form a colored product. The color intensity should be measured within twenty-four hours after sampling, since it is not stable. Results are calculated on a basis of 70 percent conversion of nitrogen dioxide to nitrite ion (the species measured by the reaction), since the exact conversion factor is disputed (Huygen, 1970).

Nitric Oxide

To measure nitric oxide, the sampling apparatus is modified by substituting an ascarite tube and a filter holder containing a chromate sulfuric acid filter in place of a prefilter assembly, as shown in Figure 41.3 (Pate, 1966). The ascarite (sodium hydroxide on asbestos) cartridge removes the nitrogen dioxide from the air sample. The chromate–sulfuric acid filter then oxidizes the nitric

Sampling: Prefilter replaced with NO assembly

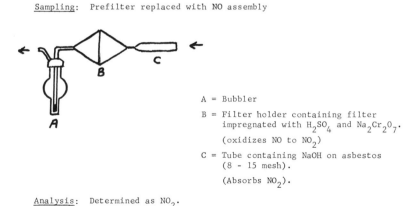

A = Bubbler

B = Filter holder containing filter
    impregnated with H_2SO_4 and $Na_2Cr_2O_7$.
    (oxidizes NO to NO_2)

C = Tube containing NaOH on asbestos
    (8 - 15 mesh).
    (Absorbs NO_2).

Analysis: Determined as NO_2.

Figure 41.3 Analysis for nitric oxide

oxide to nitrogen dioxide, which is determined. The reagents in the bubbler are the same as those for nitrogen dioxide.

Aliphatic Aldehydes

The aldehydes are removed from the air sample by a solution of methylbenzothiazole hydrazone (Altshuller and Leng, 1963). In the laboratory, the reaction product is oxidized with ferric chloride to give a colored product, which is measured. Since most aliphatic aldehydes react with this reagent to some extent, the result is indeterminate and is reported as "total aliphatic aldehydes, formaldehyde equivalent."

Hydrogen Sulfide

Two methods for hydrogen sulfide are currently being field tested. The first of these involves the suppression of the fluorescence of the analytical reagent by hydrogen sulfide (Axelrod et al., 1969). The sample is bubbled through a sodium hydroxide solution in the field. A fluorescent compound, fluorescein mercuric acetate, is added and the decrease in fluorescence is measured. When properly standardized, the decrease in fluorescence is proportional to the amount of sulfide in the sample. As with all methods of this type, any material which fluoresces or which suppresses the fluorescence of the reagent will give erroneous results.

The other method that has been used is the modified methylene blue method reported by Bamesberger (Bamesberger and Adams, 1969). The hydrogen sulfide reacts with cadmium hydroxide suspended in the collecting solution in a bubbler. The solution is acidified, N,N-dimethylparaphenylenediamine and ferric chloride added, and, in the presence of sulfide, methylene blue is formed, which is measured colorimetrically. In the past the method has been subject to loss of collected hydrogen sulfide and lack of sensitivity of the method. The modifications by Bamesberger and Adams (1969) seem to have eliminated these problems, although the sensitivity is barely sufficient for the needs of sampling in remote areas.

Hydrocarbons

For hydrocarbon analysis, samples of air are collected in evacuated stainless steel canisters, or, more commonly, in the gastight syringes manufactured by Precision Sampling Corporation (Baton Rouge, Louisiana) shown in Figure 41.4. The syringe is made of stainless steel, glass, and Teflon and is able to be locked once a sample has been collected. The syringes are carried or sent by air mail to NCAR, where the sample is injected directly into a gas chromatograph with a Porapak Q column. This is an organic polymer that separates hydrocarbon materials. When the component fractions are eluted in the helium carrier gas, they pass through a flame ionization detector connected to a recorder. The location of each peak and the area beneath the peak are related to those given by known concentrations of specific hydrocarbons under identical conditions. In an actual air sample, the qualitative

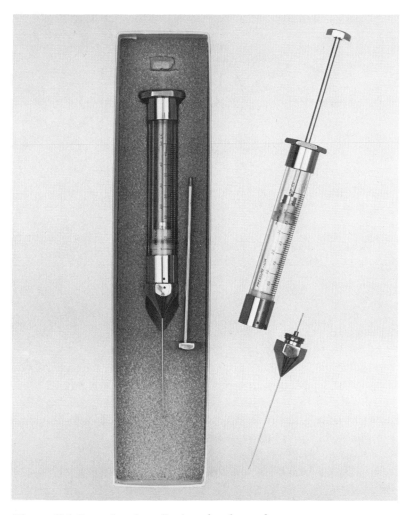

Figure 41.4 Gas syringe for collection of grab sample

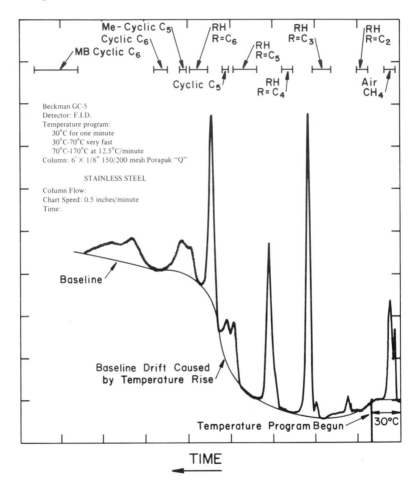

Figure 41.5 Typical gas chromatogram of an air sample from Albrook Forest, Panama

identification of as many as twenty-five peaks in each sample is a major and time-consuming problem. The peaks are usually identified on an interim basis as tentative or "presumptive" species. The chromatogram is used as a "fingerprint" characterizing the sample, as shown in Figure 41.5.

Nitrous Oxide

Nitrous oxide, the other gas analyzed chromatographically, is sufficiently stable to resist reaction when the sample is concentrated. The system used to collect the sample is shown in Figure 41.6.

Using a calibrated hypodermic needle as a disposable critical orifice, the air sample is drawn through a tube containing ascarite (sodium hydroxide on asbestos) and anhydrous calcium sulfate, to remove carbon dioxide and water, respectively. The air sample then passes through a tube of activated molecular sieve (Linde Type 5A—Synthetic Zeolite) that absorbs the nitrous oxide from the sample. The tube is sealed and returned to the laboratory. The sample of molecular sieve is transferred to a flask connected to a vacuum handling system. The system is evacuated, and the sample is heated to drive off the nitrous oxide, which is collected in a coil of stainless steel tubing immersed in liquid nitrogen. This "sample loop" is then connected to the gas chromatograph and allowed to warm to room temperature, and the sample is swept into the gas chromatograph by helium carrier gas. A Porapak Q column is used to separate the nitrous oxide from the traces of oxygen,

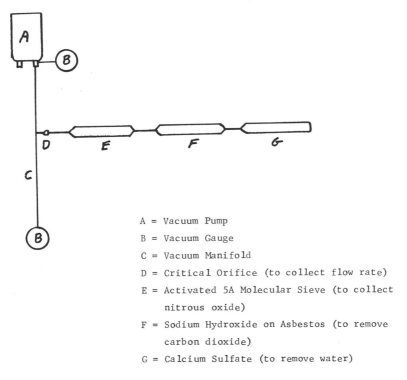

A = Vacuum Pump

B = Vacuum Gauge

C = Vacuum Manifold

D = Critical Orifice (to collect flow rate)

E = Activated 5A Molecular Sieve (to collect
    nitrous oxide)

F = Sodium Hydroxide on Asbestos (to remove
    carbon dioxide)

G = Calcium Sulfate (to remove water)

Figure 41.6 Schematic diagram of the apparatus used for the collection for nitrous oxide air samples

nitrogen, water, and other components. The eluted nitrous oxide passes through a thermal conductivity detector, and the recorder trace is interpreted by reference to a standard run under identical conditions (LaHue, Pate, and Lodge, 1970).

Ozone

The Komhyr oxidant, or ozone, meter is one of the few instruments that is suitable for measuring trace gas concentrations in the unpolluted atmosphere. The instrument uses the reaction of potassium iodide with ozone to give iodine (Komhyr, 1969). The iodine formation is measured in a concentration cell. The cell, which contains 1.5 percent potassium iodide solution in the cathode chamber and a saturated potassium iodide solution in the anode chamber, is shown in Figure 41.7. The anode and cathode are connected by a filter bridge. When the air sample is bubbled through the cell, electrons are released in the chemical reaction as iodide is converted to iodine and iodine is converted back to iodide. The resultant current is amplified and read directly on a small recorder.

The oxidant meter contains a small Teflon pump that

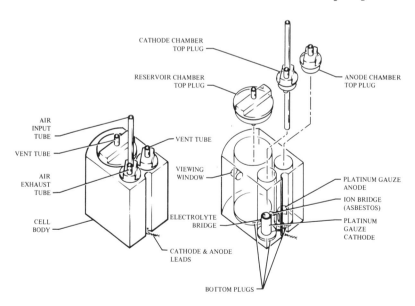

Figure 41.7 Details of construction of electrochemical concentration cell for gas analysis (ESSA Photo)

samples the air at a rate of 200 ml per minute through the cell, a small current amplifying circuit, a meter, and cadmium-nickel batteries. It is completely portable, and the batteries are recharged after about sixteen hours' use. The iodine reaction measures other strong oxidizing agents besides ozone. However, ozone is the major strong oxidant in an unpolluted atmosphere, and thus the instrument can successfully monitor ozone levels in remote areas.

Particles

Atmospheric particles may vary in size from grains of sand to short-lived aggregates of a few molecules. Between these limits (at least seven orders of magnitude in size) lie the particles that obscure visibility, change the distribution of heat in the atmosphere, fall out of the atmosphere as dust or dirt, serve as nuclei for moisture to condense upon, and play an important role in the formation of some types of pollution. Most of the particles to be measured are extremely small, requiring either the accumulation of large samples over long periods of time, or the use of extremely sensitive analytical techniques. Accordingly, we have grouped our discussion of methods into the treatment of these two types of sample.

Bulk Collection of Particles

Accumulating large samples of particles makes the final analysis easier, but it can lead to erroneous results. Upon collection, particles tend to react with one another and with the trace gases sorbed on the particles or in the air stream passing over the accumulation of particles during the sampling process.

In bulk samples, identity of the individual particles is lost. These samples are used to determine the mass collected from the volume of air sampled. The mass of a particular constituent of interest may also be determined for this type of sample. Since the quantities of material involved are so very small, the primary objective for this type of sample is to collect as large a quantity as possible so that the limit of sensitivity of the analytical balance or analytical method to be used in the analysis is reached.

Total Mass—Filter Samples

In air pollution work, a fan-type blower (to provide a large sample volume) and a glass fiber filter (to provide efficient particle collec-

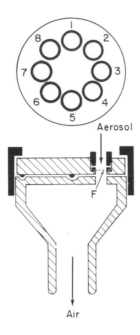

Figure 41.8 Filter holder for nuclepore filters

tion with low flow resistance) are used. For sampling periods of as much as twenty-four hours, an unpolluted atmosphere does not provide sufficient mass to be determined within an order of magnitude. For this purpose we use Nuclepore filters. This filter is made from a sheet of polycarbonate plastic approximately 10 μm thick. When the plastic is placed in a nuclear reactor and bombarded with fission fragments, each bombarding fragment leaves a track of radiation damage that can subsequently be etched out. The final etched holes are round, straight, and extremely uniform in size; depending on the etching process, the diameters can range from a few tenths of a micron to about 10 μm. The filter material is uniform in weight and is not very hygroscopic. We are able to obtain samples for quite short periods on small filter disks, which have been weighed in advance. Using a microbalance, we can accurately determine the increase in weight due to the collected particles (Spurny et al., 1969). Figure 41.8 shows details of the filter holder used for sampling with eight preweighed filter disks.

Sulfate Content of Bulk Samples

Since the weighing of a filter is a nondestructive technique, it would seem that the samples used to determine mass loading could thereafter be used for analysis for specific constituents. Unfortunately, the constituent desired is usually present in the glass filter material at a concentration much higher than that of the same constituent in the sample. Since this was true for sulfate, which was considered to be the most important constituent which should be measured in the particles, polystyrene fiber filter mats are prepared at NCAR by the procedure developed by Cadle and Thuman (1960). These filters were used in sampling for sulfate aerosol in the stratosphere and upper troposphere (Blifford and Ringer, 1969; Cadle et al., 1969) as well as for surface measurements and in the lower troposphere. Extreme care is needed at all steps of the production in order to avoid contamination of the filters. A commercial polystyrene filter of similar characteristics proved in the past to be too contaminated for use in these determinations. The manufacturer (Delbag-Luftfilter GmbH, Berlin, Germany) has improved the quality of these filters, and so they are now replacing the NCAR-made filters in some of the sampling programs.

When used for surface sampling in remote areas, 4″ circles or 8 × 10″ sheets of filter material are used: Any one of the standard high-volume samplers used in air pollution work is used to take samples for a period of one to twenty-four hours. In all cases, line voltage is required because the samplers are not suitable for use with storage batteries. Since the samplers produce quantities of carbon dust which can contaminate particulate samples being collected in an unvented area (such as under the forest canopy), care must be exercised in the siting of these samplers.

Miscellaneous Bulk Samples

Almost the entire array of particulate sampling devices developed for industrial hygiene and air pollution work can be and is used to collect bulk samples for special analyses. Probably most used is the cascade impactor, illustrated in Figure 41.9 (Lippmann, 1961). Illustrative of the considerations governing the size of the sample and limitation imposed by the analytical method to be used are samples collected with this impactor. A 15-minute to 1-hour sample usually provides sufficient sample to be used for x-ray diffraction

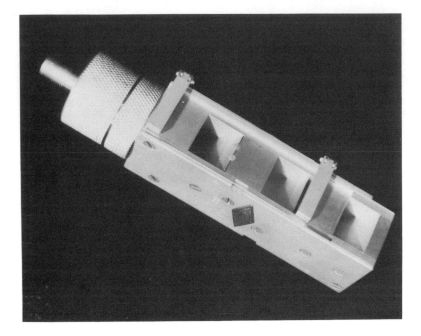

Figure 41.9 Cascade impactor

using the camera method. The predominant crystalline species can thus be identified. On the other hand, the sample is not sufficiently large to provide material for elemental analysis by x-ray spectroscopy nor for x-ray diffraction using a diffractometer.

Individual Particle Analysis

One way to minimize the interaction of particles during collection is to collect few enough to limit the probability of two particles landing on the same spot. Clearly such "thin" samples cannot be analyzed by ordinary means—a sample of this sort may well contain less than one nanogram of total collected material. Special techniques thus become necessary.

Morphology

Many kinds of particles can be recognized at sight by their appearance in the optical or electron microscope. Particles for examination may be collected either on filters or impactor slides.

The impactor provides a convenient means of collecting particles for examination with the electron microscope. The special specimen grids used for electron microscopy are mounted directly on the slide mounted on the impactor, as shown in Figure 41.9. For analysis, the grids are removed and examined with the electron microscope after shadowing. For these samples, only the fourth stage of the impactor is used, thus collecting only the smaller particles in the sample. The percentage of the total number of particles examined is calculated for each type of particle. No attempt is made to calculate a concentration in air, since an indeterminate portion of a nonuniform deposit is collected on a particular specimen grid. The appearance and characteristics of the various types of particles encountered in remote and unpolluted areas have been reported (Lodge and Pate, 1966; Sheesley, Pitombo, Pate, Wartberg, Frank, and Lodge, 1970; and Sheesley, Pate, Frank, and Lodge, 1970).

Recently we have developed particle identification techniques for use with the Nuclepore as well. Unlike membrane filters that were previously used, thin samples collected on Nuclepore filters can be used for either optical or electron microscopy. The deposit is uniform and the total deposit is collected on the active filter surface, thus allowing the concentration of specific types of particles to be calculated.

Size and Count—Total

If a sample is saturated with water vapor and then rapidly expanded, the resultant cooling will produce a supersaturation of water vapor, and the cloud droplets will form on the particles present. If the supersaturation is very slight, only the largest particles will grow into water drops. If the supersaturation is much greater, much smaller particles will also grow. This phenomenon was used as the basis for the development of the "condensation nucleus counter" (Ohta, 1959; Rich, 1955). In the instrument (shown schematically in Figure 41.10), the cloud is formed and its density is measured photoelectrically; the results are read from a meter calibrated directly in particle concentration.

In the field this instrument is useful not only for obtaining particle concentrations but also to detect the presence of sources of contamination from human activity, pumps, smoking, engines, and so on. It thus is used as a monitor for the area as well as a particle counter.

In some areas, freezing nuclei counts would be of direct in-

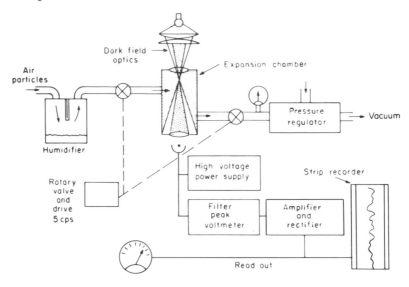

Figure 41.10 Schematic of "condensation nuclei counter"

terest. For these, the field method developed by Langer and Rosinski (1967) is useful.

Size and Count—Specific Components

Many times it is desirable to obtain information on the size distribution and number concentrations of particles containing particular species, such as chloride or sulfate. These samples are collected on membrane filters (Lodge, 1954). The filter may be impregnated with a specific reagent prior to sampling (as in the case of sulfate described next) or may be a standard membrane filter that is treated with reagents in the laboratory during analysis (as in chloride, as described later). The filter holder used for sampling is shown in Figure 41.11. The base permits insertion of a critical orifice to control the flow rate.

For sulfate-containing particles, a membrane filter is impregnated in the laboratory with insoluble barium rhodizonate, which colors it bright pink. Any particle containing sulfate will make a nearly colorless spot on a pink background, the size of the spot being proportional to the amount of sulfate in the particle (Lodge and Parbhakar, 1963; Lodge and Frank, 1966). The samples are returned to the laboratory and the number of reaction spots counted and sized. The size distribution is calculated, and the

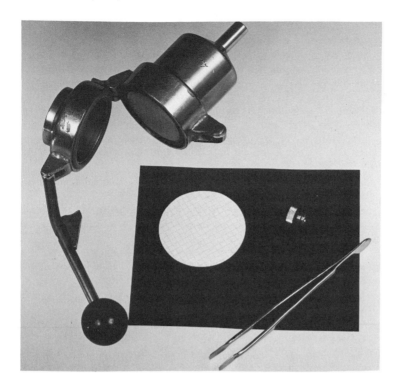

Figure 41.11 Lauterback filter holder, membrane filter and critical orifice

geometric mean and geometric standard deviation of the sample is reported together with the total concentration of sulfate-containing particles.

To determine chloride-containing particles, a portion of the membrane filter on which the sample was collected is floated on a slightly acidic solution of mercurous fluorosilicate, chloride-containing particles reacting to form pale blue spots of mercurous chloride. The reaction spots are counted and the concentrations calculated accordingly. The remainder of the filter can be analyzed for ammonium ion, calcium, nitrate, and so on.

References

Altshuller, A. P., and Leng, L. J., 1963. Application of the 3-methyl-2-benzothiazolone hydrazone method for atmospheric analysis of aliphatic aldehydes, *Analytical Chemistry, 35:* 1541.

Axelrod, H. A., Cary, J. H., Bonelli, J. E., and Lodge, J. P., Jr, 1969. Fluorescence determination of sub-parts per billion hydrogen sulfide in the atmosphere, *Analytical Chemistry, 41:* 1856.

Bainbridge, A. E., and Heidt, L. E., 1966. Meaurements of methane in the troposphere and lower stratosphere, *Tellus, 18:* 221.

Bamesberger, W. L., and Adams, D. F., 1969. Improvements in the collection of hydrogen sulfide in cadmium hydroxide suspension, *Environmental Science and Technology, 3:* 258.

Blifford, I. H., Jr., and Ringer, L. D., 1969. The size and number distribution of aerosols in the continental troposphere, *Journal of the Atmospheric Sciences, 26:* 716.

Byers, R. L., and Davis, J. W., 1970. Sulfur dioxide absorption and desorption on various filter material, *Journal of the Air Pollution Control Association, 20:* 236.

Cadle, R. D., and Thuman, W. C., 1960. Filters from submicron-diameter organic fibers, *Industrial and Engineering Chemistry, 52:* 315.

Cadle, R. D., Bleck, R., Shedlovsky, J. P., Blifford, I. H., Rosinski, J., and Lazrus, A. L., 1969. Trace constituents in the vicinity of jet streams, *Journal of Applied Meteorology, 8:* 348.

Dütsch, H. V., 1966. *Two Years of Regular Ozone Soundings Over Boulder, Colorado,* NCAR Tech. Note No. 10 (Boulder, Colorado: National Center for Atmospheric Research).

Huygen, I. C., 1970. Reaction of nitrogen dioxide with Griess type reagents, *Analytical Chemistry, 42:* 407.

Komhyr, W. D., 1969. Electrochemical concentration cells for gas analysis, *Annals of Geophysics, 25:* 203.   •

LaHue, M. D., Pate, J. B., and Lodge, J. P., Jr., 1970. Atmospheric nitrous oxide concentrations in the humid tropics, *Journal of Geophysical Research, 75:* 2922.

Langer, G., 1969. Evaluation of NCAR Ice and Cloud Condensation Nuclei Counters, in *Proceedings of the 7th International Conference on Condensation and Ice Nuclei* (Prague and Vienna), pp. 215–219.

Langer, G., and Rosinski, J., 1967. A continuous ice nucleus counter and its application to tracking in the troposphere, *Journal of Applied Meteorology, 6:* 114.

Lauterback, K. E., 1954. New design of filter holder for dust sampling, *American Industrial Hygiene Association Quarterly, 15:* 78.

Lippmann, M., 1961. A compact cascade impactor for field sampling, *Journal of the American Industrial Hygiene Association, 22:* 348.

Lodge, J. P., Jr., 1954. Analysis of micron-sized particles, *Analytical Chemistry, 26:* 1829.

Lodge, J. P., Jr., and Parbhakar, K. J., 1963. An improved method for the detection and estimation of micron-sized sulfate particles, *Analytica Chimica Acta, 29:* 372.

Lodge, J. P., Jr., and Frank, E. R., 1966. An improved method for the detection and estimation of micron-sized sulfate particles: correction, *Analytica Chimica Acta, 35:* 270.

Lodge, J. P., Jr., Pate, J. B., Ammons, B. E., and Swanson, G. A., 1966. The use of hypodermic needles as critical orifices in air sampling, *Journal of the Air Pollution Control Association, 16:* 197.

Lodge, J. P., Jr., and Pate, J. B., 1966. Atmospheric gases and particulates in Panama, *Science, 153:* 408.

Lodge, J. P., Jr., Pate, J. B., Baspagill, W., Swanson, G. A., Hill, K. C., Lorange, E., and Lazrus, A. L., 1968. *Chemistry of the United Precipitation* (Boulder, Colorado: National Center for Atmospheric Research), August.

Martell, E. A., 1968. Tungsten radioisotope distribution and stratospheric transport processes, *Journal of the Atmospheric Sciences, 25:* 113.

Martell, E. A., 1970. High altitude air sampling with a rocket-borne cryocondensor, *Journal of Applied Meteorology, 9:* 1970.

Ohta, S., 1959. Investigation on the formation and disappearance of fog in an expansion fog chamber: its application to the measurement of concentration of condensation nuclei with Pollack photoelectric nucleus counter, *The Geophysical Magazine, 29* (2): 229, Tokyo.

Pate, J. B., 1966. Determination of nitric oxide in air, presented at the American Industrial Hygiene Association Conference, Denver, Colorado, October 1, 1966.

Pate, J. B., Ammons, B. E., Swanson, G. A., and Lodge, J. P., Jr., 1965. Nitrite interference in spectrophotometric determination of atmospheric sulfur dioxide, *Analytical Chemistry, 37:* 942.

Pate, J. B., Lodge, J. P., Jr., Sheesley, D. C., and Wartburg, A. W., 1970. Atmospheric chemistry of the tropics, *Proceedings of the Introductory Symposium on Integrated Research into Ambient Factors in the Amazonian Region* (Manaus, Brazil: Instituto Nacional Pesquisas Amazonas), forthcoming.

Rich, T. A., 1955. A photo-electric nucleus counter with size distribution, *Geofisica Pura e. Applicata, 31:* 60.

Saltzman, B. E., 1954. Colorimetric microdetermination of nitrogen dioxide in the atmosphere, *Analytical Chemistry, 26:* 1949.

Sheesley, D. C., Pate, J. B., Frank, E. R., and Lodge, J. P., Jr., 1970. Measurement of particles in the non-urban atmosphere, *Proceedings of the Introductory Symposium on Integrated Research into Ambient Factors in the Amazonian Region* (Manaus, Brazil: Instituto Nacional Pesquisas Amazonas), forthcoming.

Sheesley, D. C., Pitombo, L. R. M., Pate, J. B., Wartburg, A. F., Frank, E. R., and Lodge, J. P., Jr., 1970. Atmospheric sampling (chemical) in Amazonas III. Preliminary results of particle sampling, *Proceedings of the 2nd Symposium on Integrated Research into Ambient Factors in the Amazonian Region* (Manaus, Brazil: Instituto Nacional Pesquisas Amazonas), forthcoming.

Spurný, K. R., Lodge, J. P., Jr., Frank, E. R., and Sheesley, D. C., 1969. Aerosol filtration by means of nuclepore filters structural and filtration properties, *Environmental Science and Technology, 3,* Part II, 453.

Tetlow, J. A., and Wilson, A. L., 1964. An absorptiometric method for determining ammonia in boiler feed-water, *Analyst, 89:* 453.

Wartburg, A. L., Pate, J. B., and Lodge, J. P., Jr., 1969. An improved gas sampler for air pollution analysis, *Environmental Science and Technology, 3:* 767.

West, P. W., and Gaeke, G. C., 1956. Fixation of sulfur dioxide as disulfitomercurate (II) and subsequent colorimetric estimation, *Analytical Chemistry, 28:* 1816.

42

Some Considerations   Seymour Edelberg
in the Use of Remote
Environmental Monitoring

Introduction

The impact of man's activities on the global environment can be estimated by computation and by data gathering. The former has been discussed elsewhere in this volume. The latter, both *in situ* and remote data gathering, is also discussed in this volume. This chapter is a discussion of the advantages and drawbacks of remote sensing.

In situ measurements have been used quite successfully for many years in urban pollution control (Morgan, Ozolins, and Tabor, 1970) and in atmospheric science. Two excellent examples of the latter involve CO_2 (Pales and Keeling, 1965) and particulate matter. The two major disadvantages of *in situ* monitoring are (1) the necessary sample "handling" may lead to distortion of the data; (2) for monitoring on a global scale "point" spatial resolution may lead to an unduly large amount of accumulated data which could be avoided only by assuming a spatial distribution of the measured parameters.

Remote measurements, as discussed by Philip Hanst et al. in Chapter 43, can often avoid these pitfalls. However other difficulties arise in many measurements, such as limited spatial resolution, accuracy, and sensitivity. Remote measurements require new techniques, some just now being developed and others newly conceived, to overcome these difficulties. It is the purpose here to discuss these techniques, which could be used to monitor global or regional changes in certain atmospheric gases and aerosols, vertical temperature distributions, and cloud characteristics, which will lead to a better understanding of radiative and energy balance. The measurements could involve ground-based, airborne, or satellite platforms with active or passive instruments operating in the visible, infrared, or microwave portions of the electromagnetic spectrum.

Prepared during SCEP and expanded after SCEP.

Monitoring Atmospheric Gases

Carbon Dioxide

The *in situ* CO_2 measurements (Pales and Keeling, 1965) are very well developed, although an improvement in accuracy to ± 0.1 ppm is desired (SCEP, 1970), and this appears to be feasible. However this accuracy or even an accuracy of 1 percent (3 ppm), which will not register the mean annual increase of CO_2 (0.7 ppm), will be difficult to obtain from a satellite measurement. The reasons will be discussed below.

A passive (no onboard energy transmitter) measurement can be made from a satellite, looking down, using either emitted radiation or absorbed radiation. A major disadvantage in using emitted radiation is the need for a knowledge of the vertical temperature distribution, which is a difficult measurement to make accurately and should be avoided if possible when only CO_2 total levels are desired.

An absorption measurement using the sun as a source is preferred for the determination of atmospheric CO_2 levels. This measurement is made in a wavelength region where the earth's emitted radiation is a minimum, and which avoids the complexity of a second source term in the data interpretation process. Figure 42.1, due to Robinson (1970), indicates that the earth's radiation decreases to negligible levels at wavelengths below about 4 μ. Furthermore, below 4 μ, the solar radiation increases so that it could be used as an energy source for absorption spectroscopy.

The sun is a broadband source, as seen in Figure 42.1, which implies the use of a monitoring system with sophisticated filtering. Such a system has been fabricated by Convair (Ludwig, Bartle, and Griggs, 1969) and is described in Chapter 43 under the subsection entitled "A Remote Sensing System Based on Gas Filter Correlation Principles." This instrument, looking down, senses the solar radiation which passes down through the atmosphere, is reflected by the earth, and then passes up through the attenuating atmosphere. Ludwig (private communication) has performed a set of calculations which indicates that a CO_2 concentration change of 1 ppm can be detected from satellites using this instrument with a signal-to-noise ratio of 100 or better.

The Ludwig calculations are based on an assumed "one-slab"

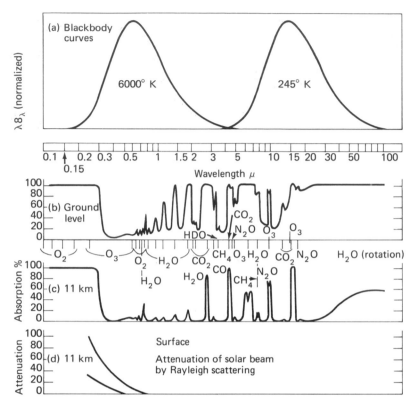

Figure 42.1 (a) Blackbody emission for 6000°K and 245°K, being approximate emission spectra of the sun and earth, respectively (since inward and outward radiation must balance the curves have been drawn with equal areas—though in fact 40 percent of solar radiation is reflected unchanged); (b) atmospheric absorption spectrum for a solar beam reaching the ground; (c) the same for a beam reaching the tropopause in temperate latitudes; (d) attenuation of the solar beam by Rayleigh scattering, at the ground and at the temperate tropopause.

Source: Robinson, 1970.

model of the atmosphere, implying that vertical density and temperature contours within the instrument's field of view need not be known. If the gas is not "black," so that it is optically thin or semitransparent, then this assumption is valid. This is a great advantage since the aforementioned contours need not be obtained as a prerequisite for the determination of the total amount of CO_2 in the vertical column, below the satellite, within the instrument's field of view. The wavelength that was selected to satisfy the optically thin criterion is one of the overtones of CO_2, at 1.6 μ. There does not appear to be any interfering radiation from atmospheric molecules in this wavelength band. Changes in ground reflectivity are monitored with a narrow-band radiometer operating in an atmospheric window at 1.5575 μ.

The calculations indicate that a 1 ppm change in atmospheric CO_2 (about 320 ppm) could be detected over land with a signal-to-noise ratio of 130, which is quite adequate. However, over the ocean the signal-to-noise ratio drops to about 7, which is not the basis for a confident measurement. This decrease is due to a smaller albedo, by a factor of approximately 20, of ocean compared to land surface. If the spectral bandwidth is made wider to accept adjacent CO_2 radiation, then perhaps a factor of 2 or 3 will be gained in signal-to-noise ratio, which is probably a marginal increase. It should be noted that the spectral bandwidth cannot be made too wide without incurring the difficulties of interference from other molecules.

The field of view of the correlation instrument is 0.1 ster. For a satellite in a 600 nautical mile orbit, the linear ground resolution is 200 n.m. Herein lies the greatest difficulty in making an accurate measurement of CO_2 concentration. Within this field of view will lie cloud distributions with different albedos, sizes, numbers, and heights. Thus there will be, within the instrument's view and along incident solar rays, some absorptive paths that are ground-reflected, while others are cloud reflected with different total path lengths. Thus it becomes very difficult to determine the volume of atmosphere which has been sensed for CO_2, and the resulting accuracy will be poor except in cloud-free regions.

A solution to this problem would be to determine the cloud positions and to decrease the field of view so that the absorbing paths are not obstructed by clouds. The field of view is decreased by increasing the size of the instrument's optics and by decreasing the size of the detector resolution element. However this cannot be done without at the same time decreasing the system's signal-to-noise ratio, an unacceptable possibility. Thus with this geometrical configuration it appears necessary to accept the large field of view with its cloud problem. It may be possible to make corrections for the effects of clouds by determining their spatial distributions and albedos. This will be discussed later.

Since the sun is used as a source in the preceding measurements, daylight operation is necessary. The use of a laser source could permit data gathering at night. However the wavelength of the available laser radiation does not match that of the CO_2 overtone, and a tunable laser of sufficient output power is not available for this measurement, as yet.

The cloud problem could be minimized by viewing the atmosphere horizontally, instead of vertically. Methods have been developed (Gray, 1969) which enable an optical instrument onboard a satellite to have its field of view stabilized very accurately on the earth's limb so that precise and stable measurements could be made with a viewing aspect which is parallel to the earth's surface. Absorption spectroscopy could use the solar source, as above. This measurement implies that the field of view lies in the cloud-free region above the troposphere. Since CO_2 mixes very well in the atmosphere, a measurement in the stratosphere will give useful data.

Operation at night is also easier with this geometrical configuration. Instead of the sun as a source, a strongly radiating star could be used. There are a number of stars that radiate in this wavelength region. Otherwise a tunable laser source operating in a "mother-daughter" satellite system might be possible.

When the earth-limb or occultation configuration is used, the system sensitivity is much greater than that in the downward-looking mode because the losses in effective source radiation due to albedo (especially the ocean albedo, 3 percent) incurred with the latter mode are not present. Therefore the field of view can be decreased, thus significantly increasing the spatial resolution in the vertical direction.

Other Gases

The gas filter correlation analyzer is also a very useful instrument, especially in the earth-limb configuration, for the determination of mean mixing ratios of other gases. For example, Ludwig (private communication) has estimated that CO could be sensed to 10^{-4} ppm at an altitude of 20 km or about 10^{-2} ppm at 70 km, with a vertical height resolution of 2 km and a signal-to-noise ratio of about 100.

The gas filter correlation analyzer is used when radiation absorbed by one, or several, specific gases is to be sensed. The specific gases must be chosen a priori. However there are situations when the gaseous components in the atmosphere are not well known and should be determined. For example, SST flights in the lower stratosphere will generate in their gaseous exhausts various forms (presently not well known) of sulfates and nitrates as well as water vapor, and hydrocarbons. It would be useful to identify and analyze the gaseous constituents in those stratospheric regions where SSTs fly, using techniques of absorption spectroscopy in order to be sure that there are no "surprise" constituents which would otherwise be missed. This calls for wavelength scanning across the entire infrared band where molecules absorb radiation. Since many of the absorbing wavelength bands are closely spaced or are even overlapping, very high spectral resolution is required. Interference filters or grating will not, in general, give the required resolution.

This problem might be avoided or minimized by using newly developed, very narrow-band diode laser sources (Hinkley, 1970). Such a source, using a lead-tin-telluride diode, can yield spectral resolutions of about 50 kHz or 1.7×10^{-6} cm^{-1}, which is the linewidth of the laser's output radiation. This laser can be tuned over a 40 cm^{-1} range, by merely varying the dc diode current. These lasers can be fabricated to emit in the 7 to 30-μ region. Thus, the device can be used in many wavelength regions not covered by the gas filter analyzer. However it suffers from the disadvantage of low power output. Under cw operation, output power levels of 1 mW have been obtained, with several watts for pulsed operation. This implies absorptive paths in the 1 km range, or the use of airborne platforms rather than satellites.

The diode laser appears to be limited to the longer infrared wavelengths. However tunable dye lasers (Snavely, 1969) and Raman lasers (Patel and Shaw, 1970) may also be applicable in other wavelength regions. The problem of low power output (implying a limitation in the maximum path length which can be monitored) remains.

Absorption spectroscopy can be replaced, for some gaseous monitoring purposes, by laser radar backscatter measurements. This has the advantage of placing the receiver in the same position as the transmitter, unlike the configuration used for absorption spectroscopy. It has been shown (Fiocco and Dewolf, 1968) that it is possible to separate through laser radar Doppler processing, the relatively narrow-band Doppler returns due to droplets and aerosols from the wider-band returns from the molecules. However, the molecular return does not identify which specific molecular type caused the return. This can be done by laser Raman probing of the atmosphere (Cooney, 1970), again in the radar backscatter configuration.

The Raman component of radiation backscattered by a specific gas is frequency-shifted, relative to the frequency of the incident radiation, by an amount which depends uniquely on the molecular structure of the gas. This shift therefore provides a unique signature and identifies the gas. Furthermore, the transmitted waveform is a short pulse, thereby permitting the return to be resolved in range. If the pulse is t-seconds in duration and the velocity of light is c, then the range resolution is $(ct/2)$, that is, for a 0.1-μsec pulse, the range resolution is 15 m. When the wavelength of the transmitted radiation is 6943 Å, the Raman component of the backscattered radiation from nitrogen is centered at 8283 Å, easily resolved with 100-Å spectral resolution from the Rayleigh and aerosol returns. Cooney (1970) has obtained humidity data for the first few kilometers above ground level, with 100-m spatial resolution, by using the frequency-doubled ruby laser output at 3471.5 Å. The Raman water vapor line appeared at 3976 Å, again easily resolved. In fact, spectral resolution does not appear to be a problem. With the transmitted radiation at 3471.5 Å, the wavelength positions of various Raman lines are 3616 Å for SO_2, 3634 Å and 3647 Å for CO_2, 3670 Å for O_2, 3751 Å for CO, and 3818 Å for H_2S.

The greatest limitation to the use of laser Raman sensing of the atmosphere is its limited range. Computations by Hirschfeld and Klainer (1970) indicate that for various pollutants the maximum range may be only hundreds of meters even for a sophisticated system. The minimum detectable pollution levels in ppm at a range of 250 m for their specific system are 1.8 for NH_3; 2.2, HCl; 4.8, CO_2; 1.1, H_2S; 5.5, CO.

Another "laser Raman radar," built by Kobayasi and Inaba (1970), detected SO_2 and CO_2 molecules in a polluted atmosphere. A Q-switched ruby laser was used, which generates 0.3 J in a 30

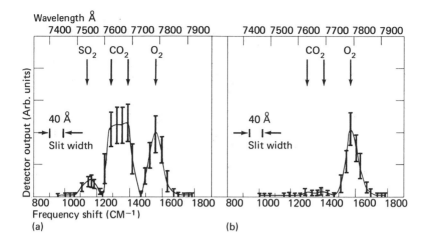

Figure 42.2 A comparison of the observed spectra of Raman backscatters from the polluted atmosphere with the ordinary one
(a) Relative detector output of Raman returns with the oil plume polluting the atmosphere as a function frequency shift; (b) The output spectrum from the ordinary atmosphere.

Source: Kobayasi and Inaba, 1970.

nsec pulse at 6943 Å. The effluence from a chimney was observed at a range of 20 meters. Figure 42.2 indicates the resulting Raman returns for CO_2, O_2, and SO_2. Since the system was calibrated, densities were obtained: 4×10^{17} cm^{-3} for the ambient CO_2 (0.015 atm), 6×10^{18} cm^{-3} for the CO_2 in the effluence (0.2 atm), and 3×10^{17} cm^{-3} for the SO_2 in the effluence (0.01 atm). It was noted (Kobayasi and Inaba, 1970) that the ambient CO_2 level was fifty times the value for the standard atmosphere, and that such a high density may be expected in winter, urban atmospheres.

The SO_2 effluence is, of course, a serious pollution problem on a regional basis, if not a global basis. A recent numerical simulation on a continental basis was made for atmospheric sulfur transport over Europe (Reiquam, 1970). The results suggest that Great Britain and Central Europe contribute to the observed concentrations of sulfur in Northern Europe. The pollutants were assumed to be contained in the lowest 3 km of the atmosphere. The sulfur might be in the form of SO_2, sulfate, or sulfuric acid. Measurements are necessary to confirm these simulations, and these could be made from aircraft using laser Raman backscatter.

On a global scale, SST gaseous exhausts are of concern, as indicated elsewhere in these volumes, and again airborne laser Raman backscatter measurements could gather useful data.

Vertical Distributions of Gases and Temperature

The preceding discussion involved the measurement of mean gas density. However for many purposes, one being to gain a better understanding of atmospheric radiative transfer, it is necessary to determine the vertical (and also the horizontal) distribution of gases. Furthermore, gas temperature distributions are also required.

Bradley (private communication) notes that it is possible to obtain atmospheric gas temperature by infrared absorption using the earth-limb (occultation) mode. He uses CO_2 as an example and observes that it is desirable to look at the absorption at the altitude where line broadening is not dominated too much by collision effects. The crossover comes when the line width due to collision broadening is approximately equal to that due to Doppler broadening. For a wavelength of about $2\ \mu$ this corresponds to an altitude of 18 km (about 0.1 atm). Measurements which are least sensitive to temperature and which yield the CO_2 molecular number density require that the effective optical thickness be in the neighborhood of unity for the excited CO_2 molecules which are to be used. A transition starting in the lowest vibrational state is chosen with an initial rotational quantum number of 22.

However, the primary consideration here is temperature determination. The use of a CO_2 molecule with an initial rotational quantum number that is much higher (about 56) is very useful as a sensitive temperature indicator. This measurement can be improved by using the isotope $^{13}C^{16}O_2$.

There is now a fairly large number of papers available on means of obtaining vertical temperature and gas density contours. A general review of some of the available literature is given in Hanel (1970). Wark (1970) indicates that 15-μ measurements made in eight channels of the SIRS instrument onboard the Nimbus III satellite frequently yielded useful temperature profiles, even in the presence of a cloud layer. This was done by inverting the radiative transfer equation. Smith (1970) obtained temperature profiles using the SIRS data from Nimbus IV as well as those from Nimbus III, by invoking a different mathematical inversion technique.

When clouds were present some profiles did not compare closely with rocketsonde temperature profiles.

Chahine (1970) uses an inversion technique which, in principle, allows him to obtain temperature profiles in the presence of clouds and to obtain, as well, the extent and height of these clouds, all with a very rudimentary first guess of the temperature profile. Chahine has used this technique to invert synthetic radiance data generated by a computer from a set of model temperature profiles in an atmosphere having a constant CO_2 mixing ratio of 462×10^{-6} by mass. The spectral interval selected to illustrate this study corresponds to thirty-three frequencies with 5 cm^{-1} spacing in the 2195–2355 cm^{-1} band (4.3 μ). An error analysis indicated that a temperature accuracy of 1°K can be expected with a 2 percent rms random error in observations, and an accuracy of 2 to 3°K can be expected with a 5 to 7 percent rms random error in observations.

Of greater interest perhaps is Chahine's analysis when the presence of clouds is taken into account. The results, shown in Figure 42.3a, are for a numerical experiment in which the synthetic radiance data were generated on a computer for several values of N (the fractional cloud cover in the field of view) and for a single layer of clouds behaving like a blackbody in equilibrium with the local ambient temperature at the 400 mb (p_c) cloud height. The figure shows different temperature profiles below the clouds for the various values of N. However above and on the clouds, the temperature profile is accurately reproduced. At the cloud level the temperature error is 1°K. Thus the cloud height can be estimated from a blackbody measurement of the cloud top temperature, T_c. Chahine concludes that a knowledge of any of the parameter combinations (p_c, N), (p_c, T_0), (T_c, N), or (T_c, T_0), where T_0 is the ground temperature, is sufficient to determine the complete temperature profile in the *clear* portion of the field of view. Also a knowledge of cloud and surface temperatures along with the radiance data allows the determination of the cloud height and the fractional cloud cover (p_c, N).

The foregoing results were obtained from "computer experiments," rather than from actual experiments. Satellite data have not been interpreted, using this approach, as yet. However the data from a balloon experiment have been so interpreted (Shaw et al., 1970). The balloon flew at an altitude of 3.5 mb with a

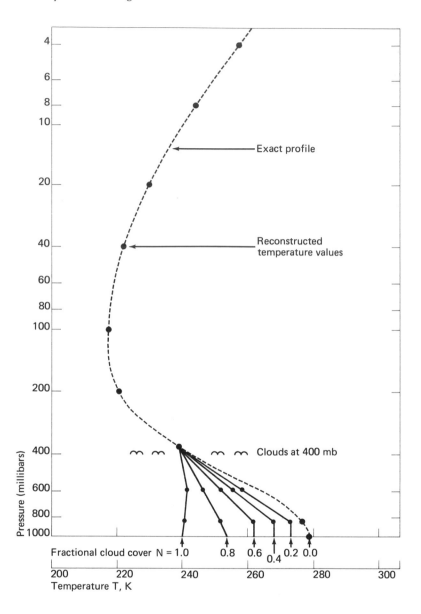

Figure 42.3a Comparison between the exact temperature profile and the reconstructed apparent-temperature values in the presence of clouds

Source: Chahine, 1970.

spectrometer which sampled 35 frequencies from the incident radiation in the spectral interval: 1975–2655 cm^{-1}. Vertical temperature contours were determined with errors not exceeding 2 to 3°K at any altitude under the worst cloud conditions. Cloud top temperatures were computed with an accuracy of about 1°K, compared with that obtained from an independent measurement. Figure 42.3b illustrates a very good comparison between the computed and measured temperature contours with 89 percent estimated cloud cover.

When a vertical temperature distribution is obtained, it is possible to obtain gas density distributions. Smith (1970) obtained water vapor profiles from SIRS-B data which appear to be accurate to factors between 2 and 4 in mixing ratio at some altitudes. (These errors may, in part, be due to temperature contours of limited accuracy.) A water vapor profile was also obtained using the IRIS instrument on Nimbus III (Conrath et al., 1970), along with vertical temperature distributions (corrected for clouds according to Smith) and total ozone. The water vapor profiles were accurate within a factor of about 2 compared with radiosonde data. Vertical ozone contours were derived from the IRIS data (Prabhakara et al., 1970). These were, generally, not accurate. However total ozone content was determined to an accuracy of about ±6 percent over a large part of the earth (Conrath et al., 1970).

Monitoring Aerosols

The importance of monitoring aersols, both in the troposphere and in the stratosphere, has been discussed elsewhere in this volume. *In situ* measurements and some remote measurements have also been discussed (Chapter 27; see also Chapter 30). Additional comments pertaining to remote measuring techniques are made in this section.

Most atmospheric aerosols are in the size range: 0.1 to 1 μ. These have their greatest effect on visible radiation. However aerosols, due to urban or industrial sources, are also in the 1 to 10-μ range and can affect infrared radiation. Since the lifetime of the larger particles is relatively short, they are not of great interest except near the source, where the source can be an urban region polluting the troposphere or an SST in the stratosphere.

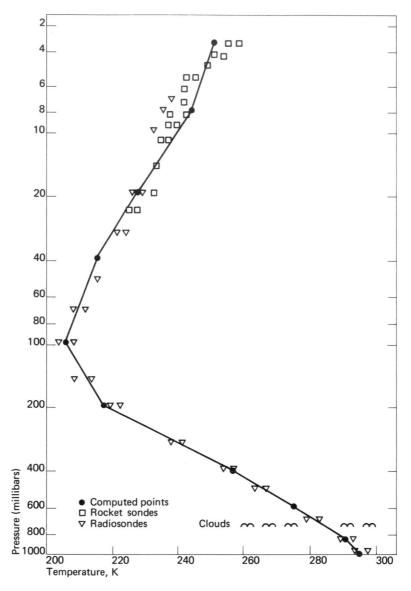

Figure 42.3b Temperature profile derived from 0505 CDT radiance data
The cloud height was estimated to be 700 mb and the cloud over 89 percent.

Source: Shaw et al., 1970.

These aerosols will absorb and scatter incident radiation, thereby affecting albedo and energy balance in the atmosphere. It is therefore necessary to measure the absorbing and scattering properties of this particulate matter. It is also desirable, in order to understand and use the data from these measurements, to determine particle size distributions and concentrations, as well as the chemical nature of the aerosol.

Turbidity (extinction of solar radiation) measurements have been made for some time (see Chapter 27). Recently the backscattered component of aerosol scattering has been measured using laser radar (lidar). Since the waveform is a pulse, backscattering can be determined as a function of altitude. The sensitivities of various systems are great enough so that measurements have been made in the stratosphere (Grams and Fiocco, 1967), even above 50 km (Grams, 1970).

The 20-km sulfate layer, first discussed by Junge (Chapter 23), has been detected by several lidars (Schuster, 1970a, b). It has been determined by these lidar measurements that the ratio of total to molecular (Rayleigh) backscatter varies from 1 to 2.5 (over the United States) with a peak usually at 20 km, the additional return presumably being due to aerosol. The estimate was made using the vertical air density contour of an assumed Standard Atmosphere.

A similar experiment was made in Australia (Bartuske, Gambling, and Elford, 1970). Here the peak return also occurred near 20 km, with the ratios of total to Rayleigh returns being somewhat smaller (about 1.75) than those measured in the Northern Hemisphere. The maximum aerosol density was determined from the data, assuming not only the Rayleigh profile but the aerosol's refractive index, shape (spherical), and size distribution.

It is obviously preferable to make as few assumptions as possible; if so, there is a need for independent measurements of the returns from aerosol and gas molecules. There are at present, two methods for this purpose. The first (Fiocco and Dewolf, 1968) makes the separation on the basis of Doppler processing. A coherent lidar is needed which will yield Doppler spectra from the received signal. The Rayleigh scattered-spectrum will be broad, and its width will depend primarily on gas temperature. The aerosol-scattered spectrum will be narrow, and will depend on the slower velocities of the particle's Brownian motion.

Another method, perhaps preferable from an experimental point of view, involves Raman backscatter (Cooney, 1970; Cooney, Orr, and Tomasetti, 1969). In this technique the return at the incident wavelength contains aerosol and Rayleigh (molecular) components. The Raman return indicates the level of return due to the molecules. These are subtracted from the return at the incident wavelength to obtain the aerosol component.

These techniques will yield the backscattered return, but not the energy scattered in other directions (diffuse scatter) or the energy absorption. The latter has been determined in a very general sense by measuring air conductivity (Cobb and Wells, 1970). This was first done almost sixty years ago and recently repeated, with an observed decrease of 20 percent (due to aerosols) over the North Atlantic, with no observed change in the South Pacific. In order to understand albedo and energy transfer changes, however, more detailed absorption and diffuse scattering data are needed. It has been suggested that earth albedo data are needed with an accuracy of a fraction of 1 percent. This indicates indirectly the quality of absorption and scattering information which is required. Since aerosols contribute only partly to albedo, absorption and scattering data due to aerosols need not be as accurate as that just stated for the earth's albedo.

Two significant papers by Robinson (1962, 1966) contain quantitative information pertaining to aerosol absorption and diffuse scattering. In the first paper two types of surface measurements are made in cloudless conditions: total solar irradiance of a horizontal surface (global radiation) and the irradiance of a horizontal surface due to scattered sunlight (diffuse radiation). A value is assumed for the solar radiation incident on the atmosphere, and its attenuation in a clean atmosphere by Rayleigh scattering and by molecular absorption (H_2O and O_3) is computed. The difference between the measured diffuse radiation and the computed downward Rayleigh scattering is attributed to downward aerosol scattering. The upward aerosol scattering is estimated from the latter and added to the sum of the computed absorption due to H_2O and O_3 and the upward Rayleigh scattering to give the expected attenuation of the global radiation. The excess of measured over expected attenuation is attributed to absorption by aerosol. In the second Robinson paper (1966), data from aircraft flights are used to obtain values of the solar constant and the characteristics of the upward and downward flux. These data verify the results of the first paper. It was found that in the altitude range between 10,000 and 20,000 feet there is absorption of about 3 percent of the solar flux and upward scatter of about 1 percent in excess of the values computed for a model clean atmosphere, these effects being attributed to particulate matter.

This method of measuring absorption due to aerosol appears to be quite adequate, even for airborne measurements to determine aerosol absorption in the lower stratosphere due to SST flights. Lasers, substituted for the solar source, would offer some advantages. The major advantage would be an experimental determination of the Rayleigh scatter and molecular absorption by using the Raman scatter technique discussed above. Also night experiments could be performed using a laser source.

It should be noted that an experiment to uniquely measure particle size distribution, density, and chemical composition has not been performed as yet. Polarized laser sources may yield aerosol shape, and wavelength diversity may give a crude measure of the size distribution.

The suggested use of lasers for atmospheric sensing has been noted many times in this volume. The question of safety also arises. The greatest danger to a human is eye damage. An energy density of 10^{-7} J/cm² incident on the cornea from a Q-switched laser is considered to be a maximum safe exposure level. An energy density of 10^{-6} J/cm² for a non-Q-switched laser (a pulse width of 1 msec), and a power density of 10^{-6} W/cm² for cw lasers are safe thresholds, all for the wavelength range: $0.4 - 1.4$ μ. For the CO_2 laser at 10.6 μ, a power density of 0.1 W/cm² is a safe threshold.

The question of the practicality of laser use on an airplane or on a satellite also arises. According to Carbone (private communication), what follows is generally the state-of-the-art, useful for experiments on these platforms:

1. Nd Glass: 1 percent efficiency; 1J output; 1.06 μ; 0.1 to 1 μsec pulse; 6 pulses/min; 9lb; 8″ × 5″ × 11″.

2. Ruby: 0.1 percent efficiency; 8 J output; 6943 Å; 3 pulses/min; nsec-μsec pulses; probably in space of 36 ft³ or less; large

range of options and energy/pulse ratio; must be tailored to volume/weight requirement.

3.   GaAs: incoherent source; small; 150 W peak; 80 nsec pulse; 1500 pulse/sec; 9100 Å (tunable); 22° beam.

4.   Dye: <1 percent efficiency; 20 mJ output; 4000–6700 Å (typical); 1 μsec pulse; 15 pulse/min.

5.   Noble Gases: 10's to 100's kW; μsec pulse; 6 pulse/min; <0.1 percent efficiency; small size.

6. CO_2:CW, pulsed, Q-switched, mode locked.

a.   Low Pressure: 5 to 10 percent efficiency CW, pulsed (μsec) <1 kW

Q-switched (100 nsec) $\sim$ 10 kW

b.   High Pressure (200 to 800 torr): 5 percent efficiency

Pulsed, Q-switched, mode locked

Pulsed (0.3–5 μsec) $\sim$ 1 to 2 J

small size, poor beam quality now

Monitoring Clouds

Cloud characteristics are of the utmost importance for understanding the earth's energy balance. Some of these characteristics are their absorbing and scattering properties, their albedo, their height distributions, size, shape, spacing, and type.

One of the most important of these is cloud height. A method of measuring this parameter from a satellite involves the determination, from multicolor radiometry, of the cloud-top temperature. The cloud height is derived from an estimate or a measurement of the vertical temperature contour (Nordberg, 1965). The latter contour can be derived from satellite data, even in the presence of clouds, as noted earlier in this chapter. Of course, the clouds are assumed to be in the monotonic portion of the temperature contour nearest the earth; otherwise a multivalued result is obtained. Furthermore the field of view of the radiometer must be narrow (about 500 m at cloud level) in order to resolve individual clouds.

It would appear that a downward-looking short pulse laser radar would be ideal for this measurement. The vertical resolution depends on the pulse length, with 0.1 μsec resolving 15 m. Furthermore a near-ground range of 500 m is resolvable from an orbital height of 600 nautical miles. This radar system would also permit the mapping of several layers of clouds if the upper layers are semitransparent, such as cirrus clouds. Various cloud structures can be made more transparent, for this purpose, by increasing the wavelength both for radiometric and radar sensing. The microwave region can be effectively used for such sensing (Shifrin, 1969).

The absorptive and scattering properties of clouds could be obtained by using the technique applied by Robinson (1962, 1966) to aerosol layers. This technique could be applied, using the solar source, to semitransparent or thin clouds where there is a detectable signal below the cloud layer. However for thick clouds and a solar source this may not be the case. It may be desirable to have a source with a smaller angular coverage in order to resolve two adjacent clouds of different types. For these reasons, a laser source may be useful, with its narrow beam and very great output power. Furthermore, a polarized source and receiver system should distinguish between spherical water drops and nonspherical ice particles, thereby giving an indication of the water-ice ratio. The measurement of earth-surface albedo from above cloud layers is extremely difficult. If enough measurements of the type, described earlier, are made so that the cloud's absorptive and scattering properties can be estimated from above, then a quantitative measurement of the range-resolved energy returned from the earth's surface may yield a computed value of the ground albedo. The accuracy of such a result is not known, at present.

In the infrared region, the emissive properties of clouds are important. These can be obtained by quantitative multicolor radiometry. The total water content in clouds can be measured by means of microwave radiometry in emission (Toong and Staelin, 1970), and absorption (Staelin, 1966), using the sun as a source. The wavelengths that were used are in the neighborhood of 1 cm. Finally, Houghton and Hunt (forthcoming) have indicated the possibility of deducing the structure of ice clouds (cirrus) by using emitted radiation in the far-infrared.

References
Bartuske, K., Gambling, D. J., and Elford, W. G., 1970. Stratospheric aerosol measurements by optical radar, *Journal of Atmospheric and Terrestrial Physics, 32:* 1535–1544.

Bradley, L. C. Private communication.

Carbone, R. J. Private communication.

Chahine, M. T., 1970. Inverse problems in radiative transfer: determination of atmospheric parameters, *Journal of Atmospheric Sciences, 27:* 960–967. Figure 42.3a reprinted with permission of the American Meteorological Society.

Cobb, W. E., and Wells, H. J., 1970. The electrical conductivity of oceanic air and its correlation to global atmospheric pollution, *Journal of Atmospheric Sciences, 27:* 814–819.

Conrath, B. J., Hanel, R. A., Kunde, V. G., and Prabhakara, C., 1970. The infrared interferometer experiment on Nimbus 3, *Journal of Geophysical Research, 75:* 5831–5857.

Cooney, J., 1970. Laser Raman probing of the atmosphere, *Laser Applications in the Geosciences,* edited by J. Gauger and F. F. Hall, Jr., Western Periodicals Co.

Cooney, J., Orr, J., and Tomasetti, C., 1969. Measurements separating the gaseous and aerosol components of laser atmospheric backscatter, *Nature, 224:* 1098–1099.

Fiocco, G., and Dewolf, J. B., 1968. Frequency spectrum of laser echoes from atmospheric constituents and determination of the aerosol content of the air, *Journal of Atmospheric Sciences, 25:* 488–496.

Grams, G., and Fiocco, G., 1967. Stratosphere aerosol layer during 1964 and 1965, *Journal of Geophysical Research, 72:* 3523–3542.

Grams, G. W., 1970. Laser radar studies of the atmosphere above 50 km, *Journal of Atmospheric and Terrestrial Physics, 32:* 729–736.

Gray, C. R., 1969. M.I.T./I.L. horizon definition experiment, *Final Report Apollo Guidance and Navigation,* M.I.T. Instrumentation Laboratory R-648, October 1969.

Hanel, R. A., 1970. Advances in satellite radiation measurements, *Advances in Geophysics, 14:* 359–390.

Hinkley, E. D., 1970. High resolution infrared spectroscopy with a tunable diode laser, *Applied Physics Letters, 16:* 351–354.

Hirschfeld, T., and Klainer, S., 1970. Remote Raman spectroscopy as a pollution radar, *Optical Spectra, 4:* 63–66.

Houghton, J. T., and Hunt, G. E., The detection of ice clouds from remote measurements of their emission in the far infrared, *Quarterly Journal of the Royal Meteorological Society,* forthcoming.

Kobayasi, T., and Inaba, H., 1970. Spectroscopic detection of SO_2 and CO_2 molecules in polluted atmosphere by laser-Raman radar technique, *Applied Physics Letters, 17:* 139–141.

Ludwig, C. B. Private communication.

Ludwig, C. B., Bartle, R., and Griggs, M., 1969. Study of air pollution detection by remote sensors, NASA CR-1380, July 1969.

Morgan, G. B., Ozolins, G., and Tabor, E. G., 1970. Air pollution surveillance systems, *Science, 170:* 289–296.

Nordberg, W., 1965. Geophysical observations from Nimbus I, *Science, 150:* 559–572.

Pales, J. C., and Keeling, C. D., 1965. The concentration of atmospheric carbon dioxide in Hawaii, *Journal of Geophysical Research, 70:* 6053–6076.

Patel, C. K. N., and Shaw, E. D., 1970. Tunable stimulated raman scattering from conduction electrons in InSb, *Physical Review Letters, 24:* 451–455.

Prabhakara, C., Conrath, B. J., Hanel, R. A., and Williamson, E. J., 1970. Remote sensing of atmospheric ozone using the 9.6 μ band, *Journal of Atmospheric Sciences, 27:* 689–697.

Reiquam, H., 1970. Sulfur: simulated long-range transport in the atmosphere, *Science, 170:* 318–320.

Robinson, G. D., 1962. Absorption of solar radiation by atmospheric aerosol, as revealed by measurements at the ground, *Archiv fur Meteorologie, Geophysik and Bioklimatologie, B12:* 19–40.

Robinson, G. D., 1966. Some determinations of atmospheric absorption by measurements of solar radiation from aircraft and at the surface, *Quarterly Journal of the Royal Meteorological Society, 92:* 263–269.

Robinson, G. D., 1970. Meteorological aspects of radiation, *Advances in Geophysics, 14:* 285–306. Figure 42.1 reprinted with permission of Academic Press, Inc.

Schuster, B. G., 1970a. Detection of the tropospheric and stratospheric aerosol layers by optical radar (lidar), *Journal of Geophysical Research, 75:* 3123–3132.

Schuster, B. G., 1970b. Lidar probing of the stratosphere in *Laser Applications in the Geosciences,* edited by J. Gauger and F. F. Hall, Jr., Western Periodicals Co.

Shaw, J. H., Chahine, M. T., Farmer, C. B., Kaplan, L. D., McClatchey, R. A., and Schaper, P. W., 1970. Atmospheric and surface properties from spectral radiance observations in the 4.3 μ region, *Journal of the Atmospheric Sciences, 27:* 773–780. Figure 42.3b reprinted with permission of the American Meteorological Society.

Shifrin, K. S., 1969. Transfer of microwave radiation in the atmosphere, NASA TTF-590.

Smith, W. L., 1970. Iterative solution of the radiative transfer equation for the temperature and absorbing gas profile of an atmosphere, *Applied Optics, 9:* 1993–1999.

Snavely, B. B., 1969. Flashlamp-excited organic dye lasers, *Proceedings of the Institute of Electrical and Electronics Engineers, 57:* 1374–1390.

Staelin, D. H., 1966. Measurements and interpretation of the microwave spectrum of the terrestrial atmosphere near 1-centimeter wavelength, *Journal of Geophysical Research, 71:* 2875–2881.

Study of Critical Environmental Problems (SCEP), 1970. *Man's Impact on the Global Environment* (Cambridge, Massachusetts: The M.I.T. Press), p. 196.

Toong, H. D., and Staelin, D. H., 1970. Passive microwave spectrum measurements of atmospheric water vapor and clouds, *Journal of the Atmospheric Sciences, 27:* 781–784.

Wark, D. Q., 1970. SIRS: an experiment to measure the free air temperature from a satellite, *Applied Optics, 9:* 1761–1766.

43

Remote Detection of Air
and Water Pollution from
Satellites and/or Airplanes

Philip Hanst, Jules Lehmann,
Henry Reichle, and
Morris Tepper

Satellites in Global Observation Systems

Satellites are uniquely suitable for the study of global environmental problems. Situated, as they are, at a distance from the earth, they command a synoptic view of the events over a large portion of the earth. They are global in concept because they are able to orbit the earth and view its every corner. Yet, through the optics of their instrumentation, they can witness the occurrences on a local scale. They reside outside of the atmosphere, away from its eroding and frictional influence and, thus, can achieve long life.

The location of the satellite at a distance from the earth permits its use not only in the direct viewing of the earth's surface, but also (1) for data relay and (2) in occultation measurements.

In the data relay mode, the satellite serves essentially to assist in communications. Because it commands such a large view of the earth, it can communicate with many sensors located in diverse areas of the globe, receive information sensed and/or recorded by the sensor, store it aboard the satellite, and yield this information on request at a data collection station. The satellite can even locate the sensor on the surface of the earth and, thus, provide information on movement.

The relative short duration of the orbit period permits the satellite to have several "sunrises" and "sunsets" each day. By viewing the sun during these periods when it is being occulted by the earth, the satellite sensor receives the solar energy after it has traversed different levels of the atmosphere. The influence of the atmospheric constituents on the solar beam can, thus, be measured as a function of altitude. This technique is particularly useful in the stratosphere above the level where clouds will intercept the solar beam. A simple variation of this technique is to replace the satellite-sun system with a mother satellite-daughter satellite system. Here, instead of receiving the energy from the sun, the satellite sensor can receive microwaves or radio transmissions from the daughter satellites. While the logistics of the system become more complicated in this latter configuration,

Prepared for SCEP and expanded during SCEP.

the selection of wavelength can decrease the interfering influence of clouds.

Manned satellites provide still another dimension. Man on board a space platform can serve as an observer, an experimenter, a technician, a repairman, a participant—in the many experiments possible from space. With his eyes, he can see unexpected phenomena, and his brain can integrate events into new patterns and evolve new relationships.

The advantages that space satellites provide in the observation of the earth carry within themselves the seeds of their primary disadvantage. This disadvantage arises from the simple geometry of the system. Since a sensor on a space platform must be situated at a distance from the feature on or near the earth that it is sensing, it can receive information only as electromagnetic transmissions—in the uv, visible, infrared, microwave, and radio. Moreover, this information comes, by and large, only from the surface of the earth phenomenon as it faces the sensor in space. Despite these constraints, remote sensing from space satellites has already proven to be a very powerful tool, as for example in meteorology.

This, then, is the essence of remote sensing. The trick is to uncover distinctive electromagnetic measurements characteristic of the particular earth feature that is of interest—its signature. Characteristic signatures are usually identified under laboratory conditions and laboratory spectra of various elements and compounds are more or less available. It is another matter to isolate these characteristic signatures under field conditions. Here the sensor receives radiation from a multitude of emitters, and the problem of signature separation becomes formidable. Moreover, the atmosphere itself and all of its constituents affect the electromagnetic transmissions through the physical processes of absorption, reflection, and scattering.

This paper discusses a number of aspects of the detection and/or monitoring of pollutants in the atmosphere and in water. The role of NASA in these fields is to demonstrate the usefulness of making remote measurements from airplanes or satellites. The following sections discuss the problems to be faced when remote measurements are considered and indicate a number of experiments that promise to provide useful data in the detection and monitoring of pollutants.

The remote detection of air pollution will generally use electromagnetic radiation as the measurement mechanism. It will be necessary to understand the absorption and the scattering properties of the pollutants and the intervening atmosphere before a measurement of the pollutant concentration and distribution can be made. In the following sections these items and several techniques for the measurement of pollution are discussed.

Interference from the Atmosphere

Remote sensing of atmospheric and water pollutants is hindered by absorbing and scattering material naturally present in the clean atmosphere. Primarily one is concerned with water vapor and water clouds, but in addition, the absorption by carbon dioxide must be considered. Scattering is a problem at the shorter wavelengths, molecular scattering being particularly strong in the blue and the ultraviolet. Particulate scattering is also strong in the blue and extends throughout the visible and into the infrared. Absorption due to ozone in the upper atmosphere cuts off sunlight and visibility in the ultraviolet and also interferes in the infrared. The water vapor and carbon dioxide absorption falls in the infrared. Emission by the atmosphere contributes background "noise," especially in the microwave region. The interferences are summarized in Table 43.1.

For sensing gaseous air pollutants the middle infrared spectral region between wavelengths of 2.5 and 20 microns is the most important. In this region there is moderate interference from scattering, absorption, and thermal emission. The interference is not so great that remote sensing is rendered excessively difficult, but neither is the interference so small that it can be overlooked. It is a limitation which the designer of remote sensing instruments and interpreter of remote sensing data must understand and reckon with. The choice of a spectral band for remote sensing of a pollutant must be based not only on a knowledge of the pollutant absorption coefficients, but also on the spectral characteristics of the source, the intensity of the background emission, the scattering of sunlight in the optical path, the absorption by other constituents of the atmosphere, and the response characteristics of the detection system. Extensive discussionss of the optical properties of the clean atmosphere are presented in Goldberg (1954),

Table 43.1 Atmospheric Effects on Remote Sensing

Spectral region / Process	0.4 μm / UV	0.4–0.7 μm / Visible	0.7–2.5 μm / Near IR	2.5–20 μm / 4000–500 cm⁻¹ / MID IR	20–1000 μm / 500–10 cm⁻¹ / FAR IR	1–15 mm / 300–20 GHz / Microwave	1.5–100 cm / 20–0.3 GHz / Microwave	1 m / 300 MHz / Radio
Molecular scattering	c	b (Blue region)	a					
Particle scattering	b (Aerosols) c (Clouds)	b (Aerosols) c (Clouds)	a (Aerosols) b (Clouds)	b (Clouds)		b (Clouds)	b (Clouds)	
Absorption	c (0.3 μm)		b (Absorption bands)	c (Absorption bands absorption)	c (Water vapor absorption)	b (Absorption bands)	a (Wing of absorption bands)	
Emission				a	a	b	b	b

Source: Tepper, 1970.
a — Minor problem
b — A problem
c — Severe problem
d — Spectral region not usable

Howard, Garing, and Walker (1965), and Migeotte, Neven, and Swensson (1956).

Spectral Characteristics of Atmospheric Pollutants

The spectrum of the polluted atmosphere is much less well known than the spectrum of the clean atmosphere. Stair and Gates have both looked at the sun through the smog of Los Angeles and observed absorption bands due to pollutants (Stair, 1955; Gates, 1955). Scott et al. (1957) observed the characteristic bands of many pollutants in the Los Angeles air in 1956 by using folded paths of several hundred meters.

The spectrum can be considered in terms of the various wavelength regions characterized by the type of physical change which takes place in the absorbing or emitting molecule, as well as by the type of instrumentation used in studying the spectrum. We consider six regions.

Ultraviolet, 0.25 μm to 0.40 μm

The ultraviolet is the region of absorption by electronic transitions in molecules. Combined with the electronic changes are vibrational and rotational changes that in a few cases produce a characteristic structure in the bands. While the ultraviolet band systems of different molecules are likely to overlap, correlation techniques have been developed for distinguishing one band system in the presence of others. The larger molecules do not have uv band structure at atmospheric pressure, and therefore the applicability of the uv in pollution monitoring is quite limited. Water pollution created by oil spillage at sea may be detectable in this region.

Visible, 0.4 μm to 0.70 μm

The visible region of the spectrum, by definition, has very few molecular absorption bands in it. NO_2 is the only colored gas which is a common pollutant. Smoke from forest fires and smokestacks has been detected from visible light photography.

Near Infrared, 0.70 μm to 2.5 μm

This is the spectral region where the overtones of the fundamental molecular vibration-rotation bands appear. The overtones are about 100 times weaker than the fundamentals, so this region is not generally useful for molecular detection and analysis. A notable exception is in the study of the atmospheres of the planets, where large concentrations of absorbing gas are viewed over extremely long paths. Thus, the overtone region is the principal source of information on the atmospheres of Mars, Venus, and Jupiter.

Middle Infrared, 2.5 μm to 25 μm

This is the spectral region where the strong fundamental vibration-rotation bands of molecules appear. Nearly every air pollutant will have a characteristic absorption band in this region. It is sometimes called the "fingerprint" region of the infrared and is used extensively in chemical analysis. In general the absorption bands in this region differ widely as to shape, location, and intensity distribution. Even the two members of pairs of similar molecules such as CO—NO and O_3—SO_2 have significantly different spectra.

Far Infrared, 25 μm to 500 μm

Rotational lines as well as some vibration-rotation bands appear in the far infrared. Unfortunately, very intense water vapor absorption blanks out the far infrared in atmospheric work.

Millimeter and Microwave

Molecular rotational lines appear in the microwave spectrum. Most large molecules have a microwave spectrum, and the region is very powerful in the study of molecular structure and other physical properties. However, the molecular gas under study must be at a very low pressure in order for the fine structure of the spectrum to be resolvable. At atmospheric pressure the spectra are so "smeared out" that distinguishing between molecules is impractical.

Atmospheric Scattering

In the atmosphere four principal types of scattering of energy need to be considered: Rayleigh scattering, Mie scattering, resonance scattering, and Raman scattering. The scattering depletes the energy of the incident beam and redistributes it in all directions with a definite intensity pattern which varies according to wavelength, type of scattering, and the composition of the scattering medium.

Rayleigh scattering occurs when the dimensions of the scatterer are much smaller than the wavelength of the light. The scattered light is concentrated along the line of flight of the pulse, with equal intensities in the forward and backward directions. Since the Rayleigh scatterers are small, they have high thermal velocities, causing the Rayleigh-scattered light to be Doppler broadened to a considerable extent. Since the intensity of Rayleigh scattering is a smooth function of wavelength, the scattered light does not identify the scatterer, and its usefulness in atmospheric analysis is limited. Gas molecules are examples of Rayleigh scatterers.

Mie scattering takes place when the dimensions of the particle are comparable to or larger than the wavelength of the light. The Mie scattered light is concentrated in the forward direction, with a much smaller intensity backwards. The relatively heavy Mie scatterers do not have high thermal velocities and hence do not broaden the laser line significantly through Doppler effect. The Mie scattering can be used to locate dusts and other aerosols in the atmosphere.

Resonance scattering occurs when the energy of the incident photon is the same as the energy difference between the ground state and an excited state of the scattering species. The resonance scattering is very strong and will have high intensity in both the forward and backward directions. The line width of the scattered radiation will be determined by the pressure, temperature, and other properties of the scatterer. In a resonance scattering experiment, the scattering species must be known in advance, and the laser must be tuned to the resonant energy difference. The tunable dye laser offers promise of applicability to resonance scattering experiments. An outstanding example of resonance scattering is the glow observed from clouds of sodium and other substances released into the sunlit upper atmosphere.

Raman scattering differs from resonance scattering in that the energy of the incident photon does not have to coincide with the energy difference between elctronic states of the scattering species. This makes the Raman scattering much weaker than resonance scattering; but with the high incident intensities available from lasers the Raman scattering is strong enough to be useful in many applications. The main distinction of Raman scattering is that the wavelength of the light is changed. The scatterer changes its energy state during the process of scattering with the increment of energy appearing in the scattered photon. These energy changes differ from one molecule to another and thus identify the scattering species. It is this Raman "signature" which makes the Raman scattering a subject of much interest in current pollution detection research.

Techniques for the Detection of Gaseous Pollutants

Lasers for the Detection of Gaseous Pollutants

Lasers of various types are being applied in remote sensing, and the applications are on the increase. It is difficult to say what the status of laser development is at the moment, because laser technology is still evolving very rapidly. The applicability of lasers in remote sensing derives mainly from their ability to produce very brief but extremely powerful bursts of more or less monochromatic radiation. These pulses can be shorter than a nanosecond (10^{-9} second) and more powerful than a gigawatt (10^9 watts). The most powerful lasers are crystals that emit specific wavelengths, such as 1.06 microns. Dye lasers are not yet as powerful as crystal lasers, but they have the ability to be tuned to any chosen wavelength over wide regions of the spectrum.

Remote probing with lasers almost always depends upon scattering of the laser energy by constituents of the atmosphere. At the present time, much of the research involving the use of lasers for the remote detection of gaseous pollutants is based upon the use of Raman scattering. Since the frequency at which the energy is scattered is determined by the composition of the scattering molecule and since the scattering frequency can be calculated in advance, this method promises to be a highly sensitive and highly specific detector of gaseous pollutants. The presently available equipment does not allow the detection of air pollutants at the concentrations at which they are found in the atmosphere. However, this situation should improve as detectors become more sensitive and lasers become more powerful.

A Remote Sensing System Based on Gas Filter Correlation Principles

For the past two years, a project has been under way at the Convair Corporation, under NASA sponsorship, to develop and test an experimental remote sensing instrument (Ludwig, Bartle, and Griggs, 1969). The first year of the contract was devoted to a study to find the best technique of surveillance of atmospheric pollution from satellites. Specifically, the following possible pollutant detection techniques were examined: (1) dispersive absorption spectroscopy; (2) matched filter absorption spectroscopy; (3) filter cell nondispersive techniques; (4) interferometric techniques; (5) laser probing techniques; and (6) microwave techniques.

From among the many detection techniques studies, a nondispersive infrared technique was selected for further development. At present, an experimental model of the proposed detection system is being prepared for testing from an airplane. The operating principle is similar to that which is used in certain commercial gas analysis systems. The radiation is alternately passed through a filter cell and a blank cell. The filter cell contains a sample of the gas whose concentration in the atmosphere is to be measured. In the NASA measurement program the source of radiation will be the earth's surface. The earth radiation has its greatest intensity in the 8- to 13-micron infrared transmission window. In addition, there is a weaker but useful radiation in the 3- to 5-micron spectral region. One can visualize the operating principle of the instrument by considering the following: (1) If the incoming radiation has no absorption lines in its spectrum, the gas-filled filter cell will remove a certain amount of energy from the incoming radiation, but the blank cell will remove none. The two cells will then have a certain transmittance ratio which by means of a neutral density filter in the reference beam may be adjusted to unity. (2) When the incoming radiation contains absorption lines that are identical to the absorption lines of the filter gas, the filter cell will remove less radiation from the beam than it does when lines are not present. The ratio of transmittances will then differ from unity by an amount proportional to the amount of absorbing gas in the light path. (3) Absorption lines of compounds other than the filter cell gas will not result in an imbalance of the transmittance ratio unless they accidentally fall on the lines of the filter gas. Since such coincidences are unlikely, the system will have a high degree of selectivity.

The instrument under development actually has two pairs of cells, each pair having a different amount of absorbing gas in its filter cell. The output signal will be the ratio of the deviations from unity by the transmittance ratios of the two pairs of cells.

The "double ratio" system will eliminate the need for measurement of the total intensity of the background emission. The instrument is diagramed in Figure 43.1. One pair of cells is shown in the cross-sectional drawing, the other pair being at right angles to the one shown.

The Correlation Interferometer for Measurement of Trace Atmospheric Constituents

The correlation interferometer concept, devised by Barringer Research Ltd., combines absorption spectroscopy with scanning interferometry for measurement of trace constituents and pollutants in the atmosphere. Development of the device will be directed initially toward measurement of atmospheric carbon monoxide (CO) in the range from 10 parts per billion to 10 parts per million. Global CO measurements with this instrument will be used in an experiment proposed by General Electric Company to identify the mechanisms or sinks which remove CO from the atmosphere.

The correlation interferometer is a passive device using near infrared solar radiation scattered from the earth and atmosphere as an energy source. The unique feature of the Barringer concept is the means for on-board correlation and identification of unknown gases. The instrument consists of a high-resolution Michelson interferometer plus a magnetic tape device and associated electronics for correlation purposes. Generally, the output of an interferometer is an interferogram that is transmitted from spacecraft to ground, where it is converted to a spectrum using Fourier transform techniques and a digital computer. This spectrum is

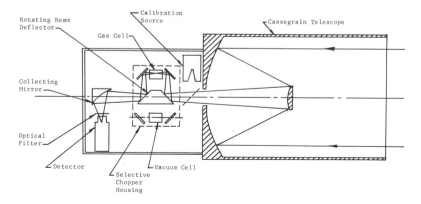

Figure 43.1 Schematic diagram of a selective chopper

compared or correlated with known absorption spectra to identify specific atmospheric constituents. This task is complex and can involve the transmission of large quantities of data if high resolution is desired.

The Barringer technique involves correlation of interferograms, representing the atmosphere and a reference gas, within the instrument. This is accomplished by recording on magnetic tape a reference interferogram for the gas of interest (that is, carbon monoxide). As the instrument views the atmosphere, an interferogram is generated and passed to a phase-sensitive detector and electronics. Simultaneously, the recorded CO interferogram is fed into the phase-sensitive detector as a reference signal. The degree of correlation between the two signals represents the presence or absence of CO in the column of atmosphere being viewed. The output of the electronics is a dc signal, whose amplitude can be related by calibration to the quantity of CO in the atmosphere.

This instrument may have the potential for measuring a variety of atmospheric constituents by providing a unique interferogram or "fingerprint" for each substance. Investigation of the applicability of the technique for this type of general measurement is being deferred pending further development of the instrument for the carbon monoxide pollution measurement.

Techniques for Detection of Particulate Pollutants

A satellite platform can be a practicable device for the detection and measurement of particulate air pollution. The observations would have two possible uses: first, the measurement of the highly diffused particulate pollution in the upper troposphere and lower stratosphere and second, for the detection and measurement of the relatively dense particulate pollution (primarily smoke) emanating from the tall exhaust stacks of large coal-fired power plants and metal smelters.

Because the size of aerosol particles to be detected is of the order of magnitude of or larger than the wavelength of light in the ultraviolet or visible portions of the spectrum, the scattering of light is the Mie scattering. At least three different techniques have been proposed for the measurement of particulate or aerosol distributions. One of these is a laser technique, and two are techniques that use the sun as a radiation source.

1.   Lasers have been used for the detection of particles in the atmosphere of the earth. The method generally used consists of aiming the laser upward and measuring the intensity of the back-scattered radiation as the pulse of laser energy traverses the atmosphere. The time history of the detected, scattered energy is compared to a computed time history for a clean (Rayleigh scattering) atmosphere. The difference between the two can be interpreted as a measure of the particulate concentration. Experiments of this type do not appear appropriate at this time for aircraft or satellite use.

2.   A second technique for the measurement of particles makes use of the fact that particles, which scatter according to the Mie law, tend to scatter energy in the forward direction very strongly as compared to gas molecules which scatter light according to the Rayleigh law. The intensity of the scattered energy is measured as a function of the angle through which the energy has been scattered. The enhanced forward scattering is interpreted in terms of the particulate concentration. An experiment of this type has been flown on a balloon by Newkirk and Eddy with good success (Newkirk and Eddy, 1969).

3.   A technique that has been proposed for the measurement of particulate distribution in the atmosphere from a satellite utilizes the fact that light scattered by particles in the atmosphere is polarized differently from light that is scattered by the gas molecules (Sekera, 1967; Hariharan, 1969). By measuring the intensity, degree of polarization, and orientation of the plane of polarization of the scattered sunlight emerging from the atmosphere one should be able to determine the total concentration, the size distribution, and the vertical distribution of the aerosols or particulates.

The instrument consists of a system of polarizing filters, neutral filters, and narrow-bandpass filters that are alternately placed in a beam of incoming atmosphere-scattered sunlight. The signal is detected by a photomultiplier tube which converts the incoming radiation to an electrical output. The intensity and degree of polarization is measured at four frequencies in the visible portion of the spectrum. Comparison of the measured parameters with those calculated for a model atmosphere should allow one to deduce the characteristics of the aerosol distribution.

Remote Detection of Water Pollution

As is the case for air pollution measurement, the detection and measurement of water pollution from an aircraft or spacecraft requires the use of electromagnetic radiation as the sensing mechanism. If the wavelength of the radiation being utilized is shorter than approximately 3.5 μm, the measurement depends upon the reflection of solar radiation by the pollutant that is being carried by the water. At wavelengths greater than approximately 3.5 μm, the measurements are utilizing the thermal emission of the water itself and are dependent upon either a change in water temperature or a change in the emissive properties of the water for the detection of pollution.

In the following discussion, water pollutants will be categorized as follows: thermal pollutants, for example, the cooling water from fossil fuel or atomic-powered electrical generating plants; pollutants carried mainly in the water volume, for example, sewage, insecticide residues, or silt; and pollutants that are carried primarily on the surface of the water, for example, oils or detergent foams. Other systems of categorizing pollutants could easily be formulated. This system is natural, however, when one considers the radiative properties of the water medium. These distinctive radiative properties will be discussed before the discussion of the detection of specific pollutant begins.

Radiative Properties of Water

First, consider the wavelength region in which the reflected solar radiation is the measuring mechanism, that is, wavelengths shorter than 3.5 μm. Figure 43.2 is a plot which indicates the depth to which incident radiation penetrates as a function of wavelength for clean water. (Figures 43.2 and 43.3 are derived from data such as that reported in Duntley, 1963, and Sullivan, 1963.) Specifically, the curve indicates the depth at which the intensity of the energy has been reduced to 5 percent of the incident value. It can be seen that in the ultraviolet region (wavelength shorter than 0.4 μm) and in the infrared region (wavelengths longer than 0.7 μm) the radiation is severely attenuated and penetrates to a depth of only a few centimeters or less. Only in the visible portion of the spectrum (wavelengths between 0.4 and 0.7 μm) does the radiation penetrate to any appreciable depth. It would appear then

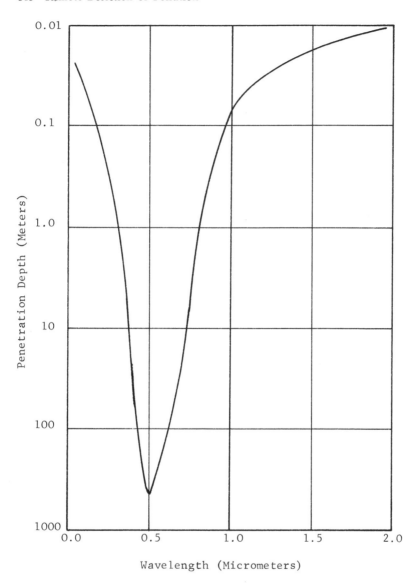

Figure 43.2 Depth at which energy is reduced to 5 percent of incident value as a function of wavelength

Sources: Duntley, 1963; Sullivan, 1963.

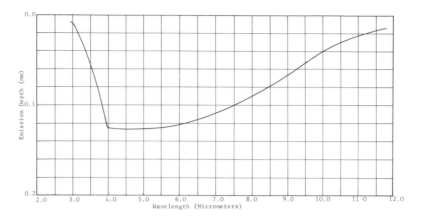

Figure 43.3 Depth above which 95 percent of energy is emitted

that only the visible portion of the spectrum would be useful for the detection of pollutants at depths greater than a few centimeters.

At wavelengths longer than approximately 3.5 μm, the amount of thermal radiation emitted by the earth (or water) becomes larger than the incident energy from the sun, and it is possible to use this emitted radiation as the sensing mechanism. The important consideration (from a pollution detection point of view) now becomes the depth of the water layer from which the energy is emitted. Figure 43.3 shows the depth above which 95 percent of the energy emitted by the surface emanates. It can be seen that the layer is very thin (of the order of tenths of a millimeter).* This is characteristic of the thermal emissions from water in both the infrared and at longer wavelengths in the microwave portion of the spectrum.

In summary then, the reflection, emission, and transmission characteristics of water impose rather severe restrictions on the measurements of pollution from a remote platform. Measurements are confined to a thin surface layer for most of the electromagnetic spectrum, only in a very narrow region (the visible) are measurements at depth possible.

* At a wavelength of 10 μm, where the intensity of the emitted radiation is the greatest, the thickness of the emitting layer is only 0.04 mm (less than the thickness of a sheet of paper).

Atmospheric Effects

Since the atmosphere (all or in part) exists between the measuring platform (the aircraft or spacecraft) and the object to be measured (the water pollution), its effects must be taken into account in the design of any pollution measuring system. As has been discussed extensively in the section on the detection and measurement of air pollution, even the clean atmosphere has regions in which it is either opaque or in which the transmission of radiation is severely reduced. In the clean atmosphere these regions of reduced transmission are produced mainly by three trace constituents: ozone, water vapor, and carbon dioxide. While water vapor and ozone are generally confined to certain altitude regions (water in the form of vapor or clouds to the lowest 10 kilometers in altitude and ozone to the region from 10 to 80 kilometers), carbon dioxide is found to be approximately uniformly mixed from the surface to altitudes of approximately 80 kilometers. The effect of these constituents (and of air pollution when it is present) is to place a variable filter in the path of the measurement and to limit severely the spectral regions that are usable for measurements of water pollution.

Thermal Water Pollution

Thermal pollution is usually generated in conjunction with the operation of factories or plants that require the dissipation of large amounts of heat. Typical examples would be steam- or atomic-powered electrical generating plants. As the demands for electrical energy increase and atomic-powered electrical generating plants become more common, it is expected that thermal pollution will become a much more serious problem. Although the addition of heat does not change the physical characteristics of the water itself appreciably (over the range of temperature considered here), it does considerably change the type of biological forms that can live in the water.

The detection and measurement of thermal pollution is carried out by measuring the increased radiation emitted by the warm water as compared to the cool water. The wavelength regions usable for remote surface temperature measurements are limited by the presence of the atmosphere. The regions available are in the infrared region from 10 to 12 μm and in the microwave region. In both of these cases the actual water whose temperature is being sensed is a very thin surface layer. Generally, the plume of warm water that is being ejected (usually at some distance below the surface) is carried to the surface by a combination of buoyancy and turbulent mixing effects, and it will be detectable there. In Figure 43.4 a Nimbus HRIR picture of the Gulf Stream is shown. It can be seen that at the resolutions presently available from satellites even as large a plume of water as this is not very distinct. As satellite equipment improves, the resolution should improve considerably. In Figure 43.5 an infrared scan of a thermal pollution plume taken from an airplane is shown. It can be seen that the definition here is much better. Fortunately since the sources of thermal pollution are of a finite nature, the airplane appears to be a highly useful tool for the measurement of thermal pollution where a precise knowledge of plume temperature and dimensions is required.

Pollutants in the Water Volume
Pollutants that are carried in the volume of the water but that do not cause a large change in the reflective or emissive properties of the water must be sensed using the visible or near infrared portions of the spectrum. Examples of pollutants of this type would be sewage, silt, or pesticide residues. The amount of work that has been done to determine the effect of various substances on the spectral transmission characteristics of water has been rather limited. An example of the spectral characteristics of water containing algae is shown in Figure 43.6. This figure shows that water containing a large amount of algae tends to scatter much more radiation back to the observer in the red end of the visible spectrum than does algae-free water. This increase is caused by scattering by the algae within the water. Similarly, changes in the spectral characteristics of pure water take place when sewage, silt, or industrial wastes are present.

This property of enhanced scattering by algae in the red end of the spectrum can be used to advantage in photographing the distribution of the algae (which is related to the distribution of pollution) with near infrared color films. Figure 43.7 shows a photograph of algae growing in the Potomac River that was taken

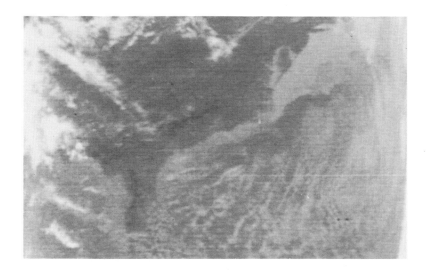

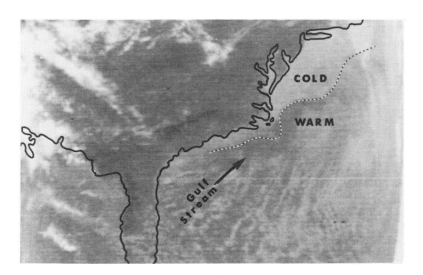

Figure 43.4 Nimbus IV infrared imagery of eastern United States and western Atlantic, April 8, 1970

from an airplane. In the original color photography the infrared energy reproduced as a red color, and the algae could be seen very vividly as a pink area against a blue water background. This distinction is still apparent in the black and white reproduction.

A second example of the use of color photographs that could be used for the measurement and detection of water pollution is shown in Figure 43.8. The figure is a black and white reproduc-

Figure 43.5 Infrared scanner picture taken from an airplane of thermal pollution plume of Los Angeles River at Long Beach Harbor

Source: Chang, 1969.

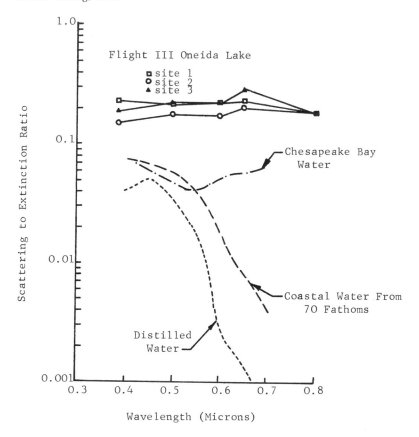

Figure 43.6 Spectral characteristics of samples of pure and algae bearing water

Sources: Hulbert, 1945; Silvestro, 1969.

Figure 43.7 Use of color infrared film to enhance algae bloom; July 1966 (Aircraft Photo, Stransberg, Itek Corporation)
Source: NASA, 1969.

tion of a color photograph. This is a photograph of the Gulf Coast area of the United States taken by Astronauts Lovell and Aldrin from Gemini XII while at an altitude of 330 kilometers. Distinct plumes of sediment-laden water are visible in the Galveston Bay, Port Arthur, and Lake Charles areas. (Smoke from a small marsh fire can also be seen in the lower part of the photograph.) Figure 43.9 shows an analysis of the photograph of Figure 43.7. Lines of constant color are shown as determined by instrument readings of the photograph. One can thereby determine the direction, rate of spreading, and relative concentration of pollutant material. Another example of similar photographs is shown in Figures 43, 44, and 45 of NASA's publication "Earth Observations Program Review" (1969).

The ability to detect pollutants at very low levels of concentration will remain limited by the sensitivity of the measurement apparatus whether it be electronic scanners or camera film combinations. Until further work is done on determining the changes in the spectral characteristics of water that are brought about by

Figure 43.8 Gemini photo S66-63034 showing the Gulf Coast area of Texas and Louisiana

Source: NASA, 1967.

the presence of various pollutants it is not possible to determine which pollutants will be detectable in this way.

Pollutants on the Water Surface

As has been discussed earlier, the energy reflected or emitted by water in the ultraviolet, infrared, and microwave portions of the spectrum interacts with a thin surface layer. Hence, a pollutant that changes the reflective or emissive properties of the surface should be detectable in these wavelength regions. Examples of surface carried pollution are oil or foams. Figure 43.10 is an in-

Figure 43.9 Analysis of Gemini photo S66-63034 shown in Figure 43.8
Source: NASA, 1967.

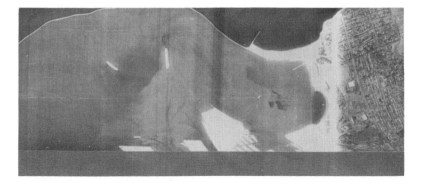

Figure 43.10 Infrared scanner picture of Los Angeles Harbor showing two ships and an oil slick
Source: Chang, 1969.

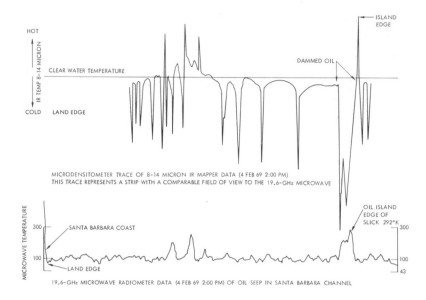

Figure 43.11 Comparison of microwave radiometer and infrared measurements of the Santa Barbara oil slick

Source: Chang, 1969.

frared scanner picture of Long Beach Harbor showing an oil slick in the area of the two ships. This picture was made using an electronic scanner that senses the radiation emitted by the surface in the wavelength region from 8 to 14 μm. Because the oil layer reduces the amount of energy emitted by the water, the oil slick appears darker in the picture. Figure 43.11 shows two traces, one of the apparent water temperature as seen in the infrared and the other of the apparent water temperature as seen in the microwave region, made as the instrument traversed the Santa Barbara oil slick. The instrument was mounted in an airplane in these flights. Measured in the infrared, the apparent temperature of the oil slick was significantly cooler than the apparent temperature of the surrounding water. Conversely, the apparent temperature of the slick was considerably higher when viewed in the microwave portion of the spectrum. Simultaneously viewing the same body of water with both instruments then allows one to separate the effects of oil slicks and water temperature.

References

Chang, Bansun, 1969. Aerial remote sensing for the earth resources program, paper presented at a seminar New horizons in color photography, American Society of Photogammetry and the Society of Photographic Scientists and Engineers, Washington, D.C., June 9–11, 1969. Figures 43.5, 43.10, and 43.11 reprinted with permission of the American Society of Photogammetry.

Duntley, Siebert Q., 1963. Light in the sea, *Journal of the Optical Society of America, 53:* 211. Figures 43.2 and 43.3 reprinted with permission of *The Journal of the Optical Society of America.*

Gates, David M., 1955. Infrared solar spectral measurements through varying degrees of smog at Los Angeles, *Proceedings of the 3rd National Air Pollution Symposium,* Pasadena, California (Los Angeles: Western Gas and Oil Association).

Goldberg, Leo, 1954. The absorption spectrum of the atmosphere, *The Earth as a Planet,* edited by G. P. Kuiper (Chicago: The University of Chicago Press).

Hariharan, T. A., 1969. An improved polarimeter for atmospheric radiation studies, *Journal of Scientific Instruments (Journal of Physics E)* Series 2, Vol. 1.

Howard, J. N., Garing, J. S., and Walker, R. G., 1965. Transmission and detection of infrared radiation, *Handbook of Geophysics and Space Environments,* edited by S. L. Valley (New York: McGraw-Hill).

Hulbert, E. O., 1945. Optics of distilled and natural water, *Journal of the Optical Society of America, 35:* 698. Figure 43.6 reprinted with permission of *The Journal of the Optical Society of America.*

Ludwig. C. B., Bartle, R., Griggs, M., 1969. *Study of Air Pollutant Detection by Remote Sensors,* NASA CR-1380.

Migeotte, M., Neven, L., and Swensson, 1956. The solar spectrum from 2.8 to 23.7 microns, part 1, photometric atlas, *Mémoirs de la Société Royale des Sciences de Liège,* Special Volume No. 1.

National Aeronautics and Space Administration (NASA), 1967. United States activities in spacecraft oceanography, The National Council on Marine Resources and Engineering Development, October 1, 1967.

NASA, 1969. Earth observations program review, Washington, D.C., NASA Headquarters, November 4–5, 1969 (unpublished).

Newkirk, Gordon, Jr., and Eddy, John A., 1964. Light scattering by particles in the upper atmosphere, *Journal of the Atmospheric Sciences, 21:* 35.

Scott, W. E., Stephens, E. R., Hanst, P. L., and Doerr, R. C., 1957. Further developments in the chemistry of the atmosphere, *Proc. A.P.I., 37* (III): 171.

Sekera, Z., 1967. Determination of atmospheric parameters from measurement of polarization of upward radiation by satellite or space probe, *Icarus, 6:* 348–359.

Silvestro, Frank B., 1969. Quantitative remote sensing of water pollution, paper presented at the 15th Annual Meeting of the Institute of Environmental Sciences, Anaheim, California, April 1969. Figure 43.6 reprinted with permission of the Institute of Environmental Sciences.

Stair, Ralph, 1955. The spectral radiant energy from the sun through varying degrees of smog at Los Angeles, *Proceedings of the 3rd National Air Pollution*

Symposium, Pasadena, California (Los Angeles: Western Gas and Oil Association).

Sullivan, Seraphin A., 1963. Experimental study of the absorption in distilled water, artificial sea water, and heavy water in the visible region of the spectrum, *Journal of the Optical Society of America, 53:* 962. Figures 43.2 and 43.3 reprinted with permission of *The Journal of the Optical Society of America.*

Tepper, Morris, 1970. Problems of remote sensing of the earth's surface, Eighth Aerospace Sciences Meeting, American Institute of Aeronautics and Astronautics, New York, January 19, 1970. Table 43.1 reproduced with permission from the American Institute of Aeronautics and Astronautics.

Though the primary focus of SCEP was on the scientific evidence that is available or must be generated to determine the true nature of critical, long-term, global problems, one group of participants addressed the set of questions which must ultimately be resolved if changes in the status quo are to be effected and if remedial action is to be taken. This Work Group on the Implications of Change was chaired by Professor Milton Katz of the Harvard Law School and included lawyers, economists, social scientists, engineers from several industries, and government officials. The summary of the report of that Work Group is reprinted here from the SCEP Report as the first paper in this series.

The next paper is a short piece written for SCEP participants by Mr. Richard Carpenter who had been a member of the Steering Committee that planned SCEP. It is a succinct bit of advice to all those who would attempt to synthesize scientific and technical material with a view toward raising the level of public discussion on important issues. Mr. Carpenter writes from the experience of six years of policy analysis in the Legislative Reference Service of the Library of Congress.

But once all available knowledge is synthesized and presented in an understandable and meaningful form to the decision maker, there is still often a large area of uncertainty with respect to the costs and benefits of any action (or inaction). Dr. Walter Spofford discusses some of the problems involved in such a case, first by reviewing the status of social science methodologies for handling these problems and then by citing some of the issues which would certainly be raised if hard decisions about reducing the amount of CO_2 in the atmosphere ever have to be made.

Should those decisions involve changes in present methods of electric power generation and private transportation—and they almost surely would—then the implications for the power and the automotive industries would be immense. Mr. Willard Crandall outlined at the Study some of the options which he saw for environmental improvement with respect to power generation. His paper is neither comprehensive nor detailed, but it does provide some idea of the complex technical issues that would have to be addressed.

It was clearly recognized at SCEP that the existence of global problems does not imply the necessity for a global solution. The sources of pollution are activities of man that can often be effectively controlled or regulated where they occur. In research and monitoring programs, however, the potential for international cooperation is high. Effective cooperation now might increase the likelihood of smooth international relations should a global problem ever demand strict international regulation or control. The final paper in this series by Professor Matthews discusses some of the implications of international cooperation on global pollution problems.

44

Implications of Change and Remedial Action

Summary of SCEP
Work Group Report

Introduction

The expansion and refinement of our knowledge and understanding are the necessary conditions for effective change in the present state of environmental management. However, these are not sufficient conditions. Even after optimal improvements have been made in our knowledge concerning the nature of key pollutants, their effects, their sources, their rates of accumulation, the routes along which they travel, and their final reservoirs, the questions will remain of how to apply our knowledge constructively and how to cope with the collateral consequences. As a practical matter, questions of environmental management will have to be faced before we have all the appropriate scientific and technical data, and this further complicates efforts of change or of remedial action.

In examining a wide range of specific problems at this Study, we have identified several aspects that are common to most of them and to many other critical environmental problems. These implications of change and remedial action are briefly discussed now.

Establishing New Priorities

Earlier in our history, the prevailing value system assigned an overriding priority to the first-order effects of applied science and technology: the goods and services produced. We took the side effects—pollution—in stride. A shift in values appears to be under way that assigns a much higher priority than before to the control of the side effects. This does not necessarily imply a reduced interest in production and consumption. When the implications of remedial action and the choices that must be made become clear, there may be second thoughts, confusion, and feelings of frustration.

In the effort to arrive at an optimal balance in specific situations, something will have to give. But the old routine assumption that it is the environment that must give has become intolerable.

Reprinted from Study of Critical Environmental Problems (SCEP), 1970. *Man's Impact on the Global Environment* (Cambridge, Massachusetts: The M.I.T. Press), pp. 31–36.

This assumption must be rejected in favor of an optimal balance to be reached from a point of departure in affixing the responsibilities for pollution.

Affixing Responsibilities

As a point of departure for taking action, we recommend a principle of presumptive "source" responsibility. While remedial measures can be attempted on the routes along which pollutants spread or in the reservoirs in which they accumulate, we believe that these measures should be generally taken at the "sources," which we define broadly to include (1) sources or the points in the processes of production, distribution, and consumption, at which the pollutant is generated, for example, factories, power plants, stockyards, bus lines; (2) protosources or earlier points that set the conditions leading to the emission of pollutants at a later stage, for example, the manufacturers of automobiles that emit pollutants when driven by motorists, or the brewers of beer sold in nonreturnable cans that are tossed aside by the consumer; and (3) secondary sources or points along the routes where pollutants are concentrated before moving on to the reservoirs, for example, sewage treatment plants or solid waste disposal centers.

The principle does not connote any element of blame or censure, nor is it intended to foreclose a judgment concerning where the financial costs of correction should ultimately be borne. It is intended, however, to indicate a point of departure for analysis and action. It rests, in part, on the basis that, if something goes wrong, it should be traced to its origin and corrected in terms of its cause; in part on a hypothesis that the source, protosource, or secondary source will typically be in the best position to take corrective measures, whether alone or with help from others; and in part on the view that the remedies available, the criteria for choice among them, and the implications of remedial action can best be appraised at the sources as here defined.

Accepting the Costs

Remedial changes will ordinarily involve financial costs, and the costs may be large in relation to the scale of the source enterprise. If the source enterprise can neither absorb the cost nor pass it on, it will be necessary to face a choice among failure of the enterprise, continuance of the pollution, or financial assistance out of public revenues. The initial change may have consequences reaching past the source enterprise to its employees, its suppliers, and its customers and beyond in widening waves of change that may engulf deep-rooted patterns of economic and social behavior. Our society is familiar with far-reaching readjustments caused by technological innovation or organizational change in the past. Comparable readjustments may be required by changes instituted to control pollution.

Assessing the Available Means for Action

The means available within the political process and legal system to encompass remedial changes include taxes designed as incentives, stimuli, or pressures, regulations, typically involving a statute, an administrative agency, and supplementary action through the courts; common-law remedies in the courts, incrementally adjusted to contemporary needs; governmental financing of research and assistance to facilitate costly adjustments to desired changes; and governmental operations, civilian and military. Governmental action in its own house can have a dual importance: in itself and as a model for others to follow.

Stimulating Effective Actions

The political, legal, and market processes of our society are profoundly affected by the nature and quantity of information available and the manner in which the information is infused into them. It is neither necessary nor feasible to postpone recommendations for action until scientific certainty can be achieved. The political process is accustomed to decisions in the face of uncertainty on the basis of a preponderance of the evidence or substantial probabilities or a reasonable consensus of informed judgment.

Thus, it is not enough for scientists and technologists to expand and refine their knowledge. They must also present their knowledge in a manner that clearly differentiates fact, assertion, and opinion and facilitates the task of relating the data to the possibilities of corrective action. But if such information is to be used, the Congress and state legislative bodies must be provided with instrumentalities and qualified staff to enable them more effectively to sort out and utilize the input of data, proposals, complaints, and suggestions that will flow into them in increasing volumes from all sectors of our society.

Developing New Professionals
In addition to general public education, we stress the special importance of some changes in scientific, technical, and professional education and training. A sensitivity to the relations between the processes of production, distribution, and consumption, on the one hand, and the processes of pollution, on the other, and a disposition to explore all the potentialities of technology and organization in the search for an optimal balance should be incorporated into their training. This applies to economists, lawyers, and social scientists as well as to scientists and engineers. Individual contributions may be undramatic now, but over time they will be critical.

Cooperating with Other Nations
Although many problems are global in nature, the solutions to these problems will generally require national as well as international action. Typically, remedial measures within one nation will need support from parallel actions within other nations. Frequently, collaborative international action will be required. The prospects for such cooperation are best for programs of collection and analysis of data. International cooperation on monitoring may also increase the likelihood of smooth relations should a global program ever demand strict international regulation or control of pollution-producing activities.

In the foreseeable future the advanced industrial societies will probably have to carry the major burden of remedial action. Developing nations are understandably concerned far more with economic growth and material progress than with second-order effects of technology. Similar attitudes were prevalent in the early stages of growth of present industrialized nations.

The challenges of international cooperation and collaboration in the critically important environmental areas studied by SCEP will be before the United Nations Conference on the Human Environment in 1972. We hope that this Report will provide useful inputs to that Conference and that the Study model furnished by SCEP will be applied to other critical problems of the environment.

45

Expectations of the Decision Maker

Richard A. Carpenter

Human beings make decisions constantly regardless of the adequacy of information and, to an extent, regardless of the penalty for being wrong. The basis for SCEP, however, is that leaders in society want to make better decisions (that is, optimum for human progress) and that science can provide approximate truths as a basis.

Decisions of individuals (freedom) can become a tyranny on the collective welfare, as in the impacts of population on the world commons. Democratic political processes seek the proper tradeoff between the common good and individual liberty. As technology and population increase, good collective decisions become more important, but that does not mean that they can or should all be made by political bodies. Through education and leadership the locus of decision can still often remain with the individual. Thus the results of SCEP are directed at both the citizen and public officials.

Both individuals and institutions have a limit as to the number of issues they can consider at any one time. The conscious attention to a problem is only one of three mechanisms—the other two being (1) a sort of autonomic nervous system of society (ecosystem) where choices are continually made for us without thought, and (2) the delegation of decision making to an established process such as the marketplace.

The setting of priorities for consideration is necessary, but the stream of events interferes. (Witness the disappearance of the "environment" from the newspapers after the Cambodian affair —eight days past Earth Day.) Thus the first expectation is that SCEP will order the critical problems of the global environment as to priority. Although beyond the scope of the Study, it should be recognized that these problems will be viewed by the decision maker in a larger context including local and regional environmental problems, world economic competition, national defense, human health care, and so on. Therefore the "criticality" of the top ranked global environmental problems must be commensurate with these other insistent demands for attention.

Prepared during SCEP.

It must be made clear why conscious decisions are necessary, why the conventional economic system is not taking care of the problem; why the common law cannot be relied upon to sort out the equities for all concerned; why the decisions of a prudent individual are inadequate to achieve the common good; why presently constituted management institutions are failing; why local or regional treatment cannot suffice.

Beyond this demonstration that the problem is worthy of special consideration most of the following criteria should also be met in order for a particular problem to be ranked in high priority:

1. Man-made sources are important relative to natural sources or background values.
2. The effect is irreversible or very difficult (costly) to reverse.
3. The effect would have great economic damage.
4. Emissions or causes are mainly from the developed countries or from technology supplied by them.
5. Emissions are susceptible to control with present technology.
6. Effects are imminent.

The second expectation is that substantial uncertainties exist and that many problems can neither be dismissed (as local or unimportant) nor given high priority on the basis of present knowledge. This is acceptable if a productive course of further study is outlined and if speculation is avoided.

Extrapolations and "if-then" scenarios are perhaps useful in arousing public interest, but they confound the decision-making process. The decision maker is used to incomplete knowledge —in fact this is always the case. The scientist is used to hypothesis and experimentation. As long as the presentation of scientific advice is careful to separate out the "do know," "don't know," and "could know," the communication will be beneficial to the politician. A pro-and-con format may be useful. Another device is the admitted weighting of evidence (on the basis of peer judgment) in order to make a definitive statement—with dissent and its reasons placed in a footnote. Consensus should not be attempted in uncertain areas, and the Study should not be "wishy-washy" about what it does not know.

The third expectation is for a listing of alternative responses to the priority problems: What can we do about it? These actions are usually suggested by the factual description of the problem. They are augmented by a study of impacts, costs, side effects, feedback mechanisms, and so on. Scientific advice to the political process stops short of advocacy even though one course of action may be obvious to the scientist. We are seeking to reserve decision making to politically responsible officials—not to establish a technocracy nor to lapse into a plebiscite of partially informed citizens on every issue which arises. Even in these complex technical matters the ability of the politician to integrate economic, social, raw political, and human intuitional inputs is valuable and should be guarded. As a citizen, the SCEP participant may opt for one or another solution, but the SCEP results should be an objective analysis of remedial alternatives as complete as possible.

The fourth expectation is recognition of the motivations of human beings which constrict the implementation of decisions. Only the archetypal ivory-towered scientist would pursue his work without consideration of human nature. The interpretation of the facts about global environmental problems should take due regard of the weakness of the flesh despite the willingness of the spirit.

For example, corrective actions will occur more easily if they can be seen to coincide with selfish motives such as direct human health effects, welfare of offspring (but not very many generations—"What has posterity ever done for me?"), damage to property, threat to livelihood, or disruption of the status quo. In contrast, appeals will go little heeded when based on goodwill toward man, voluntary reduction of standard of living to preserve the commons, stewardship of the affluent (noblesse oblige), or subjective esthetic values.

There are possibilities for clever merging of baser human desires with actions for the long-term common good. For example, ecological principles will be obeyed whether we do it willingly or not. It can be shown that a high-productivity environment is indeed a high-quality environment. Another approach is the attitude toward the developing countries. We may very well make enormous investments and outright grants in these societies in order to preserve world order. Therefore, it should be practical to underwrite whatever concessions we ask them to make to abate pollution. A third example would stress the prudence of knowing more about global systems (research and monitoring) whether or not critical problems exist, simply so we can maximize the economic exploitation of resources and the environment over a long period of time.

The decision maker can be greatly helped by the purposeful seeking out of those facts and interpretations that strengthen implementation of desired courses of action by enlisting human nature. In fact, it would appear that people will do the right thing even at some personal inconvenience if some reinforcing of their ethical armament is provided by pointing out practicality and prudence. This does not suggest any distortion of the scientific information but only that effort be directed at revealing the relevance of good environmental management to personal health and welfare. And since the ecologists seem to be on to something, that ought to be easy enough to do.

46

Decision Making under   Walter O. Spofford, Jr.
Uncertainty: The Case of
Carbon Dioxide Buildup
in the Atmosphere

Introduction

Decisions—including the lack of any action—will be made about
the quality of man's environment with or without complete information at hand. Regardless of the quality of this information,
or the process used for making these decisions, however, they will
not be made wholly on the basis of perceived present and future
damages to man resulting from a continuation of the current
practice of discharging residuals into the environment. Nor will
these decisions be made solely on the basis of the costs to society
of either reducing or eliminating these discharges. At each and
every stage in the decision process at least an implicit evaluation
of both the social costs and benefits will be made. Conceptually,
at least, society will choose that level of quality (and concomitant
damages) where the costs of eliminating the next unit of residual
discharged to the environment is equal to the benefits (that is,
damages avoided) of making this reduction. At this point, in the
language of the economist, it is said that the marginal costs equal
the marginal damages avoided (or benefits).

We do not mean to imply, however, that an explicit economic evaluation of the appropriate costs and benefits will necessarily be made (or is even needed), or that the solution will be
based on an explicit economic criterion such as economic efficiency. The nonmarket costs and benefits will, most likely, be
evaluated implicitly by a political decision-making process. Furthermore, the criterion against which various alternatives will be
judged will be established, again implicitly, through the same
decision-making process. This point will be discussed again later
on.

Where the discharge of residuals from man's production and
consumption activities is likely to cause changes in the quality of
his environment, quantitative models provide a useful tool for
making residuals management decisions. For example, when all
the pertinent cost and benefit functions are readily quantifiable

Prepared during SCEP and expanded after SCEP.

at the onset, and when interrelationships among components of the system (or systems) are accurately delineated, the numerical results of a quantitative model provide a basis for making explicit decisions. In cases where measurement is a problem (for example, where imputed or nonmarket costs and benefits are involved) and where this information is not readily available in numerical form (either in cardinal or ordinal measurements), quantitative constructs provide us with, at worst, a systematic, logical framework for organizing thought processes and, ultimately, a rationale for choosing among alternative courses of action. Some residuals management problems fall into the former category (that is, they are readily quantifiable, and hence amenable to numerical treatment) while others fall into the latter category.

The purpose of this paper is to explore the feasibility of utilizing some of the available quantitative tools and/or techniques for making better management decisions concerning those residuals that may be associated with possibly long-range, global environmental effects. In particular, we wish to examine in some detail the carbon dioxide problem. Establishing a decision framework for this residual is particularly interesting (if not overwhelming) not only because of its pervasive nature in modern society, or because of the high social costs of eliminating (or even substantially reducing) the generation of man-made carbon dioxide in the short run, or even because this problem involves decision making under uncertainty, but also because the possible environmental effects of continuing our current practices of combusting large amounts of fossil fuels are, most likely, a long way off and, hence, intergeneration transfers of capital are involved.

Admittedly, complete numerical treatment of the carbon dioxide problem is a dubious proposition before we even start our analysis. But given both the importance and the complexity of the problem, formulation in quantitative terms will, I think, provide us with a logical framework for ranking alternative courses of action.

Before continuing with an exposition of the carbon dioxide problem per se, we will discuss briefly some of the available classical and neoclassical quantitative methods used for decision making.

Some Classical and Neoclassical Decision Tools

Quantitative approaches to decision making may be grouped, for exposition purposes, into two categories: (1) deterministic, and (2) stochastic methods. For systems where either random elements do not exist (or are negligible), or if they do exist, they are not treated explicitly, deterministic methods are appropriate. If at least one element in the system is to be treated explicitly as a random variable, stochastic decision methods will be required. Assuming that explicit cost and benefit functions are available (which is not usually the case in any complex real-world situation), determination of the optimal plan (whatever the criterion) is relatively straightforward for deterministic problems. This is not to imply, however, that the deterministic problem is necessarily trivial. On the contrary, finding a solution to this problem can be quite difficult especially if many constraints are involved, if the problem is of a nonlinear form (that is, the objective function as well as the constraint set are nonlinear), and if the number of decision variables is large. For these cases, one generally relies heavily on mathematical programming methods, such as linear and nonlinear programming, and employs computers for making the necessary computations.

But what about decision making under stochastic conditions, that is, situations where the values of one or more of the system inputs (parameters, functions, and so on) is not known with certainty and may take on any one of a range of values? These "uncertain" system inputs are generally expressed as (1) a suitable statistic of the probability distribution of all possible outcomes, frequently the mean (this forces the problem back into a deterministic one), or (2) the entire probability distribution of possible outcomes (assuming it is known). In this case, we are able to establish the shape of the probability distribution function from samples of the population of possible outcomes. An example of this situation is a probability distribution of hydrologic events, for example, river flow, based on perhaps thirty or more years of record. Although we cannot say anything about the expected order of future events from this distribution, we can say something about the probability of certain events occurring sometime in the future.

A more complex situation in decision making than the one just presented is when we know we are dealing with an "uncertain" system input but have no data upon which to estimate a probability distribution or even a statistic thereof. An example of this situation would be an economic, sociologic, or demographic projection into the future where the best we can do is make an educated "guesstimate." But even this estimate is important because decisions will be made on a given issue with or without the advice of the most competent people. The economist differentiates between these two conditions by designating the first, that is, where the probability distribution is known, as *risk*, and the second, where very little (or nothing) is known a priori, as *uncertainty* (for a more complete discussion see Luce and Raiffa, 1957).

We are all quite familiar with the human mind's unbelievable capability for handling the problem of risk, at least in an intuitive fashion. We sometimes refer to this as gambling. The action taken by an individual in one of these situations will depend upon: (1) the probability (either objective or subjective) of occurrence of various states of the system, and (2) the utility the individual attaches to all possible outcomes. And in these situations an individual will usually choose that action which maximizes his (or her) *expected utility* for all possible states of the system (Chernoff and Moses, 1959). Hence, the individual's action will differ depending upon his (or her) attitude, at any one time, toward accepting substantial losses. When one decides not to risk large losses (which usually decreases the chances for massive gains), we call this being conservative, or, in the jargon of the economist, "risk aversion."

The strategies described here as conservative are all aimed at reducing the extent of possible (monetary) losses either through a reduction in the probability of occurrence of a particular outcome or through a reduction in the losses associated with a particular outcome. Reduction in either or both of these will result in a change in the expected value of losses (measured in terms of utility or disutility). This sort of "hedging" is not new to any of us. It is part of our daily lives. As we have suggested earlier, however, whether or not someone (or some nation) would be inclined to take this action depends upon his (or her) individual preferences or "utility function," or in other words, on the values he places on the levels of outcome of certain events. We emphasize these possibilities here only to point out that very often some form of action can be taken ahead of time by an individual or nation either to change the shape of the probability distribution of future outcomes or to reduce the damages should certain outcomes ever occur. Some of these possibilities will be discussed later in connection with the carbon dioxide problem.

Where decisions involving stochastic (or probabilistic) elements must be made, those of the *risk* variety are certainly less difficult to deal with than those of *uncertainty*. As already stated, alternative actions would be ranked according to the expected value (measured in terms of utility) of all possible outcomes.

Decision making under *uncertainty* is a more difficult problem (for example, see Chernoff and Moses, 1959; Luce and Raiffa, 1957; Raiffa and Schlaifer, 1968; and Von Neumann and Morgenstern, 1953). No objective probability distribution over the various states of nature is available to us. Hence, the expected value of gains (or losses) cannot be used as a means for ranking alternatives. From partial information, however, sometimes it is possible to generate an a priori probability distribution over the various states of nature and, thus, reduce the decision problem from one of uncertainty to one of risk. The a priori distribution obtained in this manner is called a *subjective* probability distribution (Luce and Raiffa, 1957; Savage, 1954). As before, alternatives would be ranked according to the expected value (based on the subjective probability distribution). An a priori distribution of this sort might be based on best "guesstimates" of the probability of all possible outcomes. For decisions dealing with long-term critical environmental problems, it would seem to me that no one is more qualified to make these estimates (perhaps along with some confidence of these estimates) than those scientists presently involved with these questions. Given this a priori (subjective) probability distribution based on the best available information, it would be a simple matter to demonstrate that the expected value of damages (valued in terms of disutility) for irreversible catastrophic events (by definition, events resulting in extremely high damages to man) may be significant even though the probability of the event ever occurring is extremely small.

The Carbon Dioxide Problem

We have been aware of the possibility of climatic change resulting from changes in the quantity of atmospheric carbon dioxide since the beginning of this century. The present concentration of carbon dioxide in the atmosphere is about 321 ppm by volume, and it has been estimated that by the year 2000, the concentration of atmospheric carbon dioxide will be 379 ppm (by volume), an 18 percent increase over the present level. It has also been pointed out that if the total recoverable fossil fuel reserve of the globe were burned, it would be within man's power in the next century to increase the carbon dioxide concentration of the atmosphere by a factor of 4 or more (Study of Critical Environmental Problems [SCEP], 1970).

Some contend that an increase in the concentration of carbon dioxide in the atmosphere would produce a "greenhouse" effect which would cause an increase in the mean global surface temperature. It has been estimated that a doubling of the carbon dioxide concentration might increase the mean annual surface temperature 2°C (SCEP, 1970). This increase in temperature might be large enough to trigger a global warming trend and thus cause a significant change in the global climate that is apt to be irreversible (at least in the short run of perhaps a few hundred years). If this were the case, the most likely scenario of events would be an eventual melting of the polar caps which, in turn, would result in a rising of the levels of the seas and oceans. A melting of the Antarctic ice cap would rise the sea level by 400 feet. It has been estimated that this process might take between 400 and 4,000 years. If 1,000 years were required, the sea level would rise about 4 feet every 10 years, or 40 feet per century. This is a hundred times greater than present worldwide rates of sea level change (President's Science Advisory Committee [PSAC], 1965).

Eventually the lower coastal lands and cities, for example, New York City, and later on, the higher lands, would be inundated. Would this necessarily be catastrophic to man if, for example, the process occurred over a relatively long period of time? The net loss of land area would probably be substantial. This would certainly cause a major shift in the spatial distribution of the world's population. But lands that are now covered over with ice and snow most of the year, or lands that presently have an extremely short growing season, might become available for extensive agricultural purposes and/or habitation by many more people than at present. New cities could (and would) be built on higher ground to replace older cities in the lower-lying areas. If the process took many, many years, in an economic sense—although the damages would most likely be enormous—the situation would probably not be catastrophic. The actual losses would depend upon, among others, the following factors: the extent, distribution, and form of climatic changes (temperature, rainfall, sunlight, and so on); the extent of rise of the oceans; and the period of time over which these changes took place. Even if we were able to predict the extent of the physical changes, the losses would be difficult to evaluate in economic terms as we have no idea of the tastes and preferences of future generations (assuming these events were to take place a few generations hence) and the values they, as a society, would place on various outcomes.

It should be pointed out, however, that if this change in the "bounties of nature" were to take place, thereby changing the relative wealth and power of nations, there would be many cataclysmic social events as well as climatological that would transpire during the transition period. The social costs of war, disintegration of the fabric of international cooperation and society, and so on, would, no doubt, be substantial.

As we mentioned previously, however, it is not at all apparent at this time that we will ever experience any measurable effect on the climate even if we were to continue our present practice of burning fossil fuels. Some, for example, are quick to point out that since 1940, the mean surface temperature of the earth has actually decreased, rather than increased, at a time when the carbon dioxide concentration of the atmosphere was increasing. The reason for this is that there are many other factors such as atmospheric particles, cloud cover, variation in energy from the sun, and water vapor, other than carbon dioxide which have an effect on the mean global temperature. In addition, negative feedback relationships do exist. Presently, we just do not know all of the elements which are involved in the radiation balance of the earth, and how they relate to one another.

How does the buildup of atmospheric carbon dioxide fit into the previous discussion of decision making under uncertainty? Here is a situation where we simply do not know for certain whether or not a given event (climatic change) of any magnitude would ever occur even if we were to continue our present practice of burning fossil fuels and thus continue to increase the concentration of carbon dioxide in the atmosphere. We have very little idea what the expected value of all possible states of nature might be or, for that matter, what the shape of the probability distribution of all possible outcomes might even look like.

Hence, in its present form, the carbon dioxide problem is clearly one of *uncertainty*. The question here is with present information on the potential future states of nature whether or not it might be possible to generate (at least) a subjective probability distribution (albeit probably not a very precise one) and thus reduce the decision problem from one of uncertainty to one of risk. Or with additional information (perhaps through the development of more sophisticated climatic models) might it be possible to generate a probability distribution over the states of nature upon which decisions could be made?

This information, together with an appropriate damage function (if one could be postulated), might at least give us a rough idea as to the magnitude of the expected damages. An admittedly rough estimate might, for example, take the form of a plot of the increase in the mean level of the ocean surface over time for an upper and lower estimate of the effect of carbon dioxide on the mean global temperature.

Decisions regarding choices of action for coping with a particular problem are ultimately based, hopefully, on the costs and benefits (in the broadest sense) of each alternative. A cursory look suggests that the social costs of substantially reducing or eliminating the discharge of man-made carbon dioxide at this time (that is, in the short run) appear to be tremendous. It would involve changes in the patterns of our lives that would affect each and every one of us.

As far as the worldwide generation of carbon dioxide from fossil fuels is concerned, it is estimated that the United States contributed 32.1 percent in 1967 and will contribute only 23.8 percent in 1980. Thus, any effort at all to curb the generation of

Table 46.1 Estimated Generation of Man-Made Carbon Dioxide on a World-wide Basis (10^9 metric tons of carbon dioxide per year)

	Year			
Emission Source	1967	1968	1980	2000
1. U.S. totals	4.3	4.62	6.2	9.6[1]
Utilities		1.07		
Industrial		1.35		
Transportation		1.27		
Domestic and commercial		.93		
2. Rest of world totals	9.1		19.8	
World totals	13.4		26.0	
U.S. percentage of world	32.1		23.8	

Source: Study of Critical Environmental Problems, 1970.
[1] Based on projected energy consumption for the year 2000 of 5.95, 16.85, and 12.35 $\times$ 10^{12} kWh(t) for coal, petroleum, and natural gas, respectively (W. E. Morrison, "Energy Resources and National Strength," presentation at Industrial College of the Armed Forces, 6 October 1970, Bureau of Mines, U.S. Department of the Interior, mimeo).

carbon dioxide within the United States will have to be coupled with extensive international cooperation (see Table 46.1).

No one (at least to my knowledge) has suggested that it was even technologically feasible to remove CO_2 from the effluent gas stream of various industrial processes or conventional fuel-fired energy sources. Consequently, if we wish to reduce the discharge of CO_2, this leaves us with two basic choices: (1) gross reduction in the consumption of energy, or (2) employing sources of energy other than fossil fuels.

Extensive reduction in the per capita consumption of energy within the United States would have such far-reaching implications involving significant changes in our present ways of life that it can, I think, be dismissed without further consideration. What, then, are our alternatives?

If we intend to maintain our present per capita levels of energy consumption, the only possibility open to us for reducing significantly the generation of man-made carbon dioxide, at least from most stationary sources, is conversion to nuclear energy. For moving sources, specifically transportation, both nuclear energy (in most cases after conversion to electrical energy) as well as the use of alternative fuels such as hydrogen, hydrazine, and ammonia are possibilities.

Let us carry our discussion of each of these, that is, stationary and moving sources of carbon dioxide, in turn, a little bit further. The ramifications throughout both the economy and society as a whole of making extensive (and rapid) changes in our processes for converting energy would be quite widespread. A detailed discussion of all the implications of these changes could fill volumes. And I think at this point, especially where we know so little about the ultimate effects of atmospheric carbon dioxide buildup on man, what we need to know is whether the social costs of various alternative courses of action are high, medium, or low. Furthermore, this whole question is so complex that without extensive resources—money, manpower, and possibly additional research—and time, nothing any more meaningful can be said about it anyway.

The first thing we should keep in mind when discussing alternatives for coping with the carbon dioxide problem is that it would be physically impossible to make all these changes in even a few years (say five years). It is not even a question of having the necessary technology, or resources available (which we do not). We just simply could not mobilize enough economic force, know-how, available technology and, most important, the necessary motivation to make such a drastic change in such a short time. The questions now become, over what time span should we attempt to make these changes (that is, the length of the transition period), and when should we start the process? I think we could safely assume that as a general rule, the longer the transition period, the less the economic disruption would be, even if the necessary technology were available. People, after a fashion, get used to changes, equipment wears out in time and must be replaced, buildings are razed and new ones put in their place, and so on. The economic system, similar to natural systems, simply is not static. But it takes a real "slug" to move the system very far in any direction very rapidly.

Some of these changes that we have in mind for reducing carbon dioxide would involve substantially eliminating a large group of existing technologies and creating a totally new set of technologies. There would certainly be much resistance to any major change so that the question of kinds of incentives required to encourage these changes would evolve. For maintaining any semblance of economic efficiency during such a transition period, maximum use of economic incentives would be required. Could such a massive change in our present way of doing things actually be brought about? I suspect so, at least partially. But even for a transition period of twenty or thirty years, the social costs will be significant.

According to Table 46.1, we estimated that stationary sources—utilities, industry, and domestic, and commercial—accounted for 72.5 percent of the man-made carbon dioxide generated in the United States for 1968. The remainder of the carbon dioxide generated in the United States came from the transportation sector. For complete elimination of carbon dioxide from the stationary sources (probably not possible much before 2020 A.D.), the only alternative would be to convert to nuclear sources of energy (for fission processes, consideration must also be given to the fission waste products such as krypton 85 and tritium) and then distribute this energy in the form of electricity to the various consumers in this category.

Whether or not complete elimination of CO_2 from this sector of the economy is what we ought to be striving for has meaning only with respect to what effect this reduction will have on the long-term buildup of carbon dioxide in the atmosphere, and ultimately, its effect on the reduction of damages to man. Hence, let us go on now to the other category of carbon dioxide sources, the transportation sector.

According to Robert Ayres (private communication) in terms of the carbon dioxide problem, there are basically two approaches to reducing emissions: (1) seek to reduce the demand for automobile transportation through (a) changes in demand for transportation (changing the physical form of cities by promoting a shift in demand from physical transportation to elaborate forms of electronic communication, or simply by imposing prohibitions or tax penalties on tripmaking) and (b) policies directed at shifting demand from the automobile to electrically powered mass transportation systems; and (2) improve the residuals production characteristics of the vehicle, and for the case of eliminating the discharge of carbon dioxide, this means a change in the present fuel. Ayres argues that the first approach, that is, reduction in demand for automobile transportation, should not be considered as a serious alternative. According to him, successful implementation in anything less than half a century or so would involve going through traumatic restructuring of our whole urban society and industrial economy. Where does that leave us then with transportation and the carbon dioxide problem?

As stated earlier, in order to completely eliminate carbon dioxide, we would have to get away from the use of fuels containing carbon. Although more expensive than petroleum products, hydrogen, hydrazine, and ammonia are all possibilities as fuels. If these fuels were synthesized in a chemical plant using nuclear energy, they would not contribute to carbon dioxide (although they would contribute to the heat problem through additional energy requirements as well as generating nuclear waste products). But systems based on this technology may be a decade or more in the future.

We have probably gone far enough to convince even the most skeptical person that eliminating man-made carbon dioxide in the short run would be impossible and that the social costs of substantially reducing it in ten, twenty, or even thirty years would be extremely high. The change will have to be made gradually. The shorter the transition period and the more complete the conversion to processes that do not generate carbon dioxide, the higher the social costs will be.

Decisions about Carbon Dioxide Buildup

Up to this point we have discussed the possible damages that future generations of people might suffer if we continue to allow a buildup of carbon dioxide in the atmosphere. We have also mentioned briefly the costs to society of reducing (or even eliminating) the discharge of man-made carbon dioxide to the environment. Now we are faced with the question of what to do about it. Should we adopt a strategy of reducing the discharge of carbon dioxide; should we put more resources into scientific research to acquire a better understanding of global systems and the ultimate effects of these discharges on man; should we collect more and better data, that is, establish a more sophisticated global monitoring network; should we put more resources into research and development of emission control devices and/or alternative energy conversion processes, or what?

Clearly, if there were any evidence at all that the discharge of a particular residual (for example, man-made carbon dioxide) might result in severe damages to man some time in the future and if the social costs of eliminating the possibility of this danger were small or negligible, the prudent country would adopt a strategy of curbing the discharge of that residual. Additional research and data collection for management purposes *alone* would be a waste of resources. That is, any additional effort would only tend to increase the costs of investigation without having any effect on the benefits (that is, avoidance of damages). As we have already demonstrated, however, this is not the case with carbon dioxide. Unfortunately for us, the carbon dioxide problem is not that simple. Not only does it involve decision making under uncertainty, but it involves another complicating feature—intergeneration transfers of income (that is, investment of one generation for the benefit of future generations).

What about this question of the time between the decision to invest and the realization of benefits (or the avoidance of damages). In the discussion thus far, we have assumed that the outcome of an event will occur in a reasonably short time after the action has been taken. And this is certainly true in the case of a card game or horse race. The stock market is another matter; the time between the initial investment and realized benefits will vary. In short, up to now we have neglected the concept of "discounting" in our analysis.

Time differentials between investments and realization of benefits, at least in a medium time frame of perhaps up to fifty years, traditionally has been handled by economists in their use of the present value concept. That is, for making benefit-cost comparisons, the costs and benefits are discounted to some base year, usually to the present time (thus the name present value). This analysis has enabled economists to make what they regard as rational decisions involving investments and returns on investments that occur during different periods of time.

But what about long-range effects such as those that might

be associated with the buildup of atmospheric carbon dioxide? Should the expected value of benefits (damages avoided) be discounted to present values? If so, at what discount rate? Most economists would agree, I think, that the present value concept is not really applicable to the situation of long-term global environmental problems. They contend the concept has more validity in capital investment theory where the intention is to compare the yields (that is, returns) of various investment alternatives. Krutilla and his coworkers, on the other hand, in their research on the value of irreplaceable natural assets have shown that discounting has meaning even when intergeneration transfers of capital are involved (Krutilla et al., 1971). They point out that as time goes on, the natural asset grows more valuable because of increasing relative scarcity. They go on further to show that if the annual value of the asset appreciates at the rate α, the total *present value PV* of the resource may be expressed as

$$PV = \gamma_0 \sum_{t=1}^{T} \frac{(1 + \alpha)^t}{(1 + i)^t} \tag{46.1}$$

where γ_0 = the initial value of the asset
   i = discount rate
   t = time index
   T = time horizon

From equation (46.1) we note that if $\alpha = i$, the total present value is equal to $\gamma_0 T$. If $\alpha > i$, the annual present value of the asset will actually increase as time goes on, whereas if $\alpha < i$, the annual present value of the asset will decrease. One might interpret the situation where $\alpha = i$ as a net discount rate of zero and where α is slightly less than i, as a very small net discount rate.

The carbon dioxide problem is analogous to their irreplaceable natural asset problem if one considers that the environment has a given "life sustaining capacity." Let us assume that carbon dioxide—or perhaps one of the other "critical" residuals—diminishes the life supporting system. The remaining life supporting capacity will therefore grow more valuable because of increasing relative scarcity, and so on. Admittedly, the carbon dioxide problem is more complex than the situation described by equation (46.1). But the approach to the problem would be identical, and, hence, at least conceptually, it appears that discounting is certainly relevant for the carbon dioxide problem as well.

It is interesting to note in this regard, however, that there is certainly enough evidence on the past behavior of people to suggest that, in general, they are not very farsighted (especially for time lengths greater than their lifetime) and that they do discount quite heavily (implicitly through their actions) the outcome of future events, even when great losses may be involved. Part of the reason for this—at least in the United States—is man's faith in technology for coping with future situations; his faith in his government; and his preoccupation with more pressing and urgent problems, such as attempting to survive from week to week.

If for decision purposes we do not employ some means of modifying costs and benefits that accrue over different periods of time, is it meaningful to make these comparisons (that is, benefits and cost) directly? Or should decisions on long-term damages and the destiny of future generations be based on other than traditional economic concepts; for example on either ethical grounds or on the rationale of preserving as many future options as possible? In the latter case, these options would not only involve those related to preserving the present quality and other characteristics of the environment but also those related to preserving petroleum (and other fossil fuels) for uses other than for energy conversion; for example, lubricants, bituments, paraffin wax, road oil, and petrochemical feedstocks for use in, among others, the manufacture of various "synthetics." The question of how much value we, as a society, might place on the preservation of future options has been discussed by others (for example, see Krutilla et al., 1971; Federal Power Commission, 1970; and Weisbrod, 1964).

These latter considerations raise a number of questions related to the mechanisms, tools, and techniques we have available for valuing the costs and potential damages associated with the generation of man-made carbon dioxide. Will these costs and damages be evaluated by social scientists, engineers, and natural scientists outside the political system, or will value judgments on potential outcomes be made by policy makers within our decision-making institutions?

Because of the difficulty of specifying meaningful, quantitative damage functions associated with the effects of a buildup of atmospheric carbon dioxide *even if we knew for certain the outcome;* the difficulty of numerically specifying the social costs of curbing (over varying transition periods) the discharge of carbon dioxide from activities of modern society; the difficulty in assessing the value of the loss of future options; and, finally, the relatively low confidence in the presently available subjective probability distribution of all possible future outcomes; the problem would seem to defy conventional numerical treatment. Does this mean then that ultimately, quantitative methods will not be employed to specify an "optimal" plan with regard to the buildup of carbon dioxide? Probably so. Nonetheless, formulating a problem of this scope and importance in quantitative terms does help, I think, to give us a better understanding of the interrelationships among the relevant components, and hence aids us in making better decisions on difficult issues.

Whether or not choices among alternatives would, or even should, be made on imputed values established through traditional economic approaches is presently a matter of contention by some political theorists. They point out that making value judgments and tradeoffs for society on nonmarket issues is within the realm of political decision making; that this is what politics is all about. Making these tradeoffs implies that some knowledge is available however.

If the undiscounted damages (either quantitatively or descriptively determined) are obviously much greater than the costs of reducing or eliminating the residual, especially if the costs are small, the prudent country would probably opt for immediate abatement. A good example in the United States of just this situation is the present concern with, and reduction of, persistent pesticides (for example, DDT) and the heavy metals (for example, mercury). Public intervention on most of these issues can be defended on the grounds that the damages are public in character or because the activity generating the residual is either not willing or in a position to bear the costs of abatement and/or future damages (Renshaw, unpublished).

On the other hand, if the costs of reducing or eliminating the residual(s) are high—as in the case of carbon dioxide—

elimination, or even reduction, of the residual on this ground alone will certainly face stiff resistance. However, aside from the costs—private as well as nonmarket—of reducing the generation of man-made carbon dioxide, at the present time there exists a lack of public concern over the issue due mainly to the following: (1) we know very little about the actual long-term effects of a buildup of carbon dioxide in the atmosphere, and (2) the damages, if they do occur, are presumably a long way off.

Let us take a moment to explore what the public reaction and government response might be in a situation involving a "critical" environmental problem—such as the carbon dioxide issue—where the elements of (1) uncertainty, and (2) a significant time lag between investment and realized benefits are involved. According to some who are in a position to observe such situations, the scenario of events—publicity, legislative and executive hearings, decision making, and the establishment of government policy—would be approximately the same for all the "critical" environmental problems. However, the time scale of occurrence of events would differ. The timing would depend essentially on two factors: (1) whether or not it could be established with certainty that an event (for example, a catastrophe) would take place if no remedial action were taken; and (2) the length of time between the initial investment in a plan of action and the realized benefits, for example, whether or not it would occur within one's own generation, one's children's generation, or many generations down the road. Included in the second part, at least implicitly, would be a factor for the lead time required for society to make the necessary transitions without unreasonable disruption.

If the foregoing considerations are any indication of the sort of public concern and timetable of events we can expect with the carbon dioxide problem, significant public action on the matter is a long way off. This is mainly because of our pervasive lack of knowledge concerning various aspects of the problem. Before society would be willing to commit large amounts of resources for reducing the quantities of man-made carbon dioxide generated in production and consumption activities, we would have to know much more about the future effects of doing nothing about it. Hence, at the present time, we should be seriously considering, as a minimum, the following issues: (1) What lead time is necessary for a relatively smooth transition from energy conversion processes that generate carbon dioxide to those that do not? (2) What additional effort should be put into monitoring and data collection in general? (3) What additional effort should be made to acquire a better understanding of the interrelationship among carbon dioxide, particles, cloud cover, weather patterns, long-term climatic change, and so on, including the development of mathematical models? (4) What additional effort should be made to acquire a better understanding of the carbon cycle, including specific routes and reservoirs? (5) What effort should be made into research and development of alternative processes for converting energy (including the automobile) that do not involve fossil fuels? The latter might include more research into, for example, breeder reactors, fusion technology, control of fission products, and processes for converting energy from other than carbon base or nuclear fuels (for example, hydrogen, ammonia, and hydrazine).

Information on the second, third, and fourth points would give us better estimates of the potential environmental changes and thus the benefits for reducing the rate of buildup of atmospheric carbon dioxide. The fifth point would provide us with a variety of technologically feasible alternatives for coping with the carbon dioxide problem (if in fact it is a problem) with enough lead time to avoid a situation that might be damaging (or even catastrophic) to future generations.

Summary
Decisions regarding possible future damages related to current practices of modern society will not be made by economists, engineers, natural scientists, the clergy, or by philosophers. They will be made by politicians within our decision-making institutions. And so in the real world of social decision making, both uncertainty as to the actual occurrence of a future catastrophic event and the unusually long times before these events are expected to occur tend to decrease the probability that any major action— except possibly for continued research—will be taken at this time. It would seem to follow from this then that if we really are concerned with our future, and if we really want our decision makers to commit large amounts of resources to remedy the situ-

ation—resources that are desperately needed elsewhere in contemporary society—we are going to have to (1) know more about the rate of buildup of carbon dioxide in the environment, including the capacity and locations of various carbon dioxide sinks; (2) know more about the effects of various ambient concentrations of carbon dioxide on the climate; (3) have a better idea of the time scale when certain environmental changes might be expected to occur; and (4) have a reasonable estimate of the necessary lead times required to (a) convert a society from processes that generate carbon dioxide as residuals to processes that do not (however, in the case of generating power by nuclear means, a similar concern must be given to the fission waste products), and (b) develop alternative processes for converting energy from other than fossil fuels. In the case of carbon dioxide, this will only be made possible through additional research and monitoring.

We have raised, unfortunately, more issues and questions than we have answered. This paper truly opens a wide area of unknowns. Nonetheless, these are all extremely important questions and ones that we are going to have to find answers for sometime in the near future.

Acknowledgments

I am indebted to my colleagues Orris Herfindahl, Allen Kneese, and Cliff Russell for their helpful suggestions during the preliminary stages of this paper. In addition, helpful comments and suggestions on the manuscript were received from Sterling Brubaker, Edwin Haefele, Orris Herfindahl, Allen Kneese, John Krutilla, and Clifford Russell, all of Resources for the Future, Inc.

References

Chernoff, H., and Moses, L. E., 1959. *Elementary Decision Theory* (New York: John Wiley and Sons, Inc.).

Federal Power Commission, 1970. Testimony of Charles J. Cicchetti in the matter of Pacific North West Power Company and Washington Public Power Supply System, Projects Numbers 2243/2273.

Krutilla, J. V., Cicchetti, G. J., Freeman, A. M., and Russell, C. S., 1971. Observations on the economics of irreplaceable assets, *Environmental Quality Analysis: Research Studies in the Social Sciences,* edited by A. V. Kneese and B. T. Bower, Resources for the Future, Inc., forthcoming.

Luce, R. D., and Raiffa, H., 1957. *Games and Decisions* (New York: John Wiley and Sons, Inc.).

President's Science Advisory Committee (PSAC), 1965. *Restoring the Quality of Our Environment*, Report of the Environmental Pollution Panel (Washington, D.C.: U.S. Government Printing Office).

Raiffa, H., and Schlaifer, R., 1968. *Applied Statistical Decision Theory* (Cambridge, Massachusetts: The M.I.T. Press).

Renshaw, E. F., The economics of survival: some notes on the theory of terminal costs, unpublished.

Savage, L. J., 1954. *The Foundation of Statistics* (New York: John Wiley and Sons).

Study of Critical Environmental Problems (SCEP), 1970. *Man's Impact on the Global Environment* (Cambridge, Massachusetts: The M.I.T. Press).

Von Neumann, J., and Morgenstern, O., 1953. *Theory of Games and Economic Behavior* (Princeton, New Jersey: Princeton University Press).

Weisbrod, Burton, 1964. Collective consumption services of individual consumption goods, *Quarterly Journal of Economics*.

47

Electric Power Generation: Willard A. Crandall
Options for Environmental
Improvement

This paper presents a very brief outline of the various options available to lessen environmental pollution from electric power generation. These involve choices of fuel, electric station and equipment design, methods of operation, and other variables. The pollutants discussed include carbon dioxide, oxides of sulfur and nitrogen, particles, waste heat, and radioactive wastes.

Carbon Dioxide

Carbon dioxide is produced as a major combustion product in the burning of any fossil fuel. Per Btu of heat released there is little difference in the quantity of CO_2 produced by the burning of coal, oil, or gas. Hence, no appreciable reduction in CO_2 emission can be made by the changing from one fossil fuel to another.

The vast quantities of "flue gas" produced by a large power plant (several million cubic feet per minute) and the elevated temperature of this gas makes the removal of CO_2 by the application of known technology not feasible in terms of equipment and space requirements, operating costs, treatment chemical requirements, and waste disposal problems. Hence the use of nuclear power as a heat source to replace fossil fuels is the only alternative available.

Obviously, existing power plants cannot be converted for the use of nuclear fuel. The installation of new nuclear power plants, under current conditions, takes many years to complete the various planning, engineering design, construction, and licensing tasks that are required. Hence, nuclear power can be considered only as a feasible alternative on a long-term basis for the construction of new electric generating facilities to meet future power demands.

Although the safety and environmental benefits of nuclear power have been well established by the record of operating plants, there are many persons who still think of nuclear power in terms of nuclear weapons and their effects. Hence, problems will often occur with respect to issues such as public acceptance and plant siting.

Prepared for SCEP.

Sulfur Dioxide

This pollutant is produced in varying quantities by the combustion of coal and oil, the amount being a function of the sulfur content of the fuel.

One alternative is the shift of fuels to gas, if available, at a cost per Btu of energy that in most cases will exceed that of the fuel replaced. The overall benefit of utilizing the supply of natural gas available in the given area to meet other energy requirements (for example, space heating and industrial needs) must also be considered.

Another alternative is the use of nuclear power. This involves again the problem cited under the control of carbon dioxide emissions.

A third alternative is the use of low-sulfur fuels—natural or desulfurized. The former (coal and oil) are in short supply and are far more costly than normally used fuels. No adequate supply of desulfurized coal is currently available due to the present state of technology in this area. Current cost estimates for coal desulfurization run as high as $15 per ton, which would result in an increase in the cost of this fuel by a factor of 2 to 3. With respect to oil, desulfurization technology has advanced to a far greater degree. It is expected that desulfurized oil, at an increased price, will become available in the near future, but the demand for this fuel will probably exceed the available supply for many years. Here again the allocation of this fuel to meet the overall utility, industrial, and domestic needs (considering the total air pollution control problems involved) must also be considered. Factors of foreign policy and national economic policy obviously have great influence in this matter of procurement and allocation of low-sulfur-content fuels.

A fourth alternative is the application of SO_2 removal equipment to treat the combustion product gas. This is an emerging technology in its formulative stage. The installed cost of equipment for this purpose runs as high as $35 per kilowatt of installed capacity for new plants and as high as $60 per kilowatt for existing plants. Increased operating costs are estimated to be in the order of $0.50 to $1.00 per ton of coal fired. Although some processes for this purpose offer the incentive of reducing overall costs by the recovery of the sulfur in a salable form, none to date have demonstrated that such practice can, in actuality, pay off the total gas treatment costs. In many areas there exists a limited market for the sulfur product (low-quality sulfuric acid, SO_2 gas, and so on) that may be recovered. Also to be considered is the fact that many of these processes produce major solid waste handling and water pollution problems.

Oxides of Nitrogen

This pollutant is produced by the combustion of any fossil fuel, the actual quantity being a function of boiler design and operation. The temperature of combustion, the quantity of excess air available, and the cooling rate of the combustion gases determine the amount of NO_x formed. Techniques are currently being developed for reducing the NO_x produced in the firing of gas and oil. They require redesign of combustion systems to permit two-stage combustion, recirculation of flue gas, and improved combustion control. The use of such techniques may reduce the NO_x emissions from 50 percent to as much as 75 percent in some cases. No estimates are currently available as to the costs involved in converting existing boilers or modifying the design of new boilers to be constructed.

Particles

Most coal-fired electric utility boilers are equipped with some type of fly ash control system. The efficiency of these systems vary. Newer boilers are being equipped with fly ash control systems having a design particulate removal efficiency of 99 percent or higher. In a few cases, systems have been used for the removal of particles produced in the combustion of oil. The technology exists for both upgrading the performance of existing systems by the installation of supplementary equipment or installing high-efficiency systems on new units. The estimated cost of high-efficiency fly ash control systems for new units run in the order to $5 per/kW of installed capacity (much higher in some areas due to space problems, local construction costs, tax structures, and so forth). The estimated cost of refitting existing boilers with high-efficiency systems range between $8 and $24 per kW.

One problem recently encountered by many utilities companies has been the decrease in efficiency of existing fly ash control systems resulting from the increased "electrical resistivity" of the fly ash produced when firing fuels having a sulfur content of 1 percent or less. This has required the upgrading of existing systems (at considerable cost) to maintain previously attained removal efficiencies.

An obvious alternative to the use of coal as a fuel and the employment of adequate fly ash control systems is the use of oil or gas. The feasibility of these alternatives is contingent upon the availability and cost of suitable oil supplies (for example, with low-sulfur content) or gas. The fly ash formed during the combustion of oil and gas is small in quantity and the particle diameter is under 1.0 micron. Hence, it does not "fall out" locally and have local environmental effects. It does go into the atmosphere, however, and therefore can have possible effects in respect to absorption of solar energy. In exploring the subject of fuel conversion, the costs of boiler conversion, fuel supply, storage problems, reliability of supply, and related problems must all be considered.

Another alternative to be considered for long-range planning is nuclear power. Concomitant to this alternative are the many problems previously mentioned.

The above discussion has been concerned with removal of all particulate matter regardless of size. It is well to point out here that it is these particles that are in both oil and gas combustion and which escape to some degree from even high-efficiency fly ash collection systems on coal-fired boilers which are of concern with respect to weather effects. The magnitude of these effects caused by particles from such sources is probably minimal in most cases.

Waste Heat

In the generation of electric power, thermal energy ranging from as low as 9,000 Btu to as high as 15,000 Btu must be released from a fuel (coal, oil, gas, or nuclear) to produce 1 kWh of electrical energy (having an equivalent energy value of 3,413 Btu). The actual value is a function of the design of the power plant and its method of operation. The waste heat (total thermal energy required less the thermal equivalent of the electric energy produced) thus amounts to about 70 percent of the thermal energy produced in electric power generation. In fossil-fuel-fired plants, this waste heat is distributed between heat released to the atmosphere by "stack losses" plus thermal radiation from the boiler and heat released to turbine condenser cooling water to the water environment. In nuclear power plants, there are no "stack losses," and the waste heat radiated from the reactor is minimal.

Due to the theoretical limitations of the power transformation "cycle efficiency" involved, there is no way known at present in which the total waste heat release can be reduced appreciably. Hence, the only procedure to be followed is to release the heat in a manner having the least adverse effect on the total environment for each alternative. Arbitrary criteria based upon data of dubious validity or upon generalized data that may not be applicable to a local area should not be used in determining the best alternative.

The alternatives to be considered in determining the best manner in which the waste heat can be released include the following:

1. The use of nuclear power to reduce stack losses. This solution is subject to the difficulties previously stated. Further, it is doubtful whether it would be preferable in any case to transfer emission of waste heat from the stack to cooling water.

2. The use of cooling towers. These are massive in size, generally considered unsightly, and can produce fog and rainfall in the local area. The estimated cost for such cooling towers may run over \$8.00/kW of installed capacity for new construction and as high as \$12.00/kW for installation at an existing plant. Further, there are operating costs involved which add to the total cost of this alternative. In the case where only salt water is available for circulation in the cooling tower, the fallout of salt water from the mist formed has had, in some cases, local adverse effects on plant life and caused some property damage.

3. The use of cooling ponds, dilution, or other means to reduce the temperature of the cooling water or the total quantity of heat carried back to the point where the cooling water reenters the water environment. These alternatives usually involve much land use (many acres for even a small power plant) and also increase the cost of power generation to a considerable degree. Studies are currently in progress, however, to try to make use of these

large quantities of low-temperature heat to raise the ground temperature in agricultural areas, increase the rate of biological action in sewage disposal plants, and other beneficial applications. It is hoped that such procedures can be practically employed in certain situations in the future.

4.   Control of reentry of cooling water to the environment. This involves the construction of facilities to insure that the warmer cooling water is dispersed into large volumes of cooler water to prevent significant local temperature rises in the water environment. Such action is possible only when there is a body of water available having adequate thermal capacity for absorbing the total heat received, without adverse effects. The costs involved with this alternative will vary in each particular case, and cost of such action will not be insignificant in respect to total power generation costs.

Radioactive Wastes

When the choice has been made to use nuclear power, the matter of radioactive wastes must be considered. The well-designed and operated nuclear power plant produces less hazardous wastes in lower quantities than is commonly believed and/or stated by critics. Technology is currently available and is being applied to reduce the level of radioactive material emitted to the environment to values approaching background radioactivity levels for naturally occuring radioisotopes. The control over the licensing and operation of nuclear power plants by the Atomic Energy Commission (AEC) and other regulatory authorities exists to insure that the emission levels during normal operation or in the event of any malfunction within the plant do not create any environmental hazards.

Research and development are currently in progress to lower further the level of radioactive waste material produced, disposing of this waste by still safer means, and further reducing the already low emissions of radioactive material to the environment.

In this area of pollution control, the choice of alternatives lies in the selection of nuclear plant design and the waste handling systems incorporated therein to insure minimal obtainable emission rates for radioactive materials. In the making of such decisions the concern of the operating utility company plus the guidelines of AEC policy and practice make the choice primarily a function of the relative effectiveness of hazard elimination systems.

The matter of nuclear fuel reprocessing must also be considered in the overall picture of environment effects. However, as this paper deals only with electric power generation, the subject of nuclear fuel reprocessing will not be discussed. Obviously, though, the same criteria for safety and environmental effects must apply here also.

Conclusion

As can be seen from this discussion of alternate methods for reducing the emission levels or effects of various pollutants, what may be the best solution to one particular pollutant problem in a given set of circumstances may also turn out to be an undesirable alternative with respect to the control of one or more other pollutant emissions. Other factors—such as both the short- and long-range power generation plans for the utility company, the methods available to finance necessary changes, the methods available to compensate for increased operating costs, the situations in respect to rate adjustment, and taxation—affect the choice between alternative actions. Other governmental regulations having little to do with environmental control by intent, may also become decisive factors in the resolution of such problems.

The options available to the electric utility industry for environmental improvement are many, but their ramifications are interrelated and their implications are often complex. It is hoped that the foregoing discussion of this subject, though brief, will aid in clarifying the perspective of others concerned with the broader aspects of environmental improvement both on a local or global basis.

48

Global Pollution and   William H. Matthews
International Cooperation

Introduction
There is growing awareness in this and other nations that the
quality of the global environment may be seriously threatened
by pollution resulting from man's activities. The advanced nations of the world are the most serious polluters of the earth's
atmosphere and oceans and are faced with the responsibility, if
not yet the necessity, for exploring ways to ameliorate these threats
and prevent the deterioration of the global environment.

Responding to this challenge will pose many difficult problems for the formulators of foreign policies, for meaningful action
may require extensive interaction and cooperation among nations.
At the first stage of cooperation, which will be gathering scientific data and learning more about the nature of the problems,
the foreign policy issues will be relatively straightforward and not
too difficult to resolve. However, if later stages are necessary that
would require international regulation or control of domestic
practices to prevent global catastrophes, the foreign policy issues
will be extraordinarily complex and perhaps even intractable.

If the likelihood of international cooperation on these issues
is to be enhanced, it will be necessary to have a clear understanding of their nature. This paper discusses several of the issues that
will ultimately have to be resolved.

Global Technologies and Global Pollution
At the most general level of analysis, global pollution can be attributed to two causes—people and technology. The combination
of the population explosion and new technological developments
has resulted in an accelerating proliferation of waste products as
the energy, food, and other needs of people have been met through
technology without centralized regulation.

The issues to be discussed in this paper are a direct consequence of the observation made by Eugene B. Skolnikoff in his
book *Science, Technology, and American Foreign Policy* (The
M.I.T. Press, 1967): "one of the aspects of the rapid advance of
science and technology is the creation of new technologies or the

Prepared for SCEP.

multiplication of the scale of application of technology, so as to *require* some form of international regulation, cooperation, and control."

In recent years, the following have become increasingly considered as "global" technologies with respect to their effects on global climate and ecology: generation of energy from fossil fuels; generation of energy from nuclear power; internal-combustion engines; industrial and transportation processes that introduce particulate matter and water vapor into the atmosphere; uses of pesticides and herbicides; uses of fertilizers; disposal techniques for chemical and human wastes; and the transport of great quantities of oil in supertankers.

In a discussion of technology and international relations, Warner R. Schilling (1968) has pointed out four characteristics that have been associated with political changes resulting from technological developments. Although pollution issues were not considered in his formulation of these characteristics, his propositions are substantiated by many aspects of pollution problems and the technologies that have helped to create them.

The first characteristic that Schilling noted is that changes result from multiple rather than single technological developments. It is clear from the list of global technologies that many technological developments have created the pollution problems and the resulting political problems. For example, the burning of fossil fuels for the production of energy would not be of such magnitude to cause global concern were it not for the myriad of technological developments and gadgets that require so much power for industrial, home, and recreational purposes and were it not for advanced techniques of locating, obtaining, and transporting the necessary fuels. The same types of relationships could be cited for nuclear power plants, automobiles, factories, and other technological developments.

Schilling has also observed that changes result from multiplicity of nontechnical, as well as technical, factors. This is now popularly characterized by "it's not the technology that is to blame, but the way it is used." Nontechnical factors determining the nature and scale of utilization of technologies include governmental prerogatives and resources, consumer preferences and needs, and the level of public and official understanding of the implications of the use of various technologies. The desires for national security, wealth, prestige; for industrial profits and expansion; for individual comfort and pleasures; and for scientific discovery and technological development can all lead to the utilization of technologies in such a way as to cause major political and social changes. The technology may make something possible, but nontechnical factors will often determine if it becomes reality. Understanding those nontechnical factors is extremely difficult.

Another characteristic of technological developments is that problems and opportunities are distributed unequally among states. Because pollution is primarily a by-product of industrialization and affluence and because the nations of the world are rather sharply divided into "haves" and "have-nots," it is obvious that all states will not view or respond to the issues raised by global pollution in the same way. The problems of irreversible damage to the earth and/or man are the concern of all nations, but the problems of ameliorating the threatening conditions will fall heavily and squarely on the developed nations, both because they are the major polluters and because they have the technology required to develop viable alternatives. Likewise the opportunities to use the pollution issues to further other national and international objectives such as image building or stabilizing relations among states will also accrue to the developed nations. The developing nations will, however, play a significant part in determining how readily or effectively the developed nations assume a sense of responsibility or leadership in this area.

Finally, Schilling hypothesized that the political consequences of technological change are largely unanticipated. Even a cursory review of the development and deployment of industrial, agricultural, and transportation technologies reveals an astonishing lack of understanding of or callous disregard for the ultimate consequences of excessive creation and emission of pollutants. Those consequences of a biological, chemical, and geophysical nature may be quite severe. The resulting political and social consequences may be equally severe. Changing polluting practices will involve changing many social practices and will surely involve both short- and long-term economic and other sacrifices. The implications for international cooperation are particularly great. Not only were these ultimate consequences largely unanticipated in the past when very few were concerned about pollution, but it is not clear that the dramatic nature of the political changes which may be required in the future are appreciated even now when environmental concern is in vogue.

The existence of these global technologies creates a number of important long-term implications for the foreign policy of this and other countries. In discussing the technologies of communications and weapons systems, Schilling (1968) has noted the double-edged effect of technology on foreign policy. This effect is also evident in the environmental area. What technology has given with one hand—by providing the means for rapid industrialization, efficient agriculture, effective transportation, and satisfaction of an ever-growing list of individual needs and comforts—it has taken with the other—by making it possible to do all of these things simultaneously and economically without having to consider the results of random and enthusiastic deployment. But perhaps in the future this can be reversed. If technology has "taken" environmental quality then perhaps it can "give" the means to regain it. As Schilling (1968) has so optimistically stated:

But just as man first used technology to overcome the limitations of his natural environment, so now, in technologically complex societies, man can turn science and technology to the task of overcoming limitations in his technical environment. Increasingly, man's values determine his technology; he can do what he wants.

There is, however, another example of this double-edged effect which may not be so easily reconciled. Even though technology may give the means for combating and abating pollution on a global level, it may have already taken away any possibility that the necessary agreements among nations could be reached. Stanley Hoffmann (1968) has suggested three factors that will seriously impede meaningful widespread international cooperation in the future—all are directly or indirectly related to technological developments: a continuing high level or military, political, and economic competition between the East and West despite a nuclear standoff; the gap between the developing and developed nations and the resulting resentments, demands, and political uncertainties; and the spread of nuclear weapons and possible destablization of the present balance of power. It will be a formidable challenge for foreign policy makers to superimpose upon or replace the present competitive international system (largely a result of military and industrial technology coupled with ideology and social structure) with a cooperative international system (which may be necessitated by excessive and unwise use of industrial technologies).

There are several consequences for nations which result from these global technologies which merit explicit recognition. Five such consequences suggested by Skolnikoff (1967) are limits to national freedom, necessity for international cooperation, strain on national capabilities, external cultural influences, and unsettling prospects for the future. It should be clear from discussions throughout this paper that the global technologies listed earlier will result in each of these consequences.

The Necessity for International Cooperation

A major foreign policy question raised by global pollution is whether nations will be willing to act in anticipation of global disaster or whether they will react only in response to specific ecological catastrophes after they have occurred. Concluding an article on pollution as an international issue, Abel Wolman (1968) made the following rather disheartening observation:

One would be rash indeed to pretend to define explicitly the position of the United States with respect to international pollution problems. What may be said confidently, however, is that the record of the last half-century indicates that the United States, like every other country (or person, for that matter), faces up to such problems only as a consequence of actual conflict or crisis. Rarely do international agreements anticipate the need. . . .

Crises or conflicts are changes in the status quo. They demand response—response which is usually aimed at reestablishing the status quo or at least establishing a new state of equilibrium. In international relations, foreign policy is formulated with respect to equilibrium among nations, not between man and nature or advanced societies and their planet. This may be an obvious point, but in developing an international environmental policy it will be a critically important, and perhaps limiting, factor.

The response to global pollution will probably demand some strong international cooperation. Indeed, George Kennan (1970) has made the following uncompromising statement:

[T]here is nothing today to give us the assurance that such efforts will be made promptly enough, or on a sufficient scale, to prevent a further general deterioration in man's environment, a deterioration of such seriousness as to be in many respects irreparable. Even to the nonscientific layman, the conclusion seems inescapable that if this objective is to be achieved, there will have to be an international effort much more urgent in its timing, bolder and more comprehensive in its conception and more vigorous in its execution than anything created or planned to date.

While this may only initially be in terms of data gathering and information exchange, it may eventually, and in the not-too-distant future, be necessary to begin seriously regulating and controlling global technologies. When one reflects on the potential issues that may arise, the extreme destabilizing effect of this regulation on the present status quo of generally laissez-faire policies regarding pollution among nations is vividly clear.

It is hardly an idle question to wonder what set of circumstances will be necessary to persuade all nations that often drastic and enormously expensive measures must be taken. And, as a practical matter, virtually all nations will probably have to agree eventually. While irreversible damage to the environment might be effectively averted if most nations, for example, discontinued burning fossil fuels or installed pollution abatement devices, the economic consequences of trade in the world market could be catastrophic to those who had undergone the time and expense necessary to save the planet as they had to compete with the few who continued to despoil the environment in order to produce cheaper goods. A nation has limits on the extent of magnanimity it can afford in these matters.

The problem of anticipatory versus responsive actions is as serious as it is difficult to resolve. Modern civilization could be fatally gored on one of the horns of the dilemma:
1.   If it chooses to anticipate and avert potential disaster but overreacts because of lack of sufficient scientific knowledge, technological options may be foreclosed which will be necessary to meet the future energy, food, and other needs of the peoples of the world; or
2.   If it cannot be compelled by scientific knowledge and probabilities to act dramatically until one or more major ecological catastrophes have occurred, effective response may be impossible because of the irreversible nature of the resulting alterations of the planet.

As with most dilemmas, a third alternative must be found—probably one between the two extremes. But, as with most dilemmas, it is not clear how or if this can be done.

The critical factors in resolving the problem of when and how to act appear to be the degree of scientific and technological understanding of these effects and how to ameliorate them, and the public conception and appreciation of the implications of this knowledge, or lack of it, on the kinds of action which may be required. Present circumstances do not allow one to be very sanguine about the state of these factors. We do not know enough, and present programs will probably not give us enough information to make wise decisions. More international cooperation in obtaining and evaluating this information is required.

The Opportunities Provided by International Cooperation

It is obvious that the direct benefits of increased international cooperation on global pollution problems could be the acquisition of needed scientific data; the development of new technologies for monitoring and abating pollution, and for replacing polluting technologies; and hopefully the preservation and enhancement of the global environment through coordinated national actions and international agreements.

The indirect benefits of such cooperation could be equally profound. Multilateral scientific and other nonmilitary activities have often been viewed by integration theorists as being very important in establishing patterns of cooperation and mutual interests. Once established, these patterns could result in a lessening of international tensions and reduce the incidents of conflict among nations. As Lincoln Bloomfield (1967) has noted:

The first category of potential communities that the United States should strive to bring into being embraces the international community of science and technology. . . . American strategy should continue to encourage proposals through the United Nations for intensive collaborative investigations and endeavors in various scientific fields that bear on the solution of the common global problems now or eventually affecting every human being.

If U.S. foreign policies are to reflect adequately some of the possibilities that are available for meaningful interaction with other nations (for example, monitoring and focused research programs), more conscious efforts must be initiated to integrate the scientific and the political considerations and constraints involved. Skolnikoff (1967) has pointed out that the "problem of how to exploit appropriate (technological) developments with taste, effectiveness, and reasonable accuracy equally requires a rare blend of competences"; he suggested that what is required is an "uncommon blend of scientific knowledge and of political/psychological common and scientific sense."

Ernst Haas is one of the leading proponents of the proposition that international conflict can be diminished by the creation and operation of international agencies that are concerned with achieving nonmilitary tasks rather than acquiring power (Haas, 1958). This concept of neofunctionalism is based on the theory that the cooperation of nations on welfare functions may "spill over" into other areas of possible international interaction to increase the likelihood of eventual integration into peaceful communities. Amitai Etzioni has taken this theory one step further and suggested that there is a "snowball" effect by which success in one area of cooperation greatly increases the probability of success in the next cooperative venture (Etzioni, 1966).

Yet, even among integration theorists there is no consensus (and there is strong disagreement between integration and conflict theorists). Karl Deutsch and his coworkers have concluded that historical evidence is inconclusive on the effects of "spill over" or "snowballing" on significant integration and that these effects have probably been overrated (Deutsch et al., 1957). Deutsch and his colleagues do, however, introduce the concept of "take-off" through which integration possibilities are enhanced when certain conditions are met and when at least one major group is committed to the success of the joint ventures.

Conflict theorists such as Raymond Aron, on the other hand, have argued that the concept of sovereignty, which is the major obstacle to integration or the establishment of a universal cooperative system, is not devalued by technological and economic dependence (Aron, 1966). If this proposition is valid, then efforts to encourage cooperation on environmental matters in order to reduce conflict could prove counterproductive.

When the issues progress from the data gathering to the regulation stage, conflict situations are likely to abound if each nation insists that it must have ultimate authority over all decisions that directly affect its domestic policies toward industry or internal governing and regulatory units. As in all national policy areas, there will be numerous competing domestic interests which must be taken into account in considering the demands and concerns of other nations.

The resulting tensions could be more destabilizing than stabilizing, especially if policy makers are not continually aware of these possibilities. As the scholars study and attempt to develop theories to explain these phenomena and hopefully make it possible to be prescriptive, foreign policy makers must nonetheless begin making decisions now that have some chance of maximizing cooperation among nations and minimizing the chances of conflict.

The problems of establishing modes for international cooperation are formidable. Even if the physical sciences were to provide definitive and apocalyptical predictions of global environmental disasters, the social sciences suggest that it is still uncertain that nations could agree whether to undertake collective rectification of the situation or agree on how the challenges should be met. While the foreign policy issues of this and other countries would then have one less uncertainty with which to deal, they would still be far from simplified.

If the theory and logic of collective action developed by Mancur Olson are valid when applied to international cooperation to attack common pollution problems, then agreements on concerted action programs will hardly be automatic even though agreement may have been been reached on the common goal. Olson (1965) has summarized the relevant portion of his theory with the statement:

If the members of a large group rationally seek to maximize their personal welfare, they will *not* act to advance their common or group objectives unless there is coercion to force them to do so, or unless some separate incentive, distinct from the achievement of the common or group interest, is offered to the members of the group individually on the condition that they help bear the costs or burdens involved in the achievement of the group objectives. Nor will such large groups form organizations to further their common goals in the absence of the coercion or the separate incentives just mentioned. These points hold true even when there is unanimous agreement in a group about the common good and the methods of achieving it.

Several years before the emergence of Olson's theory, Haas published the results of his comprehensive study of the political, social, and economic forces that acted during the 1950s to unite (and to prevent the uniting of) Europe. One of Haas's (1958) conclusions indicates that there is empirical evidence in international relations which substantiates Olson's theory: "[T]he process of community formation is dominated by nationally constituted groups with specific interests and aims, willing and able to adjust their aspirations by turning to supranational means when this course appears profitable."

It will be the responsibility of foreign policy makers to identify the numerous special interests which will be involved in extensive multilateral agreements to combat pollution and to develop ingenious policies which will make it "profitable" for nations to join together in the common interest.

The establishment of a productive union of diverse national interests, aspirations, and needs will be an extraordinary task. There must be a high degree of consensus of the governments and the peoples of at least the developed nations and probably the developing nations if "equal sacrifices" are to be calculated and agreed upon.

Karl Deutsch and co-workers have, as a result of detailed study of security communities, outlined the essential conditions which seem necessary if an amalgamated or pluralistic "community" of nations is to be formed to satisfy security, economic, or social needs (Deutsch et al., 1957). If global pollution were a credible threat to life in a region or on the planet, such a security community would probably be a necessary response.

The amalgamated community is the highest level of integration and would entail the establishment of a common government after merger. Not surprisingly, there are twelve rather stringent conditions (necessary, but perhaps not sufficient) for this to occur. On the other hand, Deutsch proposes that only three of these twelve are required for a pluralistic community in which there would be a sense of community and peaceful change but in which each government would remain intact. If either is to be reasonably hoped for in the near future, it is probably the latter.

The three conditions required for the pluralistic community

are compatibility of major values; response to each others' needs; and mutual predictability of behavior. As previously discussed, the values which must be predominant in a community organized to avert "eco-catastrophes" will have to be significantly different from those which now motivate the public and private sectors of most nations.

Internal change of these values simultaneously in most countries and external coordination of these value concepts will not be achieved without a great deal of scientific information and public education. If this compatibility of new value structures is effected, then the next two conditions should follow. The natures of the problems are such that an understanding of them and a real commitment to resolving them will leave no doubt as to the absolute interdependence of nations in undertaking meaningful action programs.

Yet, even with such commitments by many nations, there will remain the difficulties posed by countries who are unwilling to forgo quick profits or subject themselves to deprivations or retarded growth and by countries who initially agree on the principles and some of the remedial measures but who eventually balk at some new restraint or simply ignore some provision of the agreements. This is, of course, a classic problem in international law, and there is considerable disagreement among scholars and practitioners in this field on the viability of such solutions.

Those who have been frustrated at attempts to obtain "world peace through law" when issues of life and death were starkly evident, will probably shudder at the prospects of advocating "world cleanliness through law" especially if the concerns are at present only aesthetic or speculative. Agreement that pollution is harmful is certainly no more profound than agreement that war is harmful. Nations have risked and incurred substantial deprivations to fight for economic and territorial gains as well as ideological tenets. It is not unreasonable to hypothesize that they would risk at least as much to maintain present industrial practices which assure a position of economic leadership in the world market.

Illustrative of the pessimistic views of the practicality of international law through agreement are those of Raymond Aron (1966) who has cited force as the "essential imperfection of international law." He argued that since all treaties and agreements are based to some degree on force or the threat of sanctions, that the legality of such an agreement is limited in an international system much more than that of an agreement made by citizens within a country.

Stanley Hoffmann (1966), though less pessimistic in the long term, has expressed very little hope in the near future for a stable world order through law. Finally, at the other end of the spectrum is Roger Fisher who has argued that force is not the *sine qua non* for effective international law: "No more in the international than in the domestic sphere should the argument be heard that governments must be lawless because they cannot be coerced" (Fisher, 1966).

Resolution of these different views will probably only come when nations attempt to collectively regulate technologies and their uses of them and either succeed or fail. It is clear, however, that the issues are nontrivial and thus substantial thought and effort must be devoted by foreign policy planners if success is to be hoped for. If such planning begins now, perhaps courses of potential diplomatic action will be identified by the time the scientific and technological results are in.

Global technologies not only necessitate the utilization of international organizations for regulation, cooperation, and control, but they also provide an opportunity for foreign policy makers to pursue broader and more political objectives with respect to those bodies. It is very important for formulators of foreign policies which deal with international science policy to view the possible programs from several perspectives other than the purely scientific merits of any given proposal. Three additional perspectives noted by Skolnikoff (1967) are those of the effects on international organizations, on national objectives in science, and on the general interaction of the science policies and achievement of individual nations. In the environmental area, in particular, domestic and international science planning will be very closely related, and cognizance of this relationship in this country and others must always be explicit in U.S. foreign policy.

Skolnikoff has also suggested several political objectives which should be considered when developing policy toward international organizations and has listed three general ways in which science and technology can be conscious instruments of policy for contributing to these objectives: strengthening an international organization, forcing the development of new machinery, and achieving particular political objectives. Global environmental problems present many opportunities for meeting these objectives.

Implications for U.S. Foreign Policy

Advanced technology has made it possible for man to achieve extraordinary power over his natural environment and given him the means to satisfy many of his real and perceived needs. Advanced technology, coupled with population growth and lack of sophisticated and centralized regulation and control, has made it possible for man to destroy his natural environment and himself with it.

The growing realization among nations that this latter situation is indeed a possibility and that somehow it should be avoided is resulting in questions of how to avert such a catastrophe. Since the problems are universal, there is strong feeling that they will have to be tackled at the international level. Yet, the history and present status of international relations offer little comfort to those who hope that the far-reaching types of international cooperation which will be required will be easily achieved.

The challenges to U.S. and other foreign policy makers to effect meaningful and productive international interaction and cooperation will be formidable. One of the major objectives of this paper has been to suggest some of the implications of and for U.S. foreign policy which would result from the following national decisions (only the first of these decisions has thus far been made by the President and the Congress):
1.   To express openly a concern about global pollution.
2.   To support actively various cooperative scientific and technological efforts to learn more about the nature of the pollution threat.
3.   To enter seriously into negotiations with other nations to explore how harmful global technologies can be regulated and controlled.

The expression of concern may appear innocuous and can

result in an international image that portrays the U.S. as a responsible and concerned nation. Platitudinous expressions can, however, quickly result in international pressures to do something about the perceived threats. As in all "put up or shut up" situations, decisions for further action or inaction will be a complicated function of how much has been said and what the costs and benefits of concrete action are perceived to be in terms of prestige, security, wealth, knowledge, and other values.

The next logical step in operationalizing concern about environmental quality would be to actively support cooperative efforts to learn more about the environmental problems and how to solve them. This would be advantageous in at least two ways: adequate knowledge does not yet exist on which to base reasonable decisions for abatement or control; and such cooperative efforts would increase, and perhaps encourage, peaceful interaction among nations. The new knowledge which might be gained, however, could lead some nations to conclude that the problems were of sufficient gravity to warrant international regulation or control.

Thus, ultimately, the U.S. might be faced with decisions of how to balance the concerns and demands of other nations with the numerous public and private interests at home. Since this country is the major industrial country in the world and the major generator of waste products, restrictions on the emissions of wastes into the atmosphere or oceans will probably fall heavily on U.S. industrial and governmental operations. As sacrifices and social and industrial changes are demanded of this and other countries, the resulting disputes could have a severe destabilizing effect on international relations.

At the present time, "preservation of environmental quality" is a popular slogan at the local, national, and international levels. It is easy to say and difficult to disagree with. But if it is ever to be more than a slogan (and man's survival could depend on that "if"), then the steps which may be necessary to implement this objective will almost certainly be the subject of heated controversy. Some groups will have to pay, and pay heavily, for the resulting change in the status quo, and they may not do it obligingly—this does not only refer to industries but the whole spectrum from the individual consumer to the nation striving for a better position in the world's political, military, or economic markets.

Nothing less may be required than the complete restructuring of values—domestically and internationally. This will not occur easily or automatically and may not happen even if the catastrophic consequences of refusal to do it are unequivocally apparent. Responding effectively to the threat of pollution could cause such severe social and economic dislocations and disruptions that the short-term (and even the long-term) deprivations resulting from change could appear more serious than the long-term effects of global pollution.

To use an analogy, the war against global pollution may not be a "guns and butter" war—citizens of developed countries may have to forgo some comforts which they now enjoy and foreclose some future options for higher and different living standards; citizens of developing countries may have to adopt an even longer-term view of their nations' eventual industrialization and entry into the community of affluent nations; all governments may have to yield a significant degree of sovereignty to an international regulatory organization. Because present pollution problems are the result of the present state of industrial, agricultural, and governmental activities, it is clear that the problems cannot be solved by conducting "business as usual," and it might be safe to guess that the necessary changes will have profound effects on the status quo.

In spite of these implications and consequences, the problems of global pollution can be ignored only at great peril, and careful consideration must be given to initiating scientific programs to learn more about the nature of the pollution threat. Because of these implications and consequences, foreign policy makers must give careful consideration to the types of national policies and international machinery which may be required in the future if there is to be some hope for concerted action among nations in response to the threats, both physical and political, posed by global environmental pollution.

References

Aron, Raymond, 1966. *Peace and War* (Garden City, New York: Doubleday and Company).

Bloomfield, Lincoln P., 1967. *The United Nations and U.S. Foreign Policy* (Boston: Little, Brown and Company).

Deutsch, Karl, Burrell, Sidney A., Kann, Robert A., Lee, Maurice, Jr., Lichterman, Martin, Lindgren, Raymond E., Loewenheim, Francis L., and Van Wagenen, Richard W., 1957. *Political Community and the North Atlantic Area: International Organization in the Light of Historical Experience* (Princeton: Princeton University Press).

Etzioni, Amitai, 1966. The dialectics of supranational unification, *International Political Communities* (Garden City, New York: Doubleday and Company).

Fisher, Roger, 1966. Bringing law to bear on government, in R. A. Falk and S. H. Mendlovitz, *International Law* (New York: World Law Fund).

Haas, Ernst B., 1958. *The Uniting of Europe: Political, Social and Economic Forces, 1950–1957* (Stanford, California: Stanford University Press).

Haas, Ernst B., 1966. International integration: the European and the universal process, *International Political Communities* (Garden City, New York: Doubleday and Company).

Hoffmann, Stanley, 1966. International systems and international law, in R. A. Falk and S. H. Mendlovitz, *International Law* (New York: World Law Fund).

Kennan, George F., 1970. To prevent a world wasteland: a proposal, *Foreign Affairs, 48:* 3, April.

Olson, Mancur, Jr., 1965. *The Logic of Collective Action* (Cambridge, Massachusetts: Harvard University Press).

Schilling, Warner R., 1968. Technology and international relations, *International Encyclopedia of the Social Sciences* (New York: The Macmillan Company and The Free Press).

Skolnikoff, Eugene B., 1967. *Science, Technology, and American Foreign Policy* (Cambridge, Massachusetts: The M.I.T. Press).

Wolman, Abel, 1968. Pollution as an international issue, *Foreign Affairs,* October.

Name Index

Subject Index